Biosensors: Prominent Applications

Biosensors: Prominent Applications

Edited by **Marvin Heather**

CLANRYE
INTERNATIONAL

New Jersey

Published by Clanrye International,
55 Van Reypen Street,
Jersey City, NJ 07306, USA
www.clanryeinternational.com

Biosensors: Prominent Applications
Edited by Marvin Heather

International Standard Book Number: 978-1-63240-088-8 (Hardback)

Printed in the United States of America.

Contents

Preface

The prominent applications of biosensors are discussed in this state-of-the-art book. The major issue related to biosensor development is the application of biosensors for several practical tasks to maintain a steady and dependable flow of data regarding the indicators of industrial and natural processes as well as the surroundings, thus allowing optimum control and feedback. Biosensors can give important information, since the quality of life depends primarily on our knowledge about what we eat and breathe and how our bodies are able to metabolize the substances which come in our contact. This book essentially lays stress on the applications of biosensors for the purpose of monitoring the quality of food, parameters of environment and biomarkers of health.

This book unites the global concepts and researches in an organized manner for a comprehensive understanding of the subject. It is a ripe text for all researchers, students, scientists or anyone else who is interested in acquiring a better knowledge of this dynamic field.

I extend my sincere thanks to the contributors for such eloquent research chapters. Finally, I thank my family for being a source of support and help.

Editor

Biosensing Environment

Biomimetic Sensors for Rapid Testing of Water Resources

Jill M. Grimme and Donald M. Cropek

Additional information is available at the end of the chapter

1. Introduction

Environmental monitoring is a critical mission of the U. S. Army Corps of Engineers due to its commitment to support not only the soldiers, but also to maintain the environment on military lands and installations. The detection of contaminants present in the environment, specifically water resources, is vital to ensure the safety of soldiers at forward base locations and to promote the sustainable usage of military lands. Therefore, fieldable, robust, and sensitive detection mechanisms are required for screening applications that provide rapid environmental assessment. Typically, sensor and detector research focuses on new techniques and methods to characterize a sample for an analyte of interest. A metric that must be satisfied for success is the selectivity of the sensor, i.e., how well it can measure only a particular analyte without response to other closely related chemicals. Under certain conditions, it makes sense to use sensors with specificity. However, a quantitative result alone does not translate easily to a measure of toxicity to the soldier or to an assessment of environmental impact. Further, the uncertainties inherent to unfamiliar and frequently hostile environments preclude exact knowledge of a complete set of potential analytes.

Our Biomimetic Sensor for Rapid Testing of Water Resources program at the Construction Engineering Research Laboratory (ERDC-CERL) was initiated precisely to investigate and advance aspects of miniaturized microfluidic devices for toxicity sensing. This work package represents a change in the sensor and detector paradigm typical of legacy Corps of Engineer analytical research. Providing the soldier in the military environment with a rapid, universal indicator of a chemical threat may prove more powerful and vital than identifying what compound is present. Our goal in this program was to assess the property of overall toxicity, rather than acquisition of a compendium of individual chemical concentrations. The chosen way to acquire this property was to expose the water sample of interest to a population of living cells where we ascertain the harmful impact of the sample on cellular metabolism, function and communication.

Most commercially available rapid toxicity screening test kits use luminescent bacteria, photosynthetic algae, or plant enzymes to monitor toxicity via luminescence or colorimetric assay, which report a decrease in cellular respiration or enzyme turnover in relation to the toxicity of the test sample (http://www.epa.gov/nrmrl/std/etv/vt-ams.html#rtts). From a soldier health perspective, however, the use of mammalian (human) cell-based biosensors for rapid risk assessments would be more beneficial due to the direct correlation of the toxicity data to the human health effects on the soldiers. Therefore, many of the sensing mechanisms we are developing integrate more biorelevant cells for specific major organ systems into the sensing platform. Further, we maintain a more ambitious goal to move beyond homogeneous cell culture to an intelligently designed heterogeneous population of multiple cell types that recapitulates *in vivo* tissue function. Depending on the final application however, these same sensor platforms could be populated with fish and animal cells/tissues when direct environmental impact analyses are required on particular strata of an ecosystem.

Cell-based biosensors utilize immobilized living cells (either prokaryote or eukaryote) as the biorecognition element. The impact of the chemical stimulus on the cell generates changes to the intracellular or extracellular biochemistry that are most frequently detected either electrically or optically [1, 2]. While the cellular response may require an incubation time with the contaminant that increases total analysis time, the ability of cells to respond to minute changes in their complex environment imbues cell-based biosensors with high sensitivity and more importantly for our task, biorelevance. Also, the ability of the biosensor to report the bioavailability of a compound may be more critical than the analytical determination that the compound is present. Unlike some current biorecognition based sensors that utilize antibodies, aptamers, or other analyte-specific binding mechanisms, biosensors incorporating intact living cells can respond to emerging or little known chemical or biological threats and can be engaged as broad spectrum screening tools [1]. Finally, by exploiting the biochemical response of the living cell to the entire complex composition of a water sample, we can ascertain additive and synergistic effects impossible to discern using a suite of selective sensors.

Despite the advantages of using cell-based biosensors, there are non-trivial obstacles to consider when incorporating living cells onto a device to be used in austere environments. Perturbations in temperature, pH, oxygen/carbon dioxide levels, nutrients, and sterility can all have deleterious effects on the survivability of the cell population thus rendering the biosensor less reliable or simply unusable. Therefore, maintaining a well-defined and healthy cell population with adequate controls presents the most difficult challenge for the development and wide spread usage of cell-based biosensors.

Recognizing these hurdles, we formed a collaborative research effort that aimed to create sensors that instantly alert personnel to a toxic threat in water by designing, building, and testing microfluidic devices that accurately replicate the physiological response of an organ structure to contaminants in water. We are developing both electrical and optical transduction methods for the detection and translation of the cell's physiological response to a water toxicant. The ultimate goal of the project was the development of a microfluidic platform that housed a complete set of three-dimensional (3D) "organs" reconstructed from heterogeneous cell/tissue populations that worked in a physiologically realistic manner to represent the circulatory pattern of an entire organism (*e.g.*, human). With the understanding that recapitulating a truly physiologically accurate "body on a chip" is a momentous task, we have focused our research efforts to develop tools, technology and scientific knowledge that bring us closer to this vision. This chapter describes several of the resultant scientific collaborations with academic partners in this biomimetic sensor program.

A class of *in vitro* tissue models known as micro-cell culture analogs (μCCAs) has been developed by pioneering research in the laboratory of Dr. Michael Shuler at Cornell University [3-5]. Micro-CCAs are composed of a microfluidic platform that allows multiple "organs" to be combined in a physiologically realistic manner. Such a device is a tangible representation of a physiologically based pharmacokinetic (PBPK) model that translates *in vivo* conditions such as liquid to cell ratios, fluid residence times, and physiological shear stress that are realistic to each organ type into parameters that can be integrated into a microfluidic device through microfabrication. Therefore, each organ is recreated on the chip with cells housed in a precisely designed organ chamber [6] (Figure 1). The evolutionary design of the μCCA has increased from three chambers [5], to four chambers [7] and includes more complex organs and tissue types such as fat cells, which may alter the response [8]. Additionally, one of the main routes of toxicant introduction into the body is through oral ingestion and subsequent absorption from the gastrointestinal (GI) system, which represents first pass metabolism. Therefore, a critical component to the μCCA was the inclusion of the GI system mimic (Figure 2). We have developed a GI module that incorporates a mixture of cell types to better mimic realistic intestinal cell populations, which include absorptive enterocytes and mucus producing goblet-like cells [9-11]. The mixture of cells is cultured on a microporous membrane to allow introduction of the analyte of interest at the apical side, but limits passage to the basolateral side to only those analytes that successfully pass through the functional GI cell barrier layer (Figure 2C). For total system recirculation, a peristaltic pump is used with tubing connecting the analyte source and the apical GI surface (for continuous "ingestion"), and the basolateral GI surface is connected to the μCCA body chip (Figure 2A and 2B). Devices such as these are valuable not only as tools for toxicity screening but also as tools for screening drug candidates, indicating drug efficacies and side effects as well as combinatorial effects from mixtures of compounds, thus lessening the dependence on animal models.

Figure 1. Micro-CCA device with fabricated organ chambers. Image provided by M. Shuler.

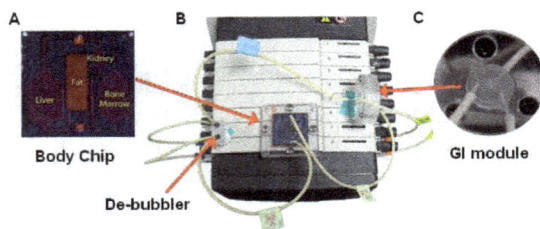

Figure 2. (A) Image of the systemic µCCA. Organ chambers and channels are fabricated to accurately scale the chamber size and the interconnecting channels so they are representative of *in vivo* circulating conditions for those organs. The other tissues of the body were represented by the external de-bubbler, which was a 200 µL reservoir. (B) Image of the systemic and GI tract µCCA experimental set-up. (C) The GI module housed in a plexiglass chamber with cells seeded on a semiporous membrane. Images provided by M. Shuler.

2. Electrochemical detection methods for cellular stress biomarkers

One type of cellular response to a toxicant is through the generation of extracellular biomarkers. In this manner, we can transform a toxicity response to a contaminant suite into a biomarker concentration. The main focus of our collaboration with Dr. Scott Banta at Columbia University is the generation of a streamlined platform for the rapid development of new biosensors that can be used to detect virtually any desired protein biomarker target. The overall selectivity and sensitivity of the biosensor are dependent not only on the recognition element, but also on the transducer used to detect the response. Therefore, our research combines generation of novel biorecognition elements with electrochemical transduction to fabricate powerful selective sensing modalities [12].

The process we have developed includes: biomarker (target) selection and immobilization, phage display to select binding peptides, peptide synthesis with a terminal thiol, quartz crystal microbalance (QCM) for *in-situ* monitoring of peptide immobilization, and analyte detection using electrochemical techniques (Figure 3A). Target selection involves several rounds of M13 phage display followed by enrichment of short unstructured peptides that bind to the desired target with increased affinity. For the sensor performance, the selected, final round peptides are chemically synthesized and immobilized on a gold surface, allowing for detection of the target using QCM (Figure 3B) or electrochemical impedance spectroscopy (EIS) (Figure 3C). The use of short unstructured peptides over more complex protein-based affinity scaffolds has several advantages: 1) peptides are stable and resistant to harsh environments, 2) peptides can be synthesized easily and inexpensively, and 3) peptides can be more amenable than antibodies to engineering at the molecular level due to their smaller size. In addition, peptide immobilization on gold electrodes has been well established.

We demonstrated the utility of this approach by creating a novel peptide-based electrochemical biosensor for the enzymes alanine transaminase (ALT), a well known biomarker of

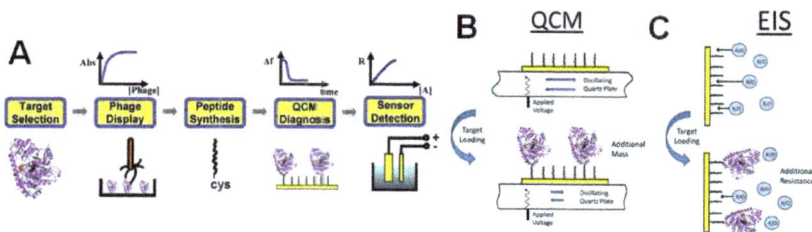

Figure 3. Schematic diagram illustrating the general process of biosensor development. (A) Work flow diagram for biosensor development: 1. Target protein biomarker selection, 2. Phage display selection, 3. Peptide synthesis, 4. QCM diagnosis, 5. Biomarker detection. B) The basic principle of QCM where the binding of the target protein to the immobilized peptides causes a frequency change in the oscillation of the quartz crystal. C) The basic principle of EIS where the binding of the target protein to the immobilized peptides causes increased resistance to the reaction of an added redox couple.

hepatotoxicity. The sensor was prepared using the selected ALT peptide that had been modified with a C-terminal cysteine to facilitate the immobilization on a gold surface. Once the peptide had been immobilized, any remaining bare gold was blocked by the addition of free cysteine to prevent non-specific binding of the target directly to the gold surface. The binding and blocking processes were monitored by QCM, cyclic voltammetry (CV) and EIS. The peptide immobilization resulted in a frequency drop in the QCM signal due to the increased mass loading, a drop in current was detectable by CV, and an impedance increase in the EIS due to the blocking of the gold surface. The packing density of the surface-bound peptide can be calculated using the Sauerbrey equation [13]:

$$\Delta f_m = -C_f x\, \Delta m \tag{1}$$

where Δf_m is the frequency change due to mass loading, Δm is the mass change at the QCM crystal surface and $C_f = 0.0566$ Hz/(ng/cm^2) for a 5 MHz AT-cut crystal at 20°C. The frequency drop of the QCM due to the immobilization of the peptide was roughly 21 Hz, which correlates to a packing density of 2×10^{-10} mol/cm^2 or 1.2×10^{14} molecule/cm^2 using a peptide molecular weight of 1851 g/mol.

Next QCM, CV and EIS were used to demonstrate the affinity and specificity of the immobilized peptide for ALT (Figures 4 and 5). A frequency drop (Df = -31 Hz) in QCM was observed upon binding of ALT versus the frequency of the sensor in buffer alone (Figure 4A). A current decrease in CV (Figure 4B) as well as a large impedance increase (Figure 4C) was measured upon addition of ALT, while buffer alone resulted in no change.

Figure 5 shows the specificity of the peptide (Au-Peptide) for ALT over streptavidin (SA) or bovine serum albumin (BSA). The greatest frequency change of -31 Hz was observed with ALT. Although a small frequency change (about -3 Hz) was seen with SA and BSA, this change may be due to a small amount of non-specific binding of the proteins to unoccupied sites on the surface or non-specific interactions with the peptide itself.

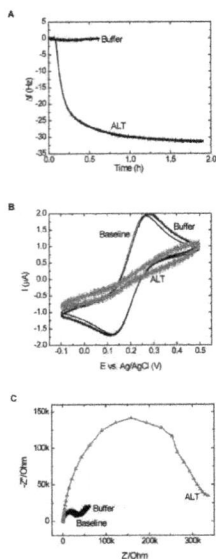

Figure 4. Sensor operation. A) QCM frequency change of Au-Peptide electrode in Buffer alone and in 10 mg/mL ALT, B) CV of Au-Peptide electrode before (Baseline) and after Buffer alone or ALT binding, C) EIS of Au-Peptide before (Baseline) and after Buffer alone or ALT binding. After incubation in Buffer or 10 mg/mL ALT, the electrodes were transferred to 1 mM solution of $Fe(CN)_6^{4-/3-}$ in 0.1 M $NaClO_4$ for the EIS measurements.

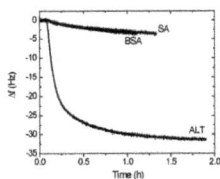

Figure 5. Specificity of the ALT biosensor. The shift in QCM frequency in response to injections of SA, BSA or ALT on the Au-Peptide modified crystal. The concentrations of all three solutions were 10 µg/mL.

For quantitative measurements, CV was not performed due to the limited detection range since the current decreases upon target binding. Therefore, we performed only QCM and EIS using a range of ALT concentrations (Figure 6). A quantitative response was observed for both QCM (Figure 6A) and EIS (Figure 6B) for the range of ALT concentrations tested. The LOD obtained by QCM is 60 ng/mL, while EIS has a LOD of 92 ng/mL. The sensitivity of the QCM system was 8.9 +/- 0.9 Hz/(mg/mL), while the sensitivity of the EIS system was 142 +/- 12 impedance percentage change %/(mg/mL). The quantitative response achieved for both QCM and EIS validates that both methods could be used as a sensing technique in quantitative ALT measurements.

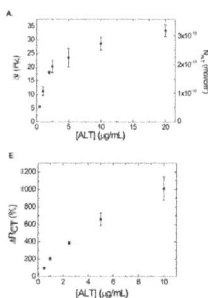

Figure 6. Response curves of ALT binding. Measurements were made using QCM (shown as absolute value of frequency change) (A) and EIS (B). Each run was performed using a single electrode with successive tests in ALT solutions from low to high concentration. The apparent amount of bound ALT (N_{ALT}) for the QCM measurements was calculated using Eqn. 1 and is shown on the right ordinate. Error bars represent standard deviations obtained from triplicate measurements.

In addition to the generation of this ALT-specific recognition element, we have developed a recognition element for the cardiac specific biomarker, troponin I (TnI) [13, 14]. Again, phage display was employed to generate the recognition peptide while QCM and EIS validated and quantified the binding events and specificity of the peptide for the target, TnI. Therefore, the generation of a peptide recognition element specific for the binding of either ALT or TnI followed by immobilization onto a gold surface for signal transduction supports this method of biorecognition element generation for the detection of any specific cellular stress biomarker. The recognition peptide provides the next step toward increasing stability over antibodies, but quantitative impact versus longevity depends upon how it is used in a more mature device. We are currently engaged in transitioning the peptide recognition schemes onto a fieldable chip. Reuse conditions for the chip must also be investigated but likely involves introduction of a release buffer with glycine-HCl.

We recognize that gold components within microfluidic devices, specifically as electrode materials, are prone to biofouling. In an effort to use an electrode material both more resistant to biofouling and possessing a larger potential range compared to noble metal electrodes [15-17], we developed a new method for the fabrication of carbon paste electrodes (CPEs) that optimizes their stability in a flow-based microfluidic device [18]. The greater stability of these CPEs results from exploiting beneficial properties of both soft and rigid composites, which are used as the binder, during CPE fabrication. As a result of our collaboration with Dr. Charles Henry at Colorado State University, we have developed a mixed binder to create on-chip CPEs for flow-based devices. The mixed binder composed of a soft composite (mineral oil) and a rigid composite (poly(dimethylsiloxane) (PDMS)) enhanced the overall CPE robustness and electrochemistry over that of a CPE fabricated with mineral oil or PDMS alone. Analysis of the CPE performance was carried out using cyclic voltammetry in static solutions or amperometry with flow injection analysis.

The CPEs are fabricated simply by filling channels molded in previously cross-linked PDMS using a method analogous to screen printing with the CPE mixture, which is composed of graphite powder and the PDMS:mineral oil binder (Figure 7). Further, we demonstrated that the PDMS: mineral oil CPEs can easily be chemically modified for the detection of biologically relevant molecules. The monitoring of thiols provides an indication of the functioning of numerous critical cellular pathways [19, 20]. Traditional carbon paste electrodes can be used to detect thiols, however, they suffer from low peak currents due to poor interactions between the carbon and the thiols. Therefore, we added the electrocatalytic reagent cobalt phthalocya-nine (CoPC), which is widely used as a reduction/oxidation mediator for lowering the overpotential of thiols [21-23], to decrease decrease the detection potential and increase the peak current obtainable with carbon electrodes. The model analyte dithiothreitol (DTT) was used to demonstrate the sensitivity of the CoPC modified CPEs for the detection of thiols. Cyclic voltammetry and amperometry were used to compare the CoPC modified CPEs to bare CPEs (Figure 8). The detectable electrocatalytic process occurs in two steps that involve the initial electrochemical oxidation of Co(II)phthalocyanine followed by the chemical oxidation of DTT as it regenerates the Co(II)phthalocyanine [23]. Figure 8A shows the cyclic voltammo-grams of a CoPC modified CPE (solid line) versus an unmodified CPE (dashed line) of 1 mM DTT in MES buffer (2-(N-morpholino)ethanesulfonic acid, 30 mM, pH 7.4). Under these conditions, the unmodified CPE failed to detect the DTT whereas the expected oxidation peak of DTT was observed for the CoPC modified CPE [24]. We evaluated the performance of the CPEs for DTT detection with respect to the amount of CoPC added to the graphite and determined that a mixture of 12% CoPC in graphite performed the best (data not shown). Also, the composition of the binder was investigated to identify the best DTT detection performance, and the ratio of 2:3 PDMS:mineral oil was optimal to maintain robustness and electrochemical performance. The secondary means to characterize the performance of the CoPC modified CPEs involved flow injection analysis using amperometric detection for the continuous detection of DTT (Figure 8B). Using a CoPC modified electrode, we observed nearly a 2-fold improvement in the electrochemical signal (4.0 ± 0.2 nA (n = 3) versus 2.5 ± 0.2 nA (n = 3)) for a bare CPE for 100 μM DTT. Additionally, the limit of detection (LOD, defined as the concen-tration yielding a 3x larger signal than baseline noise) for DTT with a CoPC modified CPE was 2.5 ± 0.2 μM (n = 5) versus 16.8 μM (n = 5) for an unmodified CPE.

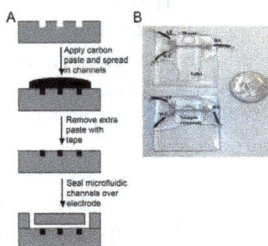

Figure 7. Flow-based microfluidic system using CPEs. (A) Schematic for electrode fabrication process (side view), (B) Photogragh of electrode alignment in a microfluidic device for the flow injection system with amperometric detection (top) and for cyclic voltammetry study (bottom). Size of electrodes: 250 μm wide and 50 μm deep. Electrode spacing (center-to-center), 300 μm. Microfluidic channel dimension: 250 μm wide, 50 μm deep and 2.5 cm long. Reproduced by permission of The Royal Society of Chemistry.

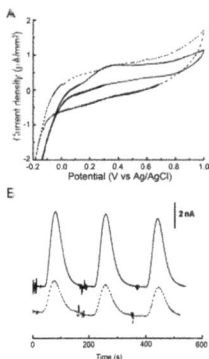

Figure 8. Comparison of working electrode performance for DTT detection between CoPC modified CPE (solid line) and unmodified CPE (dashed line). (A) Cyclic voltammograms of 1 mM DTT in 30 mM MES buffer, pH 7. (B) Flow injection analysis with amperometric detection of 100 μM DTT using 30 mM MES, pH 7, as a running buffer. Paste composition ratio is 5 : 3 : 2 (graphite:oil:PDMS). Off-chip reference and auxiliary electrodes were used. Applied potential: 0.2 V (vs. Ag/AgCl (3 M KCl)) for both types of working electrodes with Pt wire as an auxiliary electrode. Flow rate: 60 mL min⁻¹. Reproduced by permission of The Royal Society of Chemistry.

Next, we chemically modified the CPEs with multi-walled carbon nanotubes (MWCNT) for the detection of catecholamines. The importance of the ability to detect biologically relevant compounds from cells is critical not only for our current application of detection of toxic compounds in the environment, but also for the detection or monitoring of debilitating diseases. Catecholamines like dopamine are known to be involved in the progression of neurological diseases such as Parkinson's and Alzheimer's [25, 26]. Using the MWCNT modified CPEs, we were able to detect the release of dopamine from cultured rat PC12 cells. The data for the flow injection analysis for the detection of 10 μM dopamine is shown in Figure 9 using both MWCNT modified (solid line) and unmodified (dashed line) CPEs. A more significant improvement in signal was observed for the MWCNT modified CPEs when detecting catecholamines, namely dopamine, than shown above for DTT. We determined that the lowest dopamine concentration detectable by MWCNT modified CPEs was 50 nM whereas concentrations below 500 nM for unmodified CPEs were undetectable. This order of magnitude increase in detection likely results from the increased surface area of the electrode achieved with the addition of the MWCNTs and therefore, the peak currents obtained for the CPEs.

We also observed the release of catecholamines from PC12 cells upon exposure to high concentrations of potassium. At each time point, the catecholamine released was measured from potassium-stimulated PC12 cells (6 x 10⁴ cells/time point) using the flow injection analysis system with MWCNT modified CPEs. Figure 10 shows the plot of the peak area versus time for the duration of potassium exposure. At three minutes, the calculated concentration of released catecholamine was estimated to be 3.6 μM (58.3 pM/cell). Concentrations of catecholamine released by PC12 cells upon exocytotic events have been reported to be 67 μM [27, 28] or 20-160 μM dopamine from 1.3 x 10⁴ cells exposed to calcium stimulation [29]. The lower levels of catecholamines detected by our system may be the result of very low capture efficiency

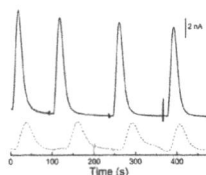

Figure 9. Flow injection analysis of dopamine. Comparison of unmodified CPE (dashed line) and MWCNT modified CPE (solid line). Sequential injections of 10 μM dopamine. Off-chip reference and auxiliary electrodes were used. Applied potential: 0.6 V (vs. Ag/AgCl (3M KCl)) with Pt wire as an auxiliary electrode. Flow rate: 60 mL min^{-1}. Reproduced by permission of The Royal Society of Chemistry.

of this non-optimized CPE design. Interestingly, it is likely that neuronal cells, like PC12s, have a process to re-uptake catecholamines, and this common theory is supported by the sharp decrease in catecholamine signal we observed with our MWCNT modified CPE system at time points after the significant release that occurs at three minutes.

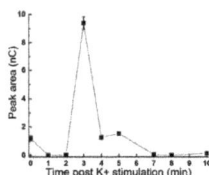

Figure 10. Effect of potassium on catecholamine release from PC12 cells. 6.15 x 10^4 cells were stimulated at each time point. Supernatant was removed and stored on ice until analysis. All experimental conditions same as in Fig. 9. Reproduced by permission of The Royal Society of Chemistry.

The ability to determine the biological consequences of known and unknown toxicants is essential for the protection of the environment and soldier health. We have generated a novel approach for the development of carbon paste electrodes that are integrated, robust, and low cost for microfluidic detection of extracellular biomarkers [18]. Figs 8B and 9 show that the CPEs are stable and reusable over many cycles, although the exact number of cycles depends upon the sample matrix and the run conditions. The CPEs can be readily modified to monitor biologically informative species, which can indicate the cellular responses to environmental stimuli. With its simple modification scheme to introduce electrode selectivity and its intricate design capability allowing for detailed electrode features, this biosensing system can be easily incorporated into many detection platforms.

3. Cell-based sensor designs

We have developed a simple and direct *in vitro* method to control the alignment and elongation of cells in order to form self-organized 3D tissues as part of a collaborative effort with Dr. Ali

Khademhosseini's laboratory at Harvard University. By harnessing the intrinsic potential of certain cells to form aligned tissues *in vivo*, we can generate functional tissues *in vitro* by confining the cells in the appropriate 3D microarchitecture [30]. Cells possessing the prerequisite ability to self-organize *in vivo* include fibroblasts, myoblasts, endothelial cells, and cardiac stem cells. The process involves embedding the cells in microengineered 3D gelatin methacrylate (GelMA) hydrogels, which are micropatterned into high aspect ratio rectangular constructs. In contrast, the cellular alignment in unpatterned microconstructs remained random. By precisely controlling this microgeometry, we can induce the controlled cellular alignment and elongation throughout the entire engineered construct in a user-defined manner. We envision this technology to enable the construction of engineered functional 3D tissues with aligned cells and anisotropic function. Additionally, we extended this technology for the development of a cell-based biosensor for the detection of cardiotoxicity in real-time by following changes in cardiomyocyte beating properties [31].

To demonstrate the control of cellular alignment and elongation properties *in vitro*, we encapsulated 3T3-fibroblasts (10×10^6 cell/mL) in a GelMA construct (800 µm (l) x 150 µm (h) x 50 µm (w)). A 1 mm border surrounded the multiple parallel microconstructs to serve as an unpatterned control (Figure 11). The GelMA macromer concentration (5% w/v, ~80% methacrylation degree) was chosen to maintain the mechanical stiffness but still preserve cell viability and allow the micropatterning of the intricate cell-laden features. After 1 day, elongation of the cells was observed mainly along the edges of the micropatterns and followed the major axis of the channel, but spread throughout the construct by 4-5 days. The lower series of panels illustrates the lack of alignment within the unstructured GelMA. Additionally, the elongation and alignment responses could be controlled by the cell density, hydrogel concentration, and culture time (data not shown) further demonstrating the ability to fine tune these cellular properties.

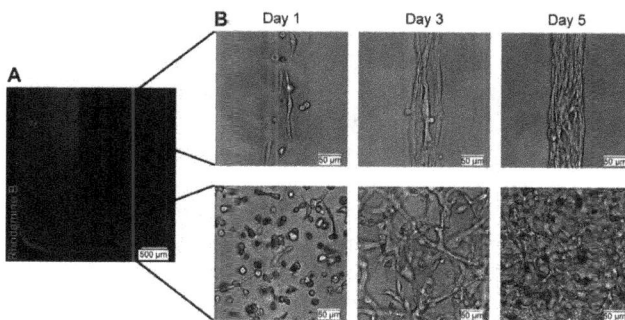

Figure 11. Cell morphology and organization as a function of time in patterned and unpatterned microconstructs. Patterning 3T3-fibroblast-laden 5% GelMA hydrogels into high aspect ratio rectangular microconstructs (50 µm (w) x 800 µm (l) x 150 µm (h)) induced cellular alignment and elongation, while cellular orientation in unpatterned hydrogels remained random. (A) Rhodamine B stained GelMA hydrogel construct shows patterned and unpatterned regions. (B) Representative phase contrast images of 3T3-fibroblasts (10×10^6 cells/mL) encapsulated in patterned (top row) and unpatterned regions of the GelMA hydrogel (bottom row) on days 1, 3 and 5 of culture show that elongation increases over time for all hydrogels, while alignment is only induced in patterned constructs.

We quantitatively analyzed the effects of microgeometry by patterning cells in GelMA constructs with a standard length and height but varying widths (800 μm (l) x 150 μm (h) x 200, 100 or 50 μm (w)) (Figure 12). We observed a strong correlation between the ability of cells to align and elongate throughout the construct with the decrease in width. Alignment was assessed by counting cells whose nuclei were within 10° of the preferred nuclear orientation as described [32]. Overall nuclear alignment and shape index were evaluated with NIH ImageJ software to evaluate cell alignment and elongation [33, 34]. The narrowest microconstruct had the best alignment with 64 ±8% of cells aligned (Figure 12A), and this increased to ~90% if the preferred nuclear orientation was defined to within 20° (Figure 12F). That alignment percentage decreased with increased channel width resulting in 40 ±6% for the 100 μm and 31 ±8% for the 200 μm channels. However, compared to the unpatterned control, the 100 μm channel was still statistically different ($p < 0.01$). The 200 μm channel was not statistically different and the unpatterned control showed only 19 ±9% alignment to the construct (Figure 12A). In comparing the nuclear shape index (Figure 12B), the results were consistent with the lowest mean nuclear shape index corresponding to the 50 μm wide rectangular microconstructs (0.807 ±0.02). The 100 μm and 200 μm wide micropatterns displayed 0.869 ±0.03 and 0.917 ±0.02 mean nuclear shape indices, respectively, while the unpatterned control had a mean nuclear shape index of 0.923 ±0.03 (Figure 12B). Figures 12G-J show representative images of the unpatterned and patterned microconstructs where increasing cell alignment and elongation is visible in constructs of decreasing width.

Next, we expanded the number of cell types used for micropatterning to include additional, self-organizing cells (human umbilical vein endothelial cells (HUVEC), rodent myoblasts (C2C12), and rodent cardiac side population cells (CSP)) and non-organizing cells (Hep-G2) to investigate the applicability of this technique to engineer a variety of tissues representative of vascular system, skeletal muscle, and myocardial tissues, respectively (Figure 13). Accurately mimicking the *in vivo* microarchitecture of these systems would require highly elongated, organized and aligned cell-ECM constructs. As a control, we selected a cell type that lacks the intrinsic potential for self alignment *in vivo* or *in vitro*, human liver carcinoma cells (Hep-G2). All cells were encapsulated in 5% (w/v) GelMA and micropatterned in 50 μm wide rectangular microconstructs. After a five day culture time, the patterned and unpatterned cells were stained with DAPI and phalloidin to visualize cell nuclei and F-actin, respectively (Figure 13A and B). Images revealed the HUVEC, C2C12, and CSP cells to be highly aligned, elongated, and interconnected along the major axis of the micropatterned hydrogels. In contrast, the unpatterned cells remained random with minimal nuclear organization observed only at the edges of the hydrogel. Significant differences were observed upon quantitative analysis for both cellular alignment and elongation (Figure 13C) as well as for mean nuclear shape index (Figure 13D) for the HUVEC, C2C12 and CSP cells when comparing the patterned to unpatterned constructs. Consistent with the lack of intrinsic potential, Hep-G2 cells did not show a significant difference in observed alignment and elongation between the patterned and unpatterned hydrogel constructs, but rather formed cell clusters of multiple cell nuclei regardless of the aspect ratio of the microconstructs. Therefore, the *in vivo* ability of cells to form aligned tissues is required to promote this alignment and elongation behavior *in vitro* via specified microgeometry.

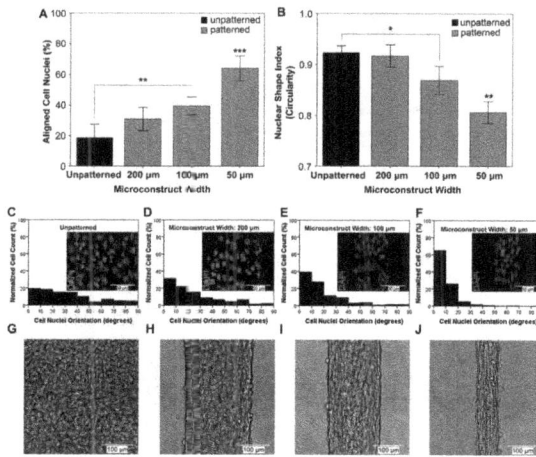

Figure 12. Cell elongation and alignment as a function of microconstruct width. Decreasing the width of patterned rectangular 5% GelMA microstructures (800 μm (l) x 150 μm (h)) increased 3T3-fibroblast alignment as well as nuclear elongation after 5 days of culture. (A) Mean percentage of aligned cell nuclei (within 10° of preferred nuclear orientation) shows 100 μm constructs were significantly more aligned than unpatterned controls, while 50 μm constructs were significantly more aligned than all other groups. (B) Mean nuclear shape index (circularity = 4*π*area/perimeter^2) shows a similar pattern with 100 μm constructs having a significantly lower index than unpatterned controls, while 50 μm constructs were significantly different than all other groups. (C), (D), (E) and (F) Histograms of the relative alignment in 10° increments demonstrates increased cellular alignment with decreased microconstruct width. (G), (H), (I) and (J) Representative phase contrast images of unpatterned, 200 μm, 100 μm and 50 μm wide microconstructs, respectively, show significantly increased cell alignment and elongation inside the constructs with decreasing width. (Error bars: ± SD; ***p < 0.001; **p < 0.01; *p < 0.01).

In the quest to form a 3D tissue construct, we tested the ability of the self-assembly properties of the microconstructs by generating a closely spaced rectangular array of cell-laden GelMA hydrogels (Figure 14). We again used 3T3-fibroblasts embedded in 5% (w/v) GelMA patterned into 50 μm wide microconstructs that were spaced 200 μm apart. After four days, the microconstructs had merged in some sections along the channel (Figure 14A, red arrows). The merging continued until day seven when the constructs had completely converged into a larger macroscale tissue construct (~ 1cm^2) (Figure 14B). F-actin staining of the macroscale tissue revealed a fiber network that was aligned in one direction along the major axis of the micro-patterns and extended across the voids between the micropillars once the constructs fully converged (Figure 14C).

Since a major goal of this research program is the development of portable devices that can monitor toxicity, we applied the ability to form a 3D microengineered hydrogel for the development of a cell-based biosensor capable of detecting cardiotoxicity using lens-free imaging [31]. We have developed a cell-based biosensor that utilizes cardiomyocytes that are imaged using a lens-free imaging technique with a complementary metal oxide semi-conductor (CMOS) imaging module extracted from a standard webcam. The cardiomyocytes are derived from mouse embryonic stem cells (ESCs) and tested for the changes in beating rates

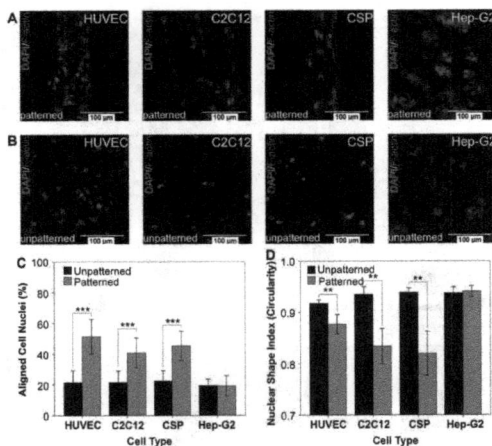

Figure 13. Cell elongation and alignment in multiple cell types. Patterning cell-laden 5% GelMA hydrogels into rectangular microconstructs (50 μm (w) x 800 μm (l) x 150 μm (h)) induced cellular alignment and elongation after 5 days of culture only in cell types possessing an intrinsic potential to organize into aligned tissues in vivo, while cellular orientation in unpatterned parts of the same hydrogels remained random. Representative z-stack overlays of DAPI/F-actin staining of (A) patterned and (B) unpatterned hydrogels laden with HUVEC, C2C12, CSP, and Hep-G2 cells respectively show patterning induced aligned and elongated cell-network formations in HUVEC, C2C12 and CSP cells and patterning independent cell-cluster formations in Hep-G2 cells. (C) Mean % of aligned cell nuclei (within 10° of preferred orientation) shows patterned HUVEC-, C2C12- and CSP-laden constructs were significantly more aligned than unpatterned controls, while in Hep-G2-laden constructs patterning failed to induce cell alignment. (D) Mean nuclear shape index (circularity = 4*π*area/ perimeter^2) similarly shows patterned HUVEC-, C2C12- and CSP-laden constructs were significantly more elongated than unpatterned controls, while in Hep-G2-laden constructs patterning failed to induce cell elongation. (Error bars: ± SD; **p < 0.01; ***p < 0.001)

and beat-to-beat variations upon exposure to known cardiotoxic compounds, isoprenaline and doxorubicin. Figure 15 shows a schematic of the experimental setup with the CMOS module positioned below the cardiomyocytes cultured in a commercially available chamber slide while the white LED light source, controlled by a pinhole, is positioned over the cell culture. The beating responses are monitored in real-time via an imaging processing program. Initial characterization of the lens-free system prior to exposing the cardiomyocytes to any drugs involved the cells being cultured in standard medium and the beating rates monitored continually for 150 minutes (Figure 16). An average of 100 beats per minute was observed and was consistent with literature reports [35, 36]. Only a 3% variation was measured in the beating rates and indicated the cells displayed normal beating patterns.

In order to measure the chronotropic effects of a drug on our cardiomyocytes-based biosensor, we selected two drugs with opposing effects: isoprenaline is a sympathomimetic β-adrenergic agonist drug that can cause an abnormally increased heart rate (tachycardia) and doxorubicin is an anti-cancer drug with side effects that can lead to an abnormally decreased heart rate (bradycardia). In the case of isoprenaline, different concentrations (1, 5, 10, and 100 nM) were dissolved in culture medium and added to the chamber slide. Measurements were collected consecutively for 20s over a period of 1 min. Figure 17A shows the increases in relative beating

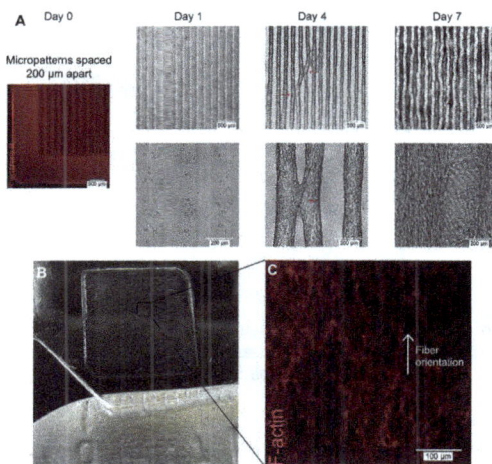

Figure 14. Self-assembly of multiple aligned microconstructs into a macroscale and aligned 3D tissue construct. 3T3-fibroblast-laden 5% GelMA hydrogels patterned into rectangular microconstructs (50 μm (w) x 800 μm (l) x 150 μm (h)) spaced 200 μm apart self-assembled into macroscale and aligned 3D tissue constructs after 7 days of culture through convergence of multiple, aligned microconstructs. (A) Rhodamine B stained hydrogel shows initial microconstruct spacing of 200 μm at day 0; representative phase contrast images of cell-laden microconstructs at days 1, 4 and 7, respectively, show focal points of contact between neighboring aligned microconstructs at day 4 (red arrows) and formation of a macroscale 3D tissue construct at day 7. (B) Image of a 1 cm x 1 cm, self-assembled 3D tissue construct at day 7. (C) Representative F-Actin staining of the middle xy-plane of a macroscale 3D tissue construct shows oriented actin fiber organization in a single direction.

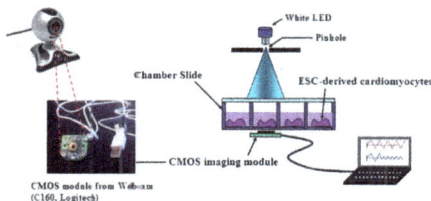

Figure 15. Schematic of the experimental setup. ESC-derived cardiomyocytes were plated on a chamber slide. The white LED and pinhole combination was used as a light source for imaging the beating cardiomyocytes. The real-time beating rates of the cardiomyocytes were measured using the CMOS imaging module which was controlled by an image processing program. Reproduced by permission of The Royal Society of Chemistry.

rates corresponding to the increase in isoprenaline concentration. After introduction of the isoprenaline, the time interval between the beats became shorter and a new pulse pattern was observed that indicated an increase in the beating rate (Figure 17B). Following a 12 min exposure, the beating rates as a function of isoprenaline concentration were compared (Figure 17C). The beating rates increased 24% and 44% for 1 and 5 nM isoprenaline, respectively, and 70% for both 10 and 100 nM. This resulted in a statistical difference between 1, 5 and 10 nM, but not between 10 and 100 nM. The patterns we observed are consistent with reported studies.

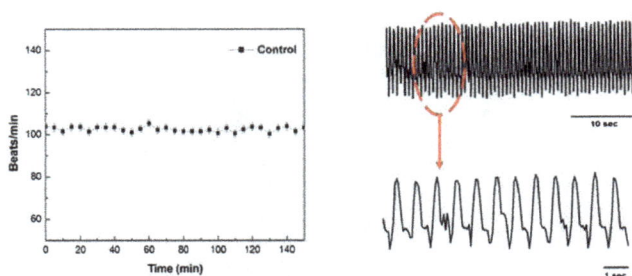

Figure 16. Characterization of the lensfree cardiotoxicity detector. Prior to injection of the drugs, the ESC-derived cardiomyocytes were cultured in normal medium and their beating rates were observed for 150 min. The beating rates were stable and the beating signal had a regular pulse pattern. Reproduced by permission of The Royal Society of Chemistry.

[36, 37]. An increasing isoprenaline concentration resulted in an increase in the beat-to-beat variation (Figure 17D).

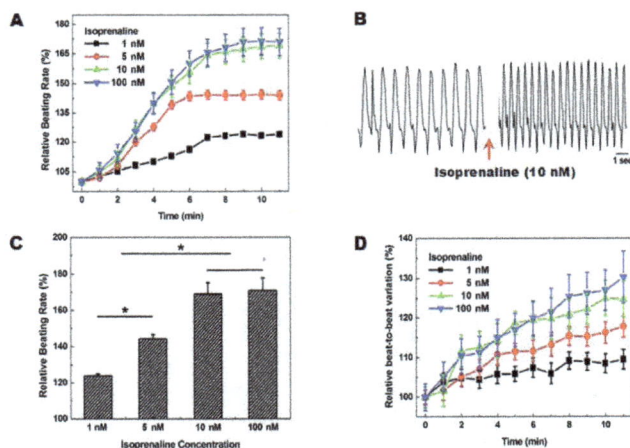

Figure 17. Chronotropic effects of cardiomyocytes under treatment of different concentrations of isoprenaline (1, 5, 10 and 100 nM). (A) The change in beating rate. The beating rates increased after isoprenaline injection. (B) A plot of the image difference value. A pulse pattern was observed. The decreased beating intervals were observed clearly after isoprenaline was injected. (C) The change in beating rates over time during the first 12 min after treatment of isoprenaline. (D) The beat-to-beat variations. Beat-to-beat variations increased under treatment of isoprenaline. * shows a significant difference in variance ($p < 0.05$). For statistical analysis, one-way ANOVA was used. Reproduced by permission of The Royal Society of Chemistry.

As expected for doxorubicin, exposure to the drug slowed beating rates in a time dependent manner (Figure 18A). Again, the drug was dissolved in culture medium at various concentrations (10, 100, 200 and 300 μM) and added to the chamber slide. The measurements were collected consecutively for 30s for a period of 5 min. The higher concentrations (200 and 300 μM) of doxorubicin resulted in a rapid decrease in the beating rate, and the beating stopped after 20 min for the highest concentration. The effects were much slower for the lower concentrations, however, a decrease in beating rate was still observed. The pulse pattern slowed as the interval between the spikes became longer (Figure 18B) indicating a decrease in the beating rates. A change in amplitude was also apparent, however, our imaging processing program did not contain an amplitude detection algorithm, and the apparent decrease could be contributed to a small difference between the reference and the live frame images. At 100 min, the beat rates as a function of doxorubicin concentration were compared and are shown in Figure 18C. A statistical difference was observed at 100 and 200 μM doxorubicin with the beating rates decreasing 26% and 54%, respectively. We also measured the beat-to-beat variations induced by the different doxorubicin concentrations (Figure 18D). The response time of the beat-to-beat variations was slower than observed for the change in beating rates. These findings are consistent with electrical measurements from a previous study [38]. Following injection of the doxorubicin, the beating of the ESC-derived cardiomyocytes became more irregular than was observed in the presence of isoprenaline.

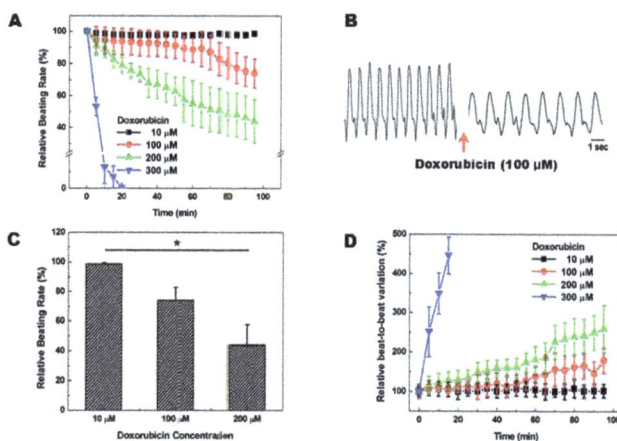

Figure 18. Chronotropic effects of cardiomyocytes under treatment with doxorubicin (10, 100, 200 and 300 μM). (A) The change in beating rate. The beating rates decreased after doxorubicin injection. (B) The increased beating intervals were clearly observed after doxorubicin injection. (C) The change in beating rates at 100 min after the treatment with doxorubicin. (D) The beat-to-beat variations. In contrast to beating rates, the beat-to-beat variations increased after treatment with doxorubicin. * shows a significant difference in variance (p < 0.05). For statistical analysis, one-way ANOVA was used. Reproduced by permission of The Royal Society of Chemistry.

The development of a cell-based biosensor using an off-the-shelf CMOS system represents a low-cost, relatively straight forward system that can be useful in monitoring the physiological response of cells to a drug or contaminant. We envision the expansion of this platform to harbor complex matrices of cells that recapitulate major organ systems and respond to toxicity via different modes of action.

Our collaborative efforts with Dr. Robert Langer at the Massachusetts Institute of Technology have yielded the development of sacrificial melt-spun interconnected microfibers as an artificial vascular system [39, 40]. The system is based on the initial experiments by Bellan, *et al.* that used standard sugar-based cotton candy to create an interconnected 3D microchannel network inside several materials, including polydimethylsiloxane (PDMS), epoxy, and polycaprolactone (PCL) [39]. Briefly, the microchannel network of cotton candy (attached to larger sticks of sugar in order to produce macrochannel interfaces) was embedded within the material of choice, which was then allowed to solidify. The embedded sugar fibers were dissolved by submerging the structure in a warm water bath, leaving behind a 3D microchannel network in the shape of the cotton candy. The resulting device supported flow of a solution of either 2 μm polystyrene spheres or heparinized whole rat blood. The channel sizes and spacing between channels were very similar to the relevant parameters for natural capillary beds thus providing an excellent means to replicate the cardiovascular network.

We next evolved the microvasculature network into a more complex 3D structure with the integration of larger, lithographically patterned microchannels [40]. This 3D structure is composed of a melt-spun sugar microfiber network "sandwiched" between a top and bottom lithographically patterned microfluidic system, and the fabrication process is detailed in Figure 19. A comparison of the filling ability was made between our hybrid device and a device with the lithographically patterned channel system connected by simple, vertical pipes. Both were filled with a fluorescent dye solution to visualize the extent of delivery into the device volume (Figure 20). Fluorescent dye was carried only to localized regions within the device lacking the 3D microfluidic network (Figure 20B). However, fluorescence was observed throughout the device volume for the hybrid device containing the microfluidic network (Figures 20C, 20D, and 20E). This indicated that the microfluidic network was well distributed throughout the device and thus greatly increased the channel surface area. We achieved an approximate five-fold increase in total volume addressed with the inclusion of the microfluidic network between the larger, lithographically patterned channel systems. Consequently, with this increased surface area comes an order of magnitude increase in the measured flow resistance ranging from 12-27 psi min mL^{-1} compared to the device lacking the microfluidic network (~0.5 psi min mL^{-1}). The drastic difference can be explained by comparing the cross-sectional area of the smallest channels of the two devices. The conventional device contains channels with the smallest cross-section of 100 μm x 50 μm, while hybrid channels are much smaller at approximately 10 μm, which is similar to a natural capillary. We also observed some device-to-device variations in the flow patterns due to the differences in the volume of sacrificial material, the fiber density, and the interfaces between the fiber network and the larger lithographic pattern. Nevertheless, a hybrid device with a combination of conventional planar microfluidics and a 3D microchannel network may provide the best option to deliver the highest volume of nutrients and oxygen to an embedded cell population while maintaining the desired channel density and interconnectivity.

We envision a hybrid device such as this containing a capillary bed-like vasculature to be a valuable tool for tissue engineering applications involving cell embedded scaffolds or biomaterials as a means to exchange oxygen and nutrients, remove waste products, and

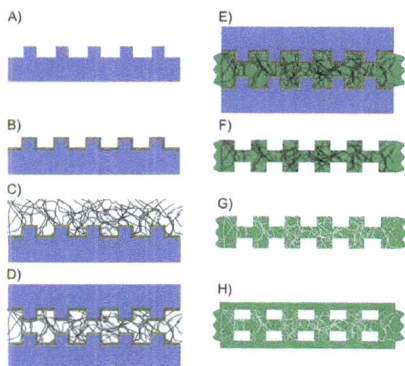

Figure 19. Illustration of hybrid device assembly process. A) A conapoxy mold is patterned from a silicone master. B) The conapoxy mold is coated with a thin layer of sugar. C) A melt-spun sugar microfiber network is adhered to the sugar layer on the conapoxy mold. D) A second sugar-coated conapoxy mold is placed on top of the microfiber network. E) The space between the two conapoxy molds is infiltrated with uncured PDMS. F) When the PDMS has cured, the conapoxy molds are removed. G) The device is placed in a water bath to remove all sugar structures. H) The top and bottom of the channel system are sealed using flat slabs of PDMS.

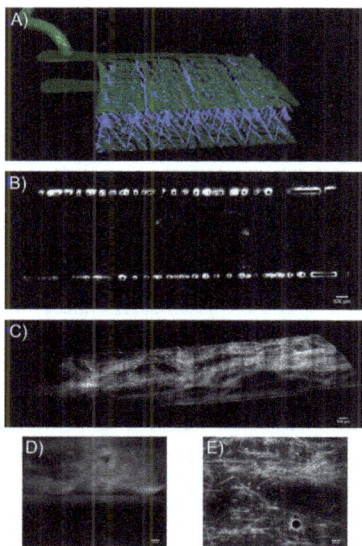

Figure 20. A) Illustration of hybrid device architecture, showing two lithographically patterned channel systems (green) connected by a 3D microfluidic network (blue) formed with sacrificial microfibers. B-D) Fluorescence microscopy images of cross-sections of devices that have been exposed to Rhodamine B dye for 45 minutes. B) A conventional two-layer microfluidic device. C) A hybrid microfluidic device containing both conventionally patterned channels and a microfiber-formed 3D channel network. D,E) Higher magnification images of a hybrid device.

reproduce the *in vivo* flow properties to achieve a physiologically realistic organ mimic. This method is rapid, inexpensive, easily scalable, and requires no toxic materials and therefore enables significant advancements in biomaterials and cell-based biosensor development. However, to produce fully biomimetic scaffolds, it must eventually be adapted for use with hydrogels and we have already made significant progress in this arena.

In collaboration with Dr. Michael Shuler's laboratory at Cornell University, we have developed a micro cell culture analog (μCCA) with 3D hydrogel-cell cultures that reproduces multi-organ interactions and enables chemical toxicity studies in a more physiologically relevant environment [41, 42]. Specific organ cell-hydrogel cultures (liver, marrow, or tumor) are maintained as discrete populations but are connected by channels mimicking blood flow within the device. The fluidic pattern of channels is precisely calculated so that upon fabrication of the chip, the resulting fluid flow mimics the properties of circulating blood. From previous studies, the μCCA was used to observe the physiological effects of naphthalene, doxorubicin, and Tegafur, which all have metabolism-dependent toxicity effects in the body [7, 41, 43]. This microfluidic network enables an accurate physical representation of pharmacokinetic-pharmacodynamic (PK-PD) models for PK profile studies of a compound (*e.g.*, toxicant, pharmaceutical, drug candidate, etc.) [44]. The body's organ systems are accurately represented as individual compartments connected by a "vascular system" of networked fluidic channels. This physical tool allows for the testing of the PK mathematical modeling that predicts the concentration and metabolite profiles of a drug from a given dose [45]. In contrast to PK, the PD of a compound relates to the pharmacological effect of that substance at a given concentration. The time course of pharmacological effects of a substance (at one concentration) can be predicted by combining the PK and PD models [46]. Our advancements of the μCCA have integrated a microfluidic system and a PK-PD modeling approach into an *in vitro/in silico* system that allows for the comparison of physical effects and mathematical predictions in order to generate a more realistic prediction of the toxicity or efficacy of a drug compound (Figure 21) [47].

The multilayered μCCA device is assembled between an aluminum frame and a plexiglass top that seals the PDMS layer that contain the fluidic channel features to a silicon gasket (0.2 mm thick) with cell culture reservoirs (Figure 21A). A polycarbonate base and an additional silicone gasket (0.5 mm thick) are also included to improve sealing of the device. The PDMS fluidic channels mimic the distributed blood flow volume observed *in vivo* for the organ systems represented on the μCCA device: liver (58%), tumor (18%) and marrow (24%) as determined by the PBPK model (Figure 21B) [41, 48]. The cell lines chosen are HepG2-C3A (liver), HCT-116 (colon cancer tumor), myeloblasts Kasumi-1 (marrow), and each population is encapsulated its own hydrogel matrix (alginate or Matrigel™). Figure 21C shows the assembled device.

As an alternative to a peristaltic pump-induced recirculation [41], we adapted the μCCA system to be a pumpless device that functions using gravity-induced flow (Figure 22A) [42]. The pumpless operation eliminates troublesome bubbles that can form and get trapped within the device channels due to use of a peristaltic pump. As shown in Figure 22B, the flow rate is linearly proportional to the height difference. However, the finite size of the reservoir likely explains why the line does not extrapolate to zero. In assembled, rocking μCCA devices, cell survivability was maintained for at least three days without replacement of the medium, but

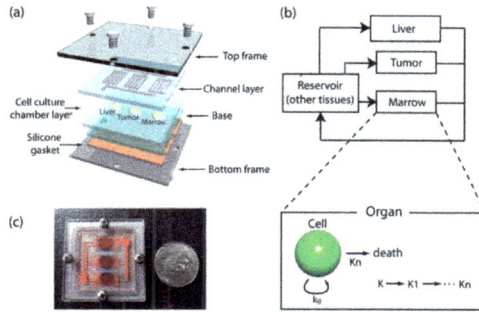

Figure 21. (a) A schematic of device components. A fluidic channel layer and a cell culture chamber layer are superimposed and sealed by top and bottom frames. A silicone gasket and a polycarbonate base are inserted for sealing. (b) A corresponding PBPK model, with the liver, tumor and marrow compartment. Below is a PD model for cell death in each compartment. Although not drawn explicitly, a PD model for each compartment exists separately, and the 'organ' can be the liver, tumor or marrow. (c) A picture of the assembled device. A red dye was used for visualization of channels, and a blue dye was mixed with the alginate. Reproduced by permission of The Royal Society of Chemistry.

viability was extended for a longer time (5 days) with regular changes of medium. For testing, the device was placed on a rocking platform with a directional change occurring every three minutes. The desired concentrations of the drug compound tested, 5-fluorouracil (5-FU) with/without uracil, was added to the cell culture medium, and after the determined exposure time, the device was disassembled for viability staining of the cells (details given in [42]).

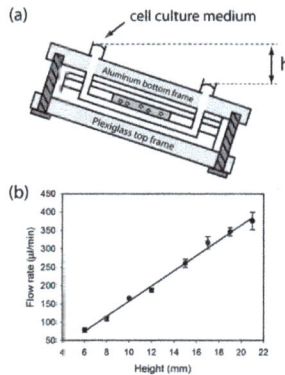

Figure 22. (a) Medium recirculation with gravity-induced flow in a mCCA. Tilting of the device causes liquid to flow from one well to the other well. In about 3 min, the rocking platform changes the angle and medium flows in the opposite direction. (b) A plot of measured flow rates against various heights. Reproduced by permission of The Royal Society of Chemistry.

The chosen cells (liver, tumor and marrow), encapsulated in a hydrogel matrix (alginate or Matrigel), were tested following a three day exposure to 5-FU (0.1 mM) in the presence or absence of uracil (0.5 mM) (Figure 23). The metabolism of 5-FU, a chemotherapeutic, is inhibited by uracil due to competition for the enzyme (DPD) that primarily metabolizes 5-FU [49, 50]. The addition of uracil allows more 5-FU to remain, and since 5-FU is toxic to cells over time, this combination should prove more toxic to the cell populations. In the case of cells encapsulated in alginate, the viabilities were not statistically different (Figure 23A). However, for the Matrigel-encapsulated cells, the tumor and marrow cells showed a significant decrease in viability with the addition of uracil (Figure 23B). Although both hydrogels are biologically compatible, they likely have different effects on the metabolic activity of the embedded cells. Cells do not readily attach to and remodel alginate because it is a negatively charged polymer, lacking functional groups that facilitate cell signaling and functioning [51]. Matrigel, a protein mixture secreted by Engelbreth-Holm-Swarm (EHS) mouse sarcoma cells, may be more suitable for eukaryotic cells to maintain their functionality and thus their toxicity response, as reported for hepatocytes [52, 53]. This may explain the viability differences we observed between the hydrogels for the cell lines tested, and this remains an area of active research.

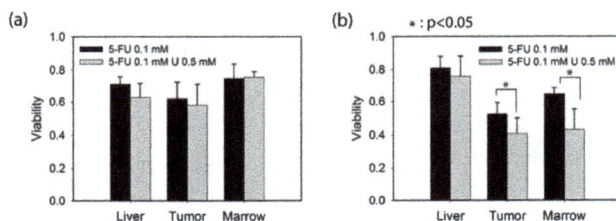

Figure 23. (a) Viability of three cell lines after 3-day treatment with 5-FU alone or 5-FU plus uracil (U) in a µCCA, encapsulated in 2% alginate. (b) Viability of three cell lines after 3-day treatment with 5-FU alone or 5-FU plus uracil (U). Cells were encapsulated in Matrigel™ in a µCCA. Reproduced by permission of The Royal Society of Chemistry.

The simplification of this rocking µCCA system lessens the requirement for specialized techniques for assembly and operation often an issue with microfluidics. We have preserved the ability to perform pharmacokinetic-based toxicity testing on multiple cell types as 3D cell-hydrogel cultures. Further, our research indicates that the physical assembly of the biosensor can be performed without significant cell damage. This supports the use of this type of cell-based biosensor for the use in toxicity monitoring applications.

4. Conclusions

We believe that the work detailed in the above sections provides key technologies and tools that bring us closer to the realization of a total "body on a chip" for applications ranging from pharmaceutical testing of drug candidates to fieldable devices that are capable of physiologically accurate toxicity responses. Additionally, the technologies developed are pursuant toward increasing the cell lifetime and survivability in the biosensor constructs. Since our focus is on military applications, a robust toxicity microanalysis system has innumerable potential uses for the soldier including toxicity devices for water resources in harsh environments, monitoring at military installations to ensure environmental sustainability, rapid screening of drinking water, and enhanced decision processing for military bases.

Acknowledgements

We would like to thank all of our collaborating academic laboratories who participated in the "In Silico Biomimetic Sensor for Rapid Testing of Water Resources" 6.2 Research Program: those represented in this chapter include Dr. Scott Banta, Columbia University; Dr. Charles Henry, Colorado State University; Dr. Ali Khademhosseini, Harvard University; Dr. Robert Langer, Massachusetts Institute of Technology; Dr. Michael Shuler, Cornell University; as well as those equally important but not described here including Dr. Paul Bohn, University of Notre Dame; Dr. William van der Schalie, U.S. Army Center for Environmental Health Research; and Dr. Yingxiao Wang, University of Illinois at Urbana-Champaign. Additional thanks goes to numerous post-doctoral researchers from those laboratories who also participated in this program. This ERDC-CERL program was supported by funds from the Army Direct Research Program.

Author details

Jill M. Grimme[1] and Donald M. Cropek[1,2*]

*Address all correspondence to: Donald.M.Cropek@usace.army.mil

1 U.S. Army Corps of Engineers, Engineer Research and Development Center – Construction Engineering Research Laboratory, Champaign, IL, USA

2 Department of Veterinary Biosciences, University of Illinois at Urbana-Champaign, Urbana, IL, USA

References

[1] Banerjee, P., Franz, B., and Bhunia, A. K. (2010) Mammalian cell-based sensor system, *Adv Biochem Eng Biotechnol*, *117*, 21-55.

[2] Wang, P. (2010) Introduction, In *Cell-Based Biosensors Principles and Applications* (Wang, P., and Liu, Q., Eds.), pp 1-12, Artech House, Norwood, MA.

[3] Sweeney, L. M., Shuler, M. L., Babish, J. G., and Ghanem, A. (1995) A cell culture ana-
     logue of rodent physiology: Application to naphthalene toxicology, *Toxicology in Vi-
     tro, 9,* 307-316.

[4] Baxter, G. T., and Freedman, R. (2004) A Dynamic In Vivo Surrogate Assay Platform
     for Cell-Based Studies, *Am Biotechnol Lab,* 1-2.

[5] Sin, A., Chin, K. C., Jamil, M. F., Kostov, Y., Rao, G., and Shuler, M. L. (2004) The de-
     sign and fabrication of three-chamber microscale cell culture analog devices with in-
     tegrated dissolved oxygen sensors, *Biotechnol Prog, 20,* 338-345.

[6] Esch, M. B., Sung, J. H., and Shuler, M. L. (2010) Promises, challenges and future di-
     rections of microCCAs, *J Biotechnol, 148,* 64-69.

[7] Viravaidya, K., Sin, A., and Shuler, M. L. (2004) Development of a microscale cell cul-
     ture analog to probe naphthalene toxicity, *Biotechnol Prog, 20,* 316-323.

[8] Viravaidya, K., and Shuler, M. L. (2004) Incorporation of 3T3-L1 cells to mimic bioac-
     cumulation in a microscale cell culture analog device for toxicity studies, *Biotechnol
     Prog, 20,* 590-597.

[9] McAuliffe, G. J., Chang, J. Y., Glahn, R. P., and Shuler, M. L. (2008) Development of a
     gastrointestinal tract microscale cell culture analog to predict drug transport, *Mol Cell
     Biomech, 5,* 119-132.

[10] Mahler, G. J., Esch, M. B., Glahn, R. P., and Shuler, M. L. (2009) Characterization of a
     gastrointestinal tract microscale cell culture analog used to predict drug toxicity, *Bio-
     technol Bioeng, 104,* 193-205.

[11]

[12] Mahler, G. J., Shuler, M. L., and Glahn, R. P. (2009) Characterization of Caco-2 and
     HT29-MTX cocultures in an in vitro digestion/cell culture model used to predict iron
     bioavailability, *J Nutr Biochem, 20,* 494-502.

[13] Wu, J., Park, J. P., Dooley, K., Cropek, D. M., West, A. C., and Banta, S. (2011) Rapid
     development of new protein biosensors utilizing peptides obtained via phage dis-
     play, *PLoS One, 6,* e24948. http://www.plosone.org/article/fetchArticle.action?arti-
     cleURI=info:doi/10.1371/journal.pone.0024948

[14] Wu, J., Cropek, D. M., West, A. C., and Banta, S. (2010) Development of a troponin I
     biosensor using a peptide obtained through phage display, *Anal Chem, 82,* 8235-8243.

[15] Park, J. P., Cropek, D. M., and Banta, S. (2010) High affinity peptides for the recogni-
     tion of the heart disease biomarker troponin I identified using phage display, *Biotech-
     nol Bioeng, 105,* 678-686.

[16] Pumera, M., Merkoci, A., and Alegret, S. (2007) Carbon nanotube detectors for micro-
     chip CE: comparative study of single-wall and multiwall carbon nanotube, and

graphite powder films on glassy carbon, gold, and platinum electrode surfaces, *Electrophoresis, 28,* 1274-1280.

[17] Shiddiky, M. J., Won, M. S., and Shim, Y. B. (2006) Simultaneous analysis of nitrate and nitrite in a microfluidic device with a Cu-complex-modified electrode, *Electrophoresis, 27,* 4545-4554.

[18] Tang, D. Y., and Xia, B. Y. (2008) Electrochemical immunosensor and biochemical analysis for carcinoembryonic antigen in clinical diagnosis, *Microchim Acta, 163,* 41-48.

[19] Sameenoi, Y., Mensack, MM., Boonsong, K., Ewing, R., Dungchai, W., Chailapakul, O., Cropek, DM., Henry, CS (2011) Poly(dimethylsiloxane) cross-linked carbon paste electrodes for microfluidic electrochemical sensing, *Analyst, 136,* 3177-3184. http://pubs.rsc.org/en/content/articlelanding/2011/an/c1an15335h

[20] Bayle, C., Causse, E., and Couderc, F. (2004) Determination of aminothiols in body fluids, cells, and tissues by capillary electrophoresis, *Electrophoresis, 25,* 1457-1472.

[21] Pastore, A., Federici, G., Bertini, E., and Piemonte, F. (2003) Analysis of glutathione: implication in redox and detoxification, *Clin Chim Acta, 333,* 19-39.

[22] Pereira-Rodrigues, N., Cofre, R., Zagal, J. H., and Bedioui, F. (2007) Electrocatalytic activity of cobalt phthalocyanine CoPc adsorbed on a graphite electrode for the oxidation of reduced L-glutathione (GSH) and the reduction of its disulfide (GSSG) at physiological pH, *Bioelectrochemistry, 70,* 147-154.

[23] Korfhage, K. M., Ravichandran, K., and Baldwin, R. P. (1984) Phthalocyanine-containing chemically modified electrodes for electrochemical detection in liquid chromatography/flow injection systems, *Analytical Chemistry, 56,* 1514-1517.

[24] Halbert, M. K., and Baldwin, R. P. (1985) Electrocatalytic and analytical response of cobalt phthalocyanine containing carbon paste electrodes toward sulfhydryl compounds, *Analytical Chemistry. 57,* 591-595.

[25] Kuhnline, C. D., Gangel, M. G., Hulvey, M. K., and Martin, R. S. (2006) Detecting thiols in a microchip device using micromolded carbon ink electrodes modified with cobalt phthalocyanine, *Analyst, 131,* 202-207.

[26] Du, M., Flanigan, V., and Ma, Y. (2004) Simultaneous determination of polyamines and catecholamines in PC-12 tumor cell extracts by capillary electrophoresis with laser-induced fluorescence detection, *Electrophoresis, 25,* 1496-1502.

[27] Westerink, R. H. (2004) Exocytosis: using amperometry to study presynaptic mechanisms of neurotoxicity, *Neurotoxicology, 25,* 461-470.

[28] Chen, T. K., Luo, G., and Ewing, A. G. (1994) Amperometric monitoring of stimulated catecholamine release from rat pheochromocytoma (PC12) cells at the zeptomole level, *Anal Chem, 65,* 3031-3035.

[29] Kozminski, K. D., Gutman, D. A., Davila, V., Sulzer, D., and Ewing, A. G. (1998) Voltammetric and pharmacological characterization of dopamine release from single exocytotic events at rat pheochromocytoma (PC12) cells, *Anal Chem, 70*, 3123-3130.

[30] Li, M. W., Spence, D. M., and Martin, R. S. (2005) A microchip-based system for immobilizing PC 12 cells and amperometrically detecting catecholamines released after stimulation with calcium, *Electroanalysis, 17*, 1171-1180.

[31] Aubin, H., Nichol, J. W., Hutson, C. B., Bae, H., Sieminski, A. L., Cropek, D. M., Akhyari, P., and Khademhosseini, A. (2010) Directed 3D cell alignment and elongation in microengineered hydrogels, *Biomaterials, 31*, 6941-6951. http://www.sciencedirect.com/science/article/pii/S0142961210006939

[32] Kim, S. B., Bae, H., Cha, J. M., Moon, S. J., Dokmeci, M. R., Cropek, D. M., and Khademhosseini, A. (2011) A cell-based biosensor for real-time detection of cardiotoxicity using lensfree imaging, *Lab Chip, 11*, 1801-1807. http://pubs.rsc.org/en/content/articlelanding/2011/lc/c1lc20098d

[33] Charest, J. L., Eliason, M. T., Garcia, A. J., and King, W. P. (2006) Combined microscale mechanical topography and chemical patterns on polymer cell culture substrates, *Biomaterials, 27*, 2487-2494.

[34] Brammer, K. S., Oh, S., Cobb, C. J., Bjursten, L. M., van der Heyde, H., and Jin, S. (2009) Improved bone-forming functionality on diameter-controlled TiO(2) nanotube surface, *Acta Biomater, 5*, 3215-3223.

[35] Charest, J. L., Garcia, A. J., and King, W. P. (2007) Myoblast alignment and differentiation on cell culture substrates with microscale topography and model chemistries, *Biomaterials, 28*, 2202-2210.

[36] Boheler, K. R., Czyz, J., Tweedie, D., Yang, H. T., Anisimov, S. V., and Wobus, A. M. (2002) Differentiation of pluripotent embryonic stem cells into cardiomyocytes, *Circ Res, 91*, 189-201.

[37] Stummann, T. C., Wronski, M., Sobanski, T., Kumpfmueller, B., Hareng, L., Bremer, S., and Whelan, M. P. (2008) Digital movie analysis for quantification of beating frequencies, chronotropic effects, and beating areas in cardiomyocyte cultures, *Assay Drug Dev Technol, 6*, 375-385.

[38] Vandecasteele, G., Eschenhagen, T., Scholz, H., Stein, B., Verde, I., and Fischmeister, R. (1999) Muscarinic and beta-adrenergic regulation of heart rate, force of contraction and calcium current is preserved in mice lacking endothelial nitric oxide synthase, *Nat Med, 5*, 331-334.

[39] Xiao, L., Hu, Z., Zhang, W., Wu, C., Yu, H., and Wang, P. (2010) Evaluation of doxorubicin toxicity on cardiomyocytes using a dual functional extracellular biochip, *Biosens Bioelectron, 26*, 1493-1499.

[40] Bellan, L. M., Singh, S. P., Henderson, P. W., Porri, T. J., Craighead, H. G., and Spector, J. A. (2009) Fabrication of an artificial 3-dimensional vascular network using sacrificial sugar structures, *Soft Matter*, 5, 1354-1357.

[41] Bellan, L. M., Kniazeva, T., Kim, E. S., Epshteyn, A. A., Cropek, D. M., Langer, R., and Borenstein, J. T. (2012) Fabrication of a hybrid microfluidic system incorporating both lithographically patterned microchannels and a 3D fiber-formed microfluidic network, *Adv Healthc Mater*, 1, 164-167. http://onlinelibrary.wiley.com/doi/10.1002/adhm.201290009/abstract

[42] Sung, J. H., and Shuler, M. L. (2009) A micro cell culture analog (microCCA) with 3-D hydrogel culture of multiple cell lines to assess metabolism-dependent cytotoxicity of anti-cancer drugs, *Lab Chip*, 9, 1385-1394.

[43] Sung, J. H., Kam, C., and Shuler, M. L. (2010) A microfluidic device for a pharmacokinetic-pharmacodynamic (PK-PD) model on a chip, *Lab Chip*, 10, 446-455. http://pubs.rsc.org/en/content/articlelanding/2010/lc/b917763a

[44] Tatosian, D. A., Shuler, M. L., and Kim, D. (2005) Portable in situ fluorescence cytometry of microscale cell-based assays, *Opt Lett*, 30, 1689-1691.

[45] Gerlowski, L. E., and Jain, R. K. (1983) Physiologically based pharmacokinetic modeling: principles and applications., *J Pharm Sci*, 72, 1103-1127.

[46] Ghanem, A., and Shuler, M. L. (2000) Combining cell culture analogue reactor designs and PBPK models to probe mechanisms of naphthalene toxicity, *Biotechnol Prog*, 16, 334-345.

[47] Derendorf, H., and Meibohm, B. (1999) Modeling of pharmacokinetic/pharmacodynamic (PK/PD) relationships: concepts and perspectives, *Pharm Res*, 16, 176-185.

[48] Sung, J. H., and Shuler, M. L. (2009) In *Methods in Bioengineering: Microdevices in Biology and Medicine* (Nahmias, Y., and Bhatia, S., Eds.), Artech House.

[49] Davies, B., and Morris, T. (1993) Physiological parameters in laboratory animals and humans, *Pharm Res*, 10, 1093-1095.

[50] Van Kuilenburg, A. B., Meinsma, R., Zoetekouw, L., and Van Gennip, A. H. (2002) Increased risk of grade IV neutropenia after administration of 5-fluorouracil due to a dihydropyrimidine dehydrogenase deficiency: high prevalence of the IVS14+1g>a mutation, *Int J Cancer*, 101, 253-258.

[51] de Bono, J., and Twelves, C. (2001) The oral fluorinated pyrimidines., *Invest New Drugs*, 19, 41-59.

[52] Lawson, M. A., Barralet, J. E., Wang, L., Shelton, R. M., and Triffitt, J. T. (2004) Adhesion and growth of bone marrow stromal cells on modified alginate hydrogels, *Tissue Eng*, 10, 1480-1491.

[53] Brandon, E. F., Raap, C. D., Meijerman, I., Beijnen, J. H., and Schellens, J. H. (2003) An update on in vitro test methods in human hepatic drug biotransformation research: pros and cons, *Toxicol Appl Pharmacol, 189*, 233-246.

[54] Castell, J. V., and Gomez-Lechon, M. J. (2009) Liver cell culture techniques, *Methods Mol Biol, 481*, 35-46.

Chapter 2

BOD Biosensors: Application of Novel Technologies and Prospects for the Development

A. Reshetilov, V. Arlyapov, V. Alferov and
T. Reshetilova

Additional information is available at the end of the chapter

1. Introduction

Household and industrial human activities cause ever increasing pollution of water bodies of rivers, lakes, water reservoirs, seas. Express assessment of the extent of pollution by organic compounds is an important and, in some cases, essential component of ecological control. Given the constantly growing list of substances released into the environment as pollutants, it can be stated that complete chemical analysis is a complex and expensive procedure. An efficient tool of analysis proves to be methods based on an integral assessment of organic components. In this context, significant attention is given to the development of biosensor methods of control that enable an integral estimate of pollution density, considerably increase operational efficiency of the analysis and reduce its cost (D'Souza, 2001).

An essential integral characteristic of the quality of water is biochemical oxygen demand (BOD), i.e., the amount of dissolved oxygen (in mg) required to oxidize all biodegradable organic compounds that occur in (1 dm^3 of) water. The BOD assessment is an empirical test in which a standardized laboratory procedure is used to determine the oxygen demand in analyzed water samples. The BOD is determined conditionally by the change of oxygen content before and after placing a water sample into a sealed flask and holding it for a certain period of time. The standard BOD determination method assumes incubation of an oxygen-saturated sample, into which activated sludge (a mixture of various microorganisms) is introduced, for 5, 7, 10 or 20 days (BOD_5, BOD_7, BOD_{10} or BOD_{20}, respectively) at 20°C(Standard Methods..., 1992). The obtained result – the amount of consumed oxygen normalized to 1 dm^3 – characterizes the total content of biochemically oxidizable organic impurities in water, as well as its capability of self-clarification. In surface waters of most water bodies, the values of BOD_5 usually change within the range of 0.5–4 mg/dm^3 and are subject to seasonal and diurnal variations. Changes of BOD_5 values vary rather significantly depending on the extent of water body pollution. Depending on the category of a water body, the value of BOD_5 is regulated as follows: it should be no more than 3 mg/dm^3 for water bodies of household water use and no more than 6 mg/dm^3 for water bodies of social-amenity and recreational water use. For seas (categories I and II of fish-husbandry water utilization) the BOD_5 at 20°C should not exceed 2 mg/dm^3. The BOD test is also widely used at wastewater treatment facilities to assess the biodegradation efficiency in wastewater purification processes. The traditional BOD test has certain advantages, is a universal means of assaying most samples of waste waters and water bodies; besides, it requires no expensive equipment. However, the test has serious limitations with respect to analysis time. This traditional technique is largely devalued by its low responsiveness. It can provoke ecologically hazardous situations, when the inflow of accidentally polluted waters to the treatment facilities or their incomplete purification in the regeneration process goes unnoticed.

Operational analysis is made possible by developing BOD assessment methods based on the use of biosensor analyzers. The biosensor is an integrated device capable of providing quantitative and semi-quantitative analytical information using a biological recognition element that is in close contact with the transducer. BOD biosensor R&D has been underway since 1970s (Karube et al., 1997b; Hikuma et al., 1979), and these systems continue to be actively developed at present (Rodriguez-Mozaz et al., 2006). It should be noted that biosensors enable a rapid determination of the BOD index (BOD_{bs}). However, BOD_{bs} is not always identical to the value of the traditional BOD_5. There is a simple explanation of this effect. The receptor element of the biosensor may contain one or several cultures. The culture(s) may have a rather broad substrate specificity, which will undoubtedly be less broad than in the cultures of activated sludge used in the standard BOD_5 method. Therefore, the oxidation of organic compounds by the culture(s) occurring in the receptor element will always be lower than by activated sludge cultures.

Recently, novel approaches to the biosensor analysis of the BOD have started to be developed; these approaches make it possible to achieve an acceptable fit of the data obtained by biosensor measurements with those determined by traditional methods and to approach the solution of many applied issues. Thus, methods of determining the BOD in water bodies of rivers are being developed, which is of extreme importance along with the assessment of the BOD in waste waters (Chee, 2011). A highly efficient approach is to couple two procedures – cleanup of polluted wastewaters from organic impurities and production of electric energy by using biofuel cells based on microbial cells (Deng et al., 2012; Du et al., 2007). Herewith, it should be noted that the general tendency of the search is to increase the correlation between the data measured by the biosensor and those determined by traditional methods. The correlation of the data obtained using a biosensor analyzer with those by the BOD_5 method can be of the order of 0.95–0.98 (Bourgeois et al., 2001). Calibration of the biosensor for BOD measurements is made using synthetic waste waters, or the biorecognition element of the BOD biosensor is made based on microorganisms capable of efficient oxidation of particular waste waters. Thus, it is expedient to develop biosensors, choose microorganisms and calibration solutions that would provide for the most efficient detection of the BOD in accordance with the particular type of waste waters, i.e., to develop specialized BOD biosensors.

The popularity of and demand for R&D of BOD-determination biosensor systems logically resulted in commercialization and industrial production of a number of models. Nevertheless, BOD-biosensor systems still have a number of limitations that impede their use. These include drawbacks in the standardization procedure, imperfections of legislation in most countries, complicated service requirements and insufficient stability of used microbial cultures with respect to heavy metals and various toxic substances (Rodriguez-Mozaz et al., 2006).

Reviews on microbial sensors (D'Souza, 2001; Liu & Mattiasson, 2002; Lei et al., 2006; Xu & Ying, 2011; Ponomareva et al., 2011), as well on the use of biosensors for analyzing objects of the environment and for ecological monitoring (Rodriguez-Mozaz et al., 2006; Baeumner, 2003) give examples of BOD sensors developed. An important role of biorecognition elements based on eukaryotic microorganisms in biosensors for solving environmental problems is noted, including for determining the BOD of water bodies (Walmsley& Keenan, 2000). Detailed information on BOD sensors based on Clark-type electrode is summed up, as well as on some commercially available biosensor systems developed prior to 2000 (Liu & Mattiasson, 2002).

In this review, we generalize information about the principles of functioning, design, analytical characteristics of BOD biosensors, the properties of biorecognition elements; the functioning parameters and characteristics of various BOD sensor types are given.

2. Operating principles of BOD biosensors

2.1. Oxygen electrode-based film biosensors

Most BOD sensors described are film-type microbial sensors based on whole cells. The principle of their operation is based on the measurement of oxygen consumption by microorganisms immobilized on the surface of the transducer. In 1977, Karube et al. (1977b) published a paper, which first described a microbial sensor for BOD_{bs} determination using microorganisms taken from activated sludge of wastewater treatment facilities. A feature of such biosensors is that between the porous (most often, cellulose) membrane and the gas permeable membrane of the oxygen electrode there is a layer of microbial film that forms the biological recognition element. Part of oxygen occurring in the layer of immobilized microorganisms is consumed in oxidation of organic compounds contained in the sample. Remaining oxygen penetrates through the gas permeable teflon membrane and is reduced at the cathode of the oxygen electrode. The strength of current in the system is directly proportional to the magnitude of oxygen reduced at the electrode. After an equilibrium is established between the diffusion of oxygen to the layer of immobilized microorganisms and the endogenous respiration rate of immobilized microorganisms, the equilibrium (background) current is recorded. When a waste water sample is introduced into a measuring cuvette, organic substances of the analyzed sample are utilized by immobilized microorganisms, as the result of which the respiratory rate of the cells increases to lead to an increase of the oxygen consumption rate. In this case, a smaller amount of oxygen penetrates through the teflon membrane to be reduced. The current will decrease until a new equilibrium is established. When the buffer solution for washing the biosensor is fed to the measuring cuvette, the microbial endogenous respiratory rate is restored and the initial equilibrium of the oxygen flows in the system is re-established. As the process is controlled by the rate of substrate diffusion to the layer of immobilized cells, the sensor signal will be proportional to the concentration of readily oxidized substrates in the sample.

Two methods of biosensor-response processing are used to obtain the analytical signal: the equilibrium method (determination by the endpoint) and the kinetic method (determination by the initial rate) (Tan et al., 1993). In the equilibrium method, the BOD_{bs} is determined using the difference between currents in two equilibrium states of the biosensor – before and after substrate is introduced. The measurement time in this case is from 15 to 30 min with the subsequent rather long recovery, which can take 1 h and more. In the kinetic method, the dependence of the rate of current strength (the first derivative of current with respect to time) on time is used as a sensor response. This rate is registered after a sample is added. This parameter reflects the rate of microbial respiration and, to a certain extent, is proportional to the concentration of substrate. In this case, the sensor response is registered within 15 to 30 s, and the recovery time of the biorecognition element is less than 10 min. It should be noted that a broader range of determined BOD values can be achieved by taking the initial rate of response as the biosensor response, at an insignificant loss in reproducibility (Yang et al., 1997). Thus, the kinetic method of biosensor signal processing is more preferable in the case when the BOD index should be constantly controlled, e.g., in the course of the purification of waste waters or in the analysis of a large number of samples (Liu et al. 2000).

At present, novel biofilm BOD sensors based on the oxygen electrode are developed (Rodriguez-Mozaz et al., 2006; Bourgeois et al., 2001; Liu & Mattiasson, 2002; Baeumner, 2003). Major attention is given to the improvement of BOD sensor parameters: increase of stability, rise of correlation of the data obtained by the biosensor and standard BOD_5 assessment methods. First and foremost, this is associated with the search for new efficient microorganisms, use of modern materials and new biomaterial immobilization methods.

2.2. Oxygen electrode-based bioreactor-type biosensor systems

The BOD_{bs} is determined using bioreactor-type sensor systems with the respirometer to constantly measure the respiratory activity of microbial suspension. Strictly speaking, in accordance with the IUPAC definition, such systems are not biosensors, because the biorecognition element is not in direct contact with the transducer. However, such systems have found wide use at wastewater treatment facilities for continuous control of the extent of purification (Iranpour & Zermeno, 2008). A common feature of all respirometric BOD sensors is the presence of bioreactors, in which activated sludge (or individual microorganisms) and readily oxidized organic substances are together (Spanjers et al., 1996). Samples of waste waters are constantly transported through the flow reactor, which has a small volume (Spanjers et al., 1991, 1993). Most often, unidentified microorganisms from waste water, e.g., activated sludge, are used in such systems as a biorecognition element. To increase the reproducibility of the results, it is proposed to use individual strains of microorganisms with a broad range of oxidized substrates that belong to the genera *Trichosporon* (Sohn et al., 1995), *Rhodococcus* and *Issatchenkia* (Heim et al., 1999).

An advantage of the bioreactor configuration of the recognition elements is that the transducer in such systems is easily replaceable. This does not disturb the activity of microorganisms. Besides, the bioreactor-type BCD sensor has more stable operational characteristics as compared with the biofilm type (Prae et al., 1995). A drawback of these devices is their stationary arrangement and impossibility of field measurements. Thus, reactor-type biosensor systems have a strictly definite purpose: continuous control of waste water purification processes at respective facilities.

2.3. Mediator-type biosensors

The BOD value determined using microbial respiration can be affected by the amount of dissolved oxygen in the sample. It is known that some synthetic compounds (artificial electron acceptors) can be reduced by certain microorganisms, i.e., are artificial acceptors of electrons (Tkac et al., 2003). If these compounds possess reversible redox properties, they can serve as carriers of electrons from the biocatalytic systems of microorganisms to the electrode.

When using mediators, the results of measurements become in practice independent of the partial pressure of oxygen in the medium, and, if the oxidation of the reduced mediator does not involve protons, the mediator electrode can be relatively insensitive to pH changes. Thus, one of the most promising trends is the development of BOD biosensors using electron transport mediators (Liu & Mattiasson, 2002; Tkac et al., 2003; Yoshida et al., 2000; Trosok et al., 2001; Yoshida et al., 2001; Nakamura et al., 2007a,b; Chen et al., 2008; Liua et al., 2010). The equilibrium state of current in such systems sets in in several seconds, which provides for a higher speed of analysis. Due to the large area of the measuring electrode, appreciable currents that can be greater than currents of the oxygen electrode are generated in mediator microbial sensors (Arlyapov et al., 2008a). An essential characteristic of biosensors is the possibility of their miniaturization. Using screen-printed electrodes, it is possible to develop cheap disposable biosensors based on microbial whole cells to extend the potential of their use by a broad range of consumers (Farré & Barceló, 2001).

One more advantage of using mediators is that the BOD can be measured under anaerobic conditions, because the microbial respiratory chain enzymes are capable of regeneration owing to the reduction of artificial electron acceptors. Pasco and co-workers (2004) proposed a fast microbial technology of BOD measurements under anaerobic conditions in the presence of co-substrate potassium hexacyanoferrate(III). Addition of substrate to the measuring cuvette increases the catabolic activity of microorganisms and leads to the accumulation of the reduced form of mediator, which is successfully re-oxidized at the working electrode; the amount of electricity is measured by a coulometric transducer.

2.4. Microbial biofuel cells as BOD sensors

Carube and coworkers (1977a) developed a sensor based on the biofuel cell (BFC)for the determination of the BOD_{bs}. The current generated in the biofueld cell is the result of the biooxidation of hydrogen or is due to the formation of products from organic compounds by way of reduction under the action of the bacteria *Clostridium butyricum* under anaerobic conditions. A significant contribution to the development of BOD sensors based on mediator-free biofuel cells was made by Korean investigators (Kim et al., 2003a, b; Moon et al., 2004; Chang et al., 2004, 2005). The long-time stability of a system based on a mediator-free fuel cell was an enormous achievement in the development of BOD sensors: the sensor was operated for 5 years without any maintenance (Moon et al., 2004). At the same time, a slow response time (about 1 h) and the fixed position of the system restricts its applications.

2.5. Optical BOD biosensors

Intensive development of fibre-optic devices at the end of the last century made it possible to produce miniature optical biosensors (Hyun et al., 1993; Karube & Yokoyama, 1993; Chee et al., 2000; Sakaguchi et al., 2003; Kwok et al., 2005; Lin et al., 2006; Jiang et al., 2006; Pang et al., 2007; Sakaguchi et al., 2007). There are two approaches to the development of optical BOD sensors: to use luminescent bacteria in the biorecognition element of the sensor or to use a luminescent support for biomaterial. In the former case, the measurement principle is based on the relation between the intensity of luminescence produced by bacteria and the cell assimilation of organic compounds from waste water samples (Hyun et al., 1993; Karube & Yokoyama, 1993; Sakaguchi et al., 2003; Sakaguchi et al., 2007). In the former case, oxygen-sensitive dyes are introduced into the material of the support, and whole microbial cells are used in this matrix as a biorecognition element. Microbial respiration intensity depends on the content of organic compounds in the analyzed sample, which are oxidized by microorganisms in the presence of oxygen. A change in the content of oxygen in the film is registered by optical methods using a dye (Chee et al., 2000; Sakaguchi et al., 2003; Kwok et al., 2005; Lin et al., 2006; Jiang et al., 2006; Pang et al., 2007). Optical biosensors possess a high sensitivity and, thus, make it possible to determine low BOD values. An important advantage of such systems is that they enable micro printed circuit boards, microsensors, on-chip biosensors (Sakaguchi et al., 2003; Sakaguchi et al., 2007).

2.6. Other types of BOD biosensors

BOD determination methods using biosensors are not limited by those described above. Thus, for instance, Vaiopoulou and co-workers (2005) developed a biosensor for on-line determination of the BOD in wastewater treatment facilities. The main operating principle of the biosensor is based on the on-line measurement of the concentration of CO_2 produced in the degradation of waste waters' carbon component by microorganisms.

An unconventional approach to BOD determination is described by Tønning et al. (2005). Samples of waste water from the a Swedish cellulose company at various degrees of purification and pure water were analyzed using an amperometric biosensor with several cells and electrodes using mathematical methods of chemometry for processing the array of obtained data (the so called biosensor language). Waste water samples were characterized by the chemical consumption of oxygen, biological consumption of oxygen, total amount of organic carbon, suppression of nitrification, inhibition of respiration and toxicity with respect to *Vibrio fischeri*, freshwater marine alga *Pseudokirchneriella subcapita* and freshwater crustacean *Daphnia magna*.

Another approach to BOD detection is based on the registration of temperature changes caused by microbial destruction of organic compounds. This approach is based on the use of calorimetric transducers: a biosensor based on this transducer is described in Mattiasson et al. (1977). In recent years, this trend has not been intensively developed.

3. Biorecognition elements of BOD sensors

3.1. Microorganisms as the basis of biorecognition

To develop biorecognition elements of BOD sensors, use is made of either pure cultures with certain consumer properties (a broad range of oxidized substrates, resistance to negative environmental factors or specificity with respect to certain waste waters), or a mixture of identified microorganisms (artificial associations), or induced consortium of microorganisms, or else activated sludge and even thermally killed bacteria. Each of these approaches has its advantages and disadvantages.

Usually BOD biosensors based on a pure culture have an advantage of biosensor system stability. At the same time, these biosensors may show a decreased value of the BOD due to the limited range of substrates oxidized by one strain. Whole cells of bacteria (*Bacillus polymyxa*, *Bacillus subtilis*, *Pseudomonas putida*) or yeasts (*Arxula adeninivorans*, *Hansenula anomala*, *Klebsiella*, *Candida*, *Trichosporon, Serratia marcescens*, *Saccharomyces cerevisiae*) are known to be used as biocatalysts. The yeasts are a more preferable biomaterial for almost all types of biosensors, as they are resistant to negative environmental factors and can function in biosensor's recognition element for a long time (Seo et al., 2009; Dhall et al., 2008). At the same time, yeast cultures are more liable to contamination than bacterial cultures.

The number of oxidized substrates is increased by using microbial associations, consisting more often of two strains, e.g., *Trichosporon cutaneum* and *Bacillus licheniformis* (Suriyawattanakulet al., 2002). As most of the described BOD sensors developed with the view to improve the convergence and operational stability, this biorecognition element makes use of a mixture of two identified strains. This led to expand the substrate specificity and to stabilize the sensor operation for a long period of time. BOD sensors based on a composite microbial population, such as in activated sludge or microbial consortia, have a high ability to detect a broad range of substrates. However, due to the instability of the composition in the consortium the sensor operation often becomes unstable in the course of time.

Biosensors based on living cells require their vital activities to be constantly sustained and need nutrients and minerals in long-time storage. BOD sensors based on thermally killed cells do not have this drawback. Cells killed by temperature can be stored in phosphate buffer for a long time at room temperature (Tan & Lim, 2005; Qian & Tan, 1999; Tag et al., 2000).

3.2. Immobilization of microorganisms

The method of immobilizing microorganisms is for each biosensor an important determining procedure, as, in fact, it sets the basic parameters of BOD biosensors. Immobilization determines their lifetime, operational stability, response, sensitivity. In this context, we should note continued studies on the introduction of new modifications of immobilization techniques (Guo et al., 2008). Microbial cells on the surface of the physico-chemical transducer are retained in most cases by simple adsorption, i.e., cells are placed, for the most part, on a porous membrane by suction or retention of water by hydrogels, a polyvinyl alcohol aqueous solution (Qian & Tan, 1999) or polycarbomoyl sulphonate (Tag et al., 2000; Chan et al., 2000). For BOD sensor miniaturization, the method of crosslinking rubber (ENT-3400) under the action of UV light was used to immobilize cells on the surface of a micro-oxygen electrode (Lehmann et al., 1999). As an alternative, disposable BOD sensors can be used, in which the biofilm should be readily replaceable. A BOD sensor was developed whose biorecognition element was prepared by mixing magnetic powder with activated sludge. Magnetized sludge was then placed on the teflon membrane of the cathode and retained due to magnetic interactions (Yang et al., 1996).

A promising modern trend of making biorecognition elements based on microbial whole cells is their immobilization in sol gel matrices (Sakai et al., 1995; Chen et al., 2002). These elements are highly permeable for analyzed samples, have a good strength and stability, as well as low toxicity for immobilized microorganisms. However, fabrication of these films is a rather complex problem, because most sol gel formation methods are based on the temperature treatment of the reagent mixture.

4. Standards used for calibration of BOD sensors

The choice of the correct standard for calibration of a BOD biosensor is one of the key factors that determine the correlation between BOD_{bs} and BOD_5. The solution of a mixture of glucose and glutamic acid (GGA) at a total concentration of 300 mg/dm^3 (glucose, 150 mg/dm^3; glutamic acid, 150 mg/dm^3), which corresponds to the BOD_5 of 205 mg/dm^3, is usually used. Although GGA is widely used as the standard for the classical method of BOD measurement (Testing Methods..., 1990; PNDF 14...., 1997; Standard Methods..., 1992), this mixture does not satisfy the conditions for the calibration of microbial BOD sensors. Firstly, GGA is unstable due to rapid microbial contamination; secondly, the rate of glutamic acid oxidation by microorganisms decreases in the presence of glucose, which does not in practice affect the 5-day analysis, but may have a strong effect on the result of an express analysis. Thirdly, GGA consists of only two readily oxidized components, and real waste waters are a complex mixture of components, predominantly with low oxidation rates, so GGA calibration may give unreliable results (Liu & Mattiasson, 2002).

Much attention at present is given to the development of calibration solutions – synthetic waste waters containing an approximate list of compounds, which are the main components of analyzed water samples (Liu et al. 2000; Sakaguchi et al., 2003; Jianbo et al., 2003; Thévenot et al., 2001; Kim & Park, 2001; Melidis et al., 2008; Tanaka et al., 1994). The most widely used are synthetic waste waters recommended by the Organization for Economic Cooperation and Development (OECD), whose basic components are peptone, meat extract, urea and various inorganic salts (Organization for Economic Corporation and Development…, 1991). As compared with the GGA calibration, the calibration using the OECD standard makes it possible to increase significantly the correlation between the BOD_{bs} and BOD_5(Liu & Mattiasson, 2002). A number of publications also describe the use of other standards for calibration of the BOD biosensor (Chee et al., 2000; Lehrmann et al., 1999; Tanaka et al., 1994); however, those compositions can simulate the composition of only certain types of waste waters and are not universal.

5. Characteristics of various types of BOD sensors

The efficiency of a biosensor is determined by its analytical and metrological characteristics and operational parameters. They include the properties of the analytical signal (response value and time) in response to the addition of an analyzed substance, reversibility of the system after the analyte is removed, stability of the biosensor, the measurement technique, operational conditions and many others. Optimization of the biosensor system is an integrated problem, because often an improvement of one property leads to a deterioration of another.

To obtain quantitative information about the content of analyzed substances in a sample, it is necessary to know the calibration characteristics of the BOD biosensor, i.e., the dependence of the analytical signal on the concentration of the tested compound. The description of the calibration should indicate under which conditions it was obtained and for which calibration solution. The linear character of the dependence of BOD biosensor responses on the concentration within a certain interval is a measure of the possibility to determine the BOD in the analysis of waste waters with various concentrations of substrates. A broad linear interval is desirable for the measurements to be correct and reliable. The linear character of BOD biosensor characteristics in steady-state measurements is lower than when using the initial rate of change of the biosensor signal. Besides the calibration dependence proper, other quantitative characteristics are also used to compare the efficiency of biosensors: the sensitivity and the detection limit (Chen et al., 2002). The sensitivity coefficient is determined as the maximal value of the derivative of the response value with respect to the concentration. An important characteristic is the detection limit. In the case of amperometric biosensors, the following regularity can be observed. The sensitivity of the sensor can be increased by increasing the amount of biomaterial. Still, as a rule, this leads to a shift of the detection limit to the region of higher concentrations of the analyzed compound. Thus, the ratio of the detection limit to the sensitivity can be an objective characteristic of a biosensor.

The linear character and quantitative characteristics of the calibration dependence are related to the design of the transducer, type of sensor and concentration of cells in the recognition element. BOD sensors with a high density of cells in the biofilm are usually more sensitive but have a narrower linear interval. These parameters are also affected by the sensitivity of the sensor with respect to certain types of organic compounds. A BOD sensor may yield dissimilar linear characteristics when using different calibration solutions and samples with different compositions of organic substrates.

As the goal of developing BOD sensors is to set up a fast alternative analytical method, the analysis by means of biosensors should be no less accurate than by the traditional BOD_5 method. The BOD_5 is determined in the 5-day test by the GGA standard solution, for which the averaged value of BOD_5 is 205 mg/l, and the standard deviation is 30.5 mg/l, which is about 15.4%. The repeatability of biofilm-type BOD sensors varies within the limits of 10–11% for a sensor based on one strain, and increases up to 15% for sensors based on a microbial association (Liu & Mattiasson, 2002).

An important consumer quality of biosensors is analysis time, which sums up from the biosensor response time and bioreceptor element activity recovery time. The BOD sensor response time varies, first and foremost, depending on the measuring technique used. The sensor signal can be registered after 5–25 min in steady-state measurements and after 15-30 s in measurements of the initial rate. In steady-state measurements, the new steady-state setup time depends on the concentration of substrate in the sample and increases significantly in the analysis of samples with high concentrations of substrates. Usually the time required for the base line to be restored is greater than the signal development time, i.e., 15–60 min in the processing of the signal by the equilibrium method and 5–10 min in the processing by the kinetic method, respectively.

It should be noted that waste waters of some productions, e.g., cereal-processing enterprises (distilleries, breweries, starch plants) are characterized by a high content of organic impurities leading to the death of the surrounding natural ecosystems. A major problem is the utilization of liquid wastes. The first step in the utilization consists in the examination of wastes for the content of organic components. For such enterprises, it is not only difficult in practice, but also inexpedient to try to develop a universal BOD sensor. It is reasonable to design biosensors and choose respective microorganisms that would provide for the most efficient detection of BOD in accordance with the particular type of waste waters, i.e., to develop specialized BOD biosensors. Thus, to control the extent of purification of waste waters in a starch works, it was proposed to use in the BOD biosensor acetobacteria *Gluconobacter oxydans* that provide for a high sensitivity to alcohols and sugars (Jung et al., 1995; Arlyapov et al., 2008b); this enabled the development of an express method for controlling the BOD of the waste waters of that works. The problem of pollution of the environment with organic compounds is the most acute in distillery industry, in particular, with respect to the utilization of distiller's spent grains, the main production waste. A biosensor method for determining the general content of readily oxidized organic compounds in the wastes of such fermentation plants was developed (Arlyapov et al., 2008a; Reshetilov et al.,. 2008). The method of analysis using the developed biosensor has a high speed, high sensitivity and selectivity. The authors note that biosensors intended for the ecological control of food-production waste waters can be also used for monitoring the fermentation processes at these productions. This will make it possible to reduce expenses for the equipment and to increase the economic returns of the enterprise.

Some characteristics and parameters of various types of BOD biosensors described in the literature are given in Table 1.

Microorganisms, immobilization	Measurement conditions	Characteristics	Reference, year
Activated sludge preparation whose cells were killed by heating at 300°C for 1.75 min	GGA		Tan & Lim, 2005
Association of microorganisms immobilized on a nylon membrane	GGA, analysis of waste water samples	Response time 90 min, stability for 400 measurement cycles, storage at 4°C, lower detection limit 1 mg O_2 l^{-1}, correlation between BOD_{bs} and BOD_5 (deviation 10%), convergence 3.39–4.45%, reproducibility 1.85–2.25%	Dhall et al., 2008
Gluconobacter oxydans cells immobilized by adsorption on a Whatman GF/A glass fibre filter	GGA, analysis of food-production waste waters	Stable operation time 12 days, sensitivity 0.28 nA×dm³/minxmg, duration of measurement 7–10 min, linear range of BOD_5 biosensor response dependence 2.0–20.3 mg/dm³	Arlyapov et al., 2008b
Microbial association	Synthetic waste waters (CECD), waste waters of a rubber-treatment plant	Biosensor response time 10–15 min, difference between the values of the standard method less than 10%	Kumlanghan et al., 2008
Saccharomyces cerevisiae cells, encapsulation in calcium alginate	GGA, analysis of waste water samples	For a 20-ppm calibration solution the correspondence between BOD_{bs} and BOD_5 is $r = 0.95$ ($n = 6$)	Seo et al., 2009
Debaryomyces hansenii cells immobilized by adsorption	GGA, analysis of communal and biotechnological waste waters	Stable operation time "/>30 days, duration of single measurement 10–17 min, linear range of BOD_5 biosensor response dependence 2.2–177 mg/dm³	Arlyapov et al., 2012
Microbial associations of *Debaryomyces hansenii*, *Pichia angusta* and *Arxula adeninovorans*	GGA, samples of fermentation mass and waste waters of waste-water treatment facilities	Stable operation time 31 days, duration of single measurement 10–15 min, linear range of BOD_5 biosensor response dependence 1–93 mg/dm³	Kamanin et al., 2012
Strain *Aeromonas hydrophila* P69.1 immobilized using a semipermeable membrane	Synthetic waste waters (CECD), waste waters from meat-processing plants	Linear range of BOD_7 biosensor response dependence 5–45 mg/dm³, biosensor lifetime 110 days, biosensor response time up to 20 min	Raud et al., 2012
Mediator-type BOD biosensors			
Yeast *Saccharomyces cerevisiae*, a two-mediator system with ferricyanide and lipophilic mediator menadione	Analysis of sea water and river water samples	Linear range within 1 µM to 10 mM concentration of hexacyanoferrate (II) ($r^2 = 0.9995$, relative standard deviation = 1.3%). For 14 days under storage conditions at 4°C, decrease of sensor response 93%	Nakamura et al., 2007b

Microorganisms, immobilization	Measurement conditions	Characteristics	Reference, year
Glass-carbon electrode modified by ferricyanide in ion-exchange polysiloxane	GGA, sea water samples	Linear interval up to 40 mg O_2 l^{-1} (r = 0.994), convergence < 3.8%, reproducibility < 7.7%, correspondence between BOD_{bs} and BOD_5 (r = 0.988)	Chen et al., 2008
Bacteria *Gluconobacter oxydans* in graphite paste with mediator ferrocene	GGA, analysis of food-production waste waters	Stable operation time 30 days, sensitivity 4 nA×dm³/mg, duration of single measurement 6–7.5 min, linear range of BOD_5 biosensor response dependence 34–680 mg/dm³	Arlyapov et al., 2008a
Klebsiella pneumoniae cells, ferricyanide mediator	GGA; synthetic waste waters (OECD), municipal waste waters	Linear range of BOD_5 biosensor response dependence 30–500 mg/l or 30–200 mg/l, using GGA and synthetic waste waters, respectively	Bonetto et al., 2011
Escherichia coli cells immobilized in PVC-polyvinyl pyridine copolymer; mediator, neutral red	GGA; synthetic waste waters (OECD); urea and real waste waters		Liu et al., 2012
Optical BOD biosensors			
Bacillus subtilis cells im-mobilized into composite sol gel of quartz and polyvinyl alcohol; oxygen-sensitive film from *tris*(4,7-diphenyl-1.10-phenanthroline)-ruthenium(II)	GGA		Kwok et al., 2005
Microorganisms *Bacillus licheniformis*, *Dietzia maris* and *Marinobacter marinus* from sea water, immobilized in polyvinyl alcohol. Sensitive film from organically modified silicate film, with em-bedded ruthenium com-plex sensitive to oxygen	GGA; sea water samples	Stable operation time up to 10 months; with GGA as a standard, the correlation coefficient (R^2) within the range of 0.3–40 mg/l BOD was 0.985; reproducibility, ±2.3%	Lin et al 2006; Jiang et al., 2006
BFC-based BOD biosensors			
	Real-time monitoring of waste waters	Stable current in 60 min after introduction of samples at various concentrations into a biofuel cell; reproducibility 10% in the determination of BOD at a concentration of 100 mg/l	Chang et al., 2004, 2005
	Synthetic and real waste waters	Linear dependence on BOD up to 350 mg/ml; stable operation time, 7 months	Di Lorenzo et al., 2009

Microorganisms, immobilization	Measurement conditions	Characteristics	Reference, year
BOD biosensors based on other registration methods			
Measurement of the concentration of CO_2, produced by the degradation of waste waters' carbon component by microorganisms. Control of CO_2 using an infrared spectrometer	Real-time determination of the BOD of waste water treatment facilities		Vaiopoulou et al., 2005
Amperometric bioelectric "language" in the group cell using mathematical data processing methods. Modification of electrodes by tyrosinase, horseradish peroxidase, acetylcholin-esterase and butyrylcholinesterase.	Samples of waste waters		Tønning et al., 2005
Activated sludge, pH-transducer. Determination of CO_2 under aerobic conditions and of NaOH under anaerobic conditions	Monitoring of the extent of pollution with organic compounds and of toxicity		Melidis et al., 2008
Saccharomyces cere-visiae, the colorimetric method in the presence of 2,6-dichlorophenol-indophenol	GGA; river water samples	Linear interval 1.1–22 mg O_2 l^{-1} ($r = 0.988$, $n = 3$), storage 36 days	Nakamura et al., 2007a

Table 1. Characteristics and parameters of BOD sensors

6. Commercial BOD biosensors

Most biosensor designs described in the literature have remained breadboard models. To date, only several models of BOD biosensors are available commercially. The first commercial model of biosensor for the BOD analysis was put on the market by Nisshin Electric Co. Ltd in 1983. Later, several commercial BOD biosensor analysers were produced by other Japanese (Central Kagaku Corp.) and European (Dr. Lange GmbH, Aucoteam GmbH, Prufgeratewerk Medingen GmbH) companies. The first commercial BOD sensors were Clark's oxygen electrode-based biosensors and, as a rule, made use of activated sludge as receptor element substrate (Liu & Mattiasson, 2002).

The method of BOD analysis using biosensors was included into the Japanese Industrial Standard in 1990 (JIS K3602). Several models of BOD biosensors are sold on the market at present: QuickBOD α1000, BOD-3300, HABS-2000 (all by Central Kagaku Corp., Japan) (http://www.aqua-ckc.jp/product2/bod.html#top). The model BOD-3300 enables the BOD determination within the range of 0–500 mg/l in 30–60 min and costs about 80 thousand US dollars. The weight of the analyzer is ~210 kg. QuickBOD α1000 is a more advanced device developed by Japanese engineers and makes it possible to determine the BOD within the range of 2–50 mg/l in 60 min; the weight of the device is 16 kg and its cost is 30 thousand US dollars. It should be noted that this analyzer, in contrast with its many precursors is based on the use of one culture (*T. cutaneum*).

7. Conclusion

Thus, the determination of the BOD by means of biosensors is a rather advanced trend of analytical biotechnology. However, BOD biosensors have a number of limitations that impede their use, so it is topical to conduct own Russian research and to perform works establishing the basis for commercial production of BOD biosensors. Biosensor BOD analyzers are robust, simple and cheap analytical tools that can be successfully used for controlling aqueous ecosystems along with the traditional BOD determination methods.

Author details

A. Reshetilov[1], V. Arlyapov[2], V. Alferov[2] and T. Reshetilova[1]

1 Federal State Budgetary Institution of Science, G.K. Skryabin Institute of Biochemistry and Physiology of Microorganisms, Russian Academy of Sciences, 5 Pr. Nauki, Pushchino, Moscow Region, Russia

2 Federal State Budgetary Educational Institution of Higher Professional Education, Tula State University, Tula, Russia

References

[1] Arlyapov, V. A., Chigrinova, E., Yu, , Ponamoreva, O. N., & Reshetilov, A. N. (2008a). Express detection of BOD in wastewaters of starch-processing industry. Starch Science and Technology, Ed. G.E. Zaikov. New York: Nova Science Publishers, 978-1-60456-950-6, 16_-175.

[2] Arlyapov, V., Kamanin, S., Ponamoreva, O., & Reshetilov, A. (2012). Biosensor analyzer for BOD index express control on the basis of the yeast microorganisms Candida maltosa, Candida blankii, and Debaryomyces hansenii. Enzyme Microb. Technol., Iss. 4-5, 0141-0229, 50, 215-220.

[3] Arlyapov, V. A., Ponamoreva, O. N., Alferov, V. A., Rogova, T. V., Blokhin, I. V., Chepkova, I. F., & Reshetilov, A. N. (2008b). Microbial biosensors for express detection of BOD in wastewaters of food enterprises. Voda: Khim. Ekol., in Russian), 2072-8158(3), 20-22.

[4] Baeumner, A. J. (2003). Biosensors for environmental pollutants and food contaminants.Anal. Bioanal. Chem., 1618-2642, 377, 434-445.

[5] Bourgeois, W., Burgess, J. E., & Stuetz, R. M. (2001). On-line monitoring of wastewater quality: a review. J. Chem. Techn. Biotechn., 0268-2575, 76, 337-348.

[6] Bonetto, M. C., Sacco, N. J., Ohlsson, A. H., & Cortón, E. (2011). Assessing the effect of oxygen and microbial inhibitors to optimize ferricyanide-mediated BOD assay.Talanta., Iss. 1, 0039-9140, 85, 455-462.

[7] Chan, C., Lehmann, M., Chan, K., Chan, P., Chan, C., Gruendig, B., Kunze, G., & Renneberg, R. (2000). Designing an amperometric thick-film microbial BOD sensor. Biosens. Bioelectron., 0956-5663, 15(7), 343-353.

[8] Chang, I. S., Jang, J. K., Gil, G. C., Kim, M., Kim, H. J., Cho, B. W., & Kim, B. H. (2004). Continuous determination of biochemical oxygen demand using microbial fuel cell type biosensor. Biosens. Bioelectron., 0956-5663, 19(6), 607-613.

[9] Chang, I. S., Moon, H., Jang, J. K., & Kim, B. H. (2005). Fluorescence and bioluminescence of bacterial luciferase intermediates. Biosens. Bioelectron., 0956-5663, 20(9), 1856-1859.

[10] Chee, G., & , J. (2011). Biosensor for the determination of biochemical oxygen demand in rivers. Environmental Biosensors, Ed. Vernon Somerset, InTech Publishers, 978-9-53307-486-3, 257-276.

[11] Chee, G., , J., Nomura, Y., Ikebukuro, K., & Karube, I. (2000). Optical fiber biosensor for the determination of low biochemical oxygen demand. Biosens. Bioelectron., 0956-5663, 15(7-8), 371-376.

[12] Chen, D., Cao, Y., Liu, B., & Kong, J. (2002). A BOD biosensor based on a microorganism immobilized on an Al_2O_3 sol-gel matrix. Anal. Bioanal. Chem., 1618-2642, 372, 737-739.

[13] Chen, H., Ye, T., Qiu, B., Chen, G., & Chen, X. (2008). A novel approach based on ferricyanide-mediator immobilized in an ion-exchangeable biosensing film for the determination of biochemical oxygen demand. Anal. Chim. Acta, 0003-2670, 612(1), 75-82.

[14] Deng, H., Chen, Zh., & Zhao, F. (2012). Energy from plants and microorganisms: progress in plant-microbial fuel cells. ChemSusChem, 5, 1006-1011, 1864-5631.

[15] Dhall, P., Kumar, A., Joshi, A., Saxsena, T. K., Manoharan, A., Makhijani, S. D., & Kumar, R. (2008). Quick and reliable estimation of bod load of beverage industrial wastewater by developing bod biosensor. Sens. Act.B, 0925-4005, 133(2), 478-483.

[16] Di Lorenzo, M., Curtis, T. P., Head, I. M., & Scott, K. (2009). A single-chamber microbial fuel cell as a biosensor for wastewaters. Water Res., Iss. 13, 0043-1354, 43, 3145-3154.

[17] D'Souza, S. F. (2001). Microbial biosensors. Biosens. Bioelectron.,0956-5663, 16, 337-353.

[18] Du, Zh., Li, H., & Gu, T. (2007). A state of the art review on microbial fuel cells: a promising technology for wastewater treatment and bioenergy. Biotechnol. Adv., 0734-9750, 25, 464-482.

[19] Farré, M., & Barceló, D. (2001). Characterization of wastewater toxicity by means of a whole-cell bacterial biosensor, using Pseudomonas putida, in conjunction with chemical analysis. Fresenius J. Anal. Chem., 0937-0633(371), 467-473.

[20] Guo, G., , M., Xin, L., , L., Wang, X., , D., Zhao, Y., & Chen, X. (2008). Study on the fluorescence characteristics of BOD sensing films immobilizing different limnetic microorganism. s. Guang Pu Xue Yu Guang Pu Fen Xi / Spectroscopy and Spectral Analysis, 1000-0593, 28, 2134-2138.

[21] Heim, S., Schnieder, I., Binz, D., Vogel, A., & Bilitewski, U. (1999). Development of an automated microbial sensor system. Biosens. Bioelectron., 0956-5663, 14, 187-193.

[22] Hikuma, M., Suzuki, H., Yasuda, T., Karube, I., & Suzuki, S. (1979). Amperometric estimation of BOD by using living immobilized yeast. Eur. J. Appl. Microbiol. Biotechnol., 0171-1741, 8, 289-297.

[23] Hyun, C., , K., Tamiya, E., Takeuchi, T., & Karube, I. (1993). A novel BOD sensor based on bacterial luminescence. Biotechnol. Bioeng., 0006-3592, 41, 1107-1118.

[24] Iranpour, R., & Zermeno, M. (2008). Online biochemical oxygen demand monitoring for wastewater process control- full-scale studies at Los Angeles Glendale wastewater plant, California. Water Environ. Res., 1061-4303, 80(4), 24-29.

[25] Jianbo, J., Tang, M., Chen, X., Qi, L., & Dong, S. (2003). Co-immobilized microbial biosensor for BOD estimation based on sol-gel derived composite material. Biosens. Bioelectron., 0956-5663, 18(8), 1023-1029.

[26] Jiang, Y., Xiao, L., , L., Zhao, L., Chen, X., Wang, X., Wong, K., & , Y. (2006). Optical biosensor for the determination of BOD in seawater. Talanta, 0039-9140, 70(1), 97-103.

[27] Jung, J., Sofer, S., & Lakhwala, F. (1995). Towards an on-line biochemical oxygen demand analyser. Biotechnol. Tech., 0095-1208X, 9(4), 289-294.

[28] Kamanin, S. S., Arlyapov, V. A., Ponamoreva, O. N., Alferov, V. A., & Reshetilov, A. N. (2012). BOD-biosensor based on yeast strains. Voda: Khim. Ekol., in Russian), 2072-8158(3), 74-81.

[29] Karube, I., & Yokoyama, K. (1993). Microbial sensors and micro-biosensors. NATO ASI Ser. E, 0016-8132X, 252, 281-288.

[30] Karube, I., Matsunaga, T., Tsuru, S., & Suzuki, S. (1977a). Biochemical fuel cell utilizing immobilized cells of clostridium butyricum. Biotechnol. Bioeng., 0006-3592, 19, 1727-1760.

[31] Karube, I., Mitsuda, S., Matsunaga, T., & Suzuki, S. (1977b). Microbial electrode BOD sensors. Biotechnol. Bioeng., 0006-3592, 19(10), 1535-1547.

[32] Kim, B. H., Chang, I. S., Gil, G. C., Park, H. S., & Kim, H. J. (2003a). Novel bod (biological oxygen demand) sensor using mediator-less microbial fuel cell. Biotechnol. Lett., 0141-5492, 25, 541-545.

[33] Kim, M., Youn, S. M., Shin, S. E., Jang, J. G., Han, S. H., Hyun, M. S., Gaddb, G. M., & Kim, H. J. (2003b). Practical field application of a novel BOD monitoring system. J. Environ. Monit., 1464-0325, 2(5), 640-643.

[34] Kim, M., , N., Park, K., & , H. (2001). Klebsiella BOD sensor. Sens. Act.B, 0925-4005, 80, 9-14.

[35] Kumlanghan, A., Kanatharana, P., Asawatreratanakul, P., Mattiasson, B., & Thavarungkul, P. (2008). Microbial BOD sensor for monitoring treatment of wastewater from a rubber latex industry. Enzyme Microb. Technol., Iss. 6, 0141-0229, 42, 483-491.

[36] Kwok, N., , Y., Dong, S., Lo, W., Wong, K., & , Y. (2005). An optical biosensor for multi-sample determination of biochemical oxygen demand (BOD). Sens. Act.B, 0925-4005, 110(2), 289-298.

[37] Lehmann, M., Chan, C., Lo, A., Lung, M., Tag, K., Kunze, G., Riedel, K., Gruendig, B., & Renneberg, R. (1999). Measurement of biodegradable substances using the salt-tolerant yeast Arxula adeninivorans for a microbial sensor immobilized with poly(carbamoyl) sulfonate (PVS): Part II: application of the novel biosensor to real samples from coastal and island regions. Biosens. Bioelectron., 0956-5663, 14, 295-302.

[38] Lei, Y., Chen, W., & Mulchandani, A. (2006). Microbial biosensors (review). Anal. Chim. Acta, 0003-2670, 568, 200-210.

[39] Lin, L., Xiao, L., , L., Huang, S., Zhao, L., Cui, J., , S., Wang, X., , H., & Chen, X. N. (2006). Novel BOD optical fiber biosensor based on co-immobilized microorganisms in ormosils matrix. Biosens. Bioelectron., 0956-5663, 21(9), 1703-1709.

[40] Liu, J., Bjornsson, L., & Mattiasson, B. (2000). Immobilised activated sludge based biosensor for biochemical oxygen demand measurement. Biosens. Bioelectron., 0956-5663, 14(12), 883-993.

[41] Liu, J., & Mattiasson, B. (2002). Microbial BOD sensors for wastewater analysis. Water Res., 0043-1354, 36, 3786-3802.

[42] Liu, L., Zhang, S., Xing, L., Zhao, H., & Dong, S. (2012). A co-immobilized mediator and microorganism mediated method combined pretreatment by TiO$_2$ nanotubes used for BOD measurement. Talanta,. 93, 314-319, 0039-9140.

[43] Liua, L., Shanga, L., Liua, C., Liua, C., Zhanga, B., & Dong, S. (2010). A new mediator method for bod measurement under non-deaerated condition. Talanta, 0039-9140, 81(4-5), 1170-1175.

[44] Mattiasson, B., Larsson, P. O., & Mosbach, K. (1977). The microbe thermistor. *Nature*, 268, 519-520, 0028-0836.

[45] Melidis, P., Vaiopoulou, E., & Aivasidis, A. (2008). Development and implementation of microbial sensors for efficient process control in wastewater treatment plants. Bioprocess Biosyst. Eng., 1615-7591, 31(3), 277-352.

[46] Moon, H., Chang, I. S., Kang, K. H., Jang, J. K., & Kim, B. H. (2004). Improving the dynamic response of a mediator-less microbial fuel cell as a biochemical oxygen demand (BOD) sensor. Biotechnol. Lett., 0141-5492, 26(22), 1717-1738.

[47] Nakamura, H., Kobayashi, S., Hirata, Y., Suzuki, K., Mogi, Y., & Karube, I. 2007a). A spectrophotometric biochemical oxygen demand determination method using 2,6-dichlorophenolindophenol as the redox color indicator and the eukaryote Saccharomyces cerevi. siae. Anal. Biochem., 0003-2697, 369(2), 168-174.

[48] Nakamura, H., Suzuki, K., Ishikuro, H., Kinoshita, S., Koizumi, R., Okuma, S., Gotoh, M., & Karube, I. (2007b). A new BOD estimation method employing a double-mediator system by ferricyanide and menadione using the eukaryote Saccharomyces cerevisiae. *Talanta*, 72(1), 210-216, 0039-9140.

[49] Organization for Economic Cooperation and Development (OECD), OECD Guidel.Testing Chem., (1991).

[50] Pang, H. L., Kwok, N. Y., Chan, P. H., Yeung, C. H., Lo, W., & Wong, K. Y. (2007). High-throughput determination of biochemical oxygen demand (BOD) by a microplate-based biosensor. Environ. Sci. Technol., 0001-3936X, 4(11), 4038-4082.

[51] Pasco, N., Baronian, K., Jeffries, C., Webber, J., & Haya, J. (2004). Micredox®- development of a ferricyanide-mediated rapid biochemical oxygen demand method using an immobilised proteus vulgaris biocomponent.Biosens. Bioelectron., 0956-5663, 20, 524-532.

[52] , P. N. D. F. 1., 1:2:3:, , & 123-9, . The method of measuring the biochemical oxygen demand after n days of incubation (BOD$_{complete}$) in surface fresh, subsurface (ground), drinking, waste and purified waste waters. Moscow, (1997). pp. (in Russian)

[53] Ponomareva, O. N., Arlyapov, V. A., Alferov, V. A., & Reshetilov, A. N. (2011). Microbial biosensors for detection of biological oxygen demand (review). Appl. Biochem. Microbiol., 0003-6838, 47, 1-11.

[54] Praet, E., Reuter, V., Gaillard, T., Vasel, J., & (1995, L. (1995). Bioreactors and biomembranes for biochemical oxygen demand estimation. Trends Anal Chem., 0165-9936, 14(7), 371-378.

[55] Qian, Z., & Tan, T. C. (1999). BOD measurement in the presence of heavy metal ions using a thermally-killed-Bacillus subtilis biosensor. Water Res., 0043-1354, 33(13), 2923-2928.

[56] Raud, M., Tenno, T., Jõgi, E., & Kikas, T. (2012). Comparative study of semi-specific Aeromonas hydrophila and universal Pseudomonas fluorescens biosensors for BOD

measurements in meat industry wastewaters. Enzyme Microb. Technol., Iss. 4-5, 2012, 0141-0229, 50, 221-226.

[57] Reshetilov, A. N., Alferov, V. A., Ledenev, V. P., & Sergeyev, V. I. (2008). A novel method of rapid test in alcohol production. Likerovodochn. Proizv. Vinodel., Iss. 99, in Russian), 3, 20-22.

[58] Rodriguez-Mozaz, S., de Alda, M. J. L., & Barcelo, D. (2006). Biosensors as useful tools for environmental analysis and monitoring. Anal. Bioanal. Chem., 1618-2642, 386(4), 1025-1041.

[59] Sakaguchi, T., Kitagawa, K., Ando, T., Murakami, Y., Morita, Y., Yamamura, A., Yokoyama, K., & Tamiya, E. (2003). A rapid BOD sensing system using luminescent recombinants of Escherichia col.Biosens. Bioelectron., 0956-5663, 19(2), 115-121.

[60] Sakaguchi, T., Morioka, Y., Yamasaki, M., Iwanaga, J., Beppu, K., Maeda, H., & Morita, Y. (2007). Rapid and onsite BOD sensing system using luminous bacterial cells-immobilized chip. Biosens. Bioelectron., 0956-5663, 22(7), 1345-1350.

[61] Sakai, Y., Abe, N., Takeuchi, S., & Takahashi, F. (1995). BOD sensor using magnetic activated sludge. J. Ferment. Bioeng., 0092-2338X, 80(3), 300-303.

[62] Seo, K. S., Choo, K. H., Chang, H. N., & Park, J. K. (2009). A flow injection analysis system with encapsulated high-density Saccharomyces cerevisiae cells for rapid determination of biochemical oxygen demand. Appl. Microbiol. Biotechnol., 0175-7598, 83(2), 217-223.

[63] Sohn, M., , J., Lee, J., , W., Chung, C., Ihn, G., , S., & Hong, D. (1995). Rapid estimation of biochemical oxygen demand using a microbial multi-staged bioreactor. Anal. Chim. Acta, 0003-2670, 313(3), 221-228.

[64] Spanjers, H., Vanrolleghem, P., Olsson, G., & Dold, P. (1996). Respirometry in control of the activated sludge process. Water Sci. Technol., Iss 3-4, 0273-1223, 34, 117-143.

[65] Spanjers, H., Olsson, G., & Klapwijk, A. (1993). Determining influent short-term biochemical oxygen demand by combined respirometry and estimation. Water Sci. Technol., Iss. 11-12, 0273-1223, 28, 401-415.

[66] Spanjers, H., & Klapwijk, A. (1991). Continuous estimation of short term oxygen demand from respiration measurements.Water Sci. Technol., Iss. 7, 0273-1223, 24, 29-32.

[67] Suriyawattanakul, L., Surareungchai, W., Sritongkam, P., Tanticharoen, M., & Kirtikara, K. (2002). The use of cc-immobilization of Trichosporon cutaneum and Bacillus licheniformis for a BOD sensor. Appl. Microbiol. Biotechnol., 0175-7598, 59(1), 40-44.

[68] Standard Methods for the Examination of Water and Wastewater(1992). Washington: Amer. Publ. Health Association, , 5.

[69] Tag, K., Lehmann, M., Chan, C., Renneberg, R., Riedel, K., & Kunze, G. (2000). Measurement of biodegradable substances with a mycelia-sensor based on the salt tolerant yeast Arxula adeninivorans LS3. Sens. Act. B, 0925-4005, 67, 142-148.

[70] Tan, T. C., Li, F., Neoh, K. G., & (1993, . (1993). Measurement of BOD by initial rate of response of a microbial sensor. Sens. Act.B, 0925-4005(10), 137-142.

[71] Tan, T. C., & Lim, E. W. C. (2005). Thermally killed cells of complex microbial culture for biosensor measurement of BOD of wastewater. Sens. Act. B, 0925-4005, 107(2), 546-551.

[72] Tanaka, H., Nakamura, E., Minamiyama, Y., & Toyoda, T. (1994). BOD biosensor for secondary effluent from wastewater treatment plants. Water Sci. Technol., 0273-1223, 30(4), 215-227.

[73] Thévenot, R. D., Toth, K., Durst, A. D., & Wilson, G. S. (2001). Electrochemical biosensors: recommended definitions and classification. Biosens. Bioelectron., 0956-5663, 16, 121-131.

[74] Testing Methods for Industrial Waste Water, JIS K3602, Japanese Industrial Standard Committee, Tokyo, 1990

[75] Tkac, J., Vostiar, I., Gorton, L., Gemeiner, P., & Sturdik, E. (2003). Application to the analysis of ethanol during fermentation. Biosens. Bioelectron., 0956-5663, 18(9), 1125-1134.

[76] Tønning, E., Sapelnikova, S., Christensen, J., Carlsson, C., Winther-Nielsen, M., Dock, E., Solna, R., Skladal, P., Nørgaard, L., Ruzgas, T., & Emnéus, J. (2005). Chemometric exploration of an amperometric biosensor array for fast determination of wastewater quality. Biosens. Bioelectron., 0956-5663, 21(4), 608-617.

[77] Trosok, S. P., Driscoll, B. T., & Luong, J. H. T. (2001). Mediated microbial biosensor using a novel yeast strain for wastewater BOD measurement. Appl. Microbiol. Biotechnol., 0175-7598, 56(3-4), 550-554.

[78] Vaiopoulou, E., Melidis, P., Kampragou, E., & Aivasidis, A. (2005). On-line load monitoring of wastewaters with a respirographic microbial sensor. Biosens. Bioelectron., 0956-5663, 21(2), 365-371.

[79] Walmsley, R. M., & Keenan, P. (2000). The eukaryote alternative: advantages of using yeasts in place of bacteria in microbial biosensor development. Biotechnol. Bioprocess Eng., 1226-8372, 5(6), 387-394.

[80] Xu, X., & Ying, Y. (2011). Microbial biosensors for environmental monitoring and food analysis (review). Food Rev. Int., 8755-9129, 27, 300-329.

[81] Yang, Z., Sasaki, S., Karube, I., & Suzuki, H. (1997). Fabrication of oxygen electrode arrays and their incorporation into sensors for measuring biochemical oxygen demand. Anal. Chim. Acta, 0003-2670, 357(1-2), 41-50.

[82] Yang, Z., Suzuki, H., Sasaki, S., & Karube, I. (1996). Disposable sensor for biochemical oxygen demand. Appl. Microbiol. Biotechnol., 0175-7598, 46(1), 10-14.

[83] Yoshida, N., Hoashi, J., Morita, T., Mc Niven, S. J., Nakamura, H., & Karube, I. (2001). Improvement of a mediator-type biochemical oxygen demand sensor for on-site measurement. J. Biotechnol., 0168-1656, 88(3), 269-275.

[84] Yoshida, N., Yano, K., Morita, T., Mc Niven, S. J., Nakamura, H., & Karube, I. (2000). A mediator-type biosensor as a new approach to biochemical oxygen demand estimation. Analyst, 0003-2654, 125(12), 2280-2284.

Chapter 3

Bacterial Sensors in Microfouling Assays

Carmenza Duque, Edisson Tello,
Leonardo Castellanos, Miguel Fernández and
Catalina Arévalo-Ferro

Additional information is available at the end of the chapter

1. Introduction

1.1. Fouling, the undesirable load

A lot of the marine invertebrates have a planktonic larval stage, in this period the larvae are dispersed and transported by currents. When larvae mature and attain the ability to meta-morphose, they start looking for suitable substrates, swimming toward the bottom and exploring the surfaces. When larvae encounter suitable substrate, they settle and metamorphose into juveniles; the survival of them is heavily dependent on where they settle. On the other hand, larval settlement and metamorphosis are influenced by local factors as salinity, temperature, light, kind of substrates, larval age, and nutritional conditions of larvae. However, one of the most important factors for settlement is the presence of chemical tracks originated from nonspecific adults and prey organisms. Microbial films are included also in those kinds of tracks and induce differentially larval settlement and metamorphosis in many invertebrate species; unfortunately, these bacterial biofilm factors have not been fully characterized [1]. The first biofilm formed on a surface, the settlement and the following steps of biological colonization are known as fouling, which could be defined (since an industrial point of view) as the undesirable accumulation of dissolved chemical compounds, microorganisms, algae and animals on submerged substrates leading to subsequent bio-deterioration of the colonized surface.

The fouling process is an ecologically complex of interactions between basibionts, surface-colonizing microbes, and fouling larvae, all mediated by chemical signaling. The assessment of fouling organism over basibionts can have severely deleterious effects on them such as inhibition of photosynthesis, blockage of filter feeding, and elevated risk of mechanical dislodgement or predation. In this scenario, competition for space represents ecological forces comparable to predation, because the space is a limited resource in the ocean; therefore, marine invertebrates have to compete for their place on a surface. Consequently, sessile invertebrates establish evolutionary weapons to colonize: when they are larvae, they must locate and colonize a surface in order to colonize and metamorphose; but when they are adults, they have to keep their own surfaces clean and ward off settlement by larvae [2].

Nowadays, the biofouling has been understood as a four-step sequential process. The initial step consists in the adsorption of organic macromolecules, it occurs almost immediately after submersion of any surface, and is characterized by the formation of a film composed by proteins, glycoproteins and polysaccharides, this film is colonized subsequently by bacteria. The second step, which occurs within an hour of surface immersion in water, is characterized by the assessment of prokaryotes and the subsequent development of a bacterial biofilm. Once the bacterial attachment to the surface has occurred, bacterial cells begin producing a matrix of extracellular polymeric substances (EPS) that is critical for maintaining adhesion and subsequent biofilm development and consolidation. Those biofilm are bacterial communities assembled coordinating different phenotypes that change with the time and the environment, and depend of different factors that influence, as well, further colonization of a surface. Recently, Quorum Sensing (QS) has been recognized as one of the main factors that determinate biofilm maturation, and this is perhaps the strongest determinant for the establishment of a proper biofilm. This phenomenon is defined as the regulation of gene expression depending on the bacterial population density and allows the synchronization of phenotypes by bacterial communication. It is important to explain briefly, how this QS regulation works (Figure 1). Basically, it works through a genetic circuit compose by a transcriptional factor (LuxR) and an acyl homoserin lactones (AHL's) synthetase (LuxI). The accumulation of AHLs in the media, due to the amount of bacterial cells preset in a culture, leads to the expression of the genes regulated by LuxR; some of them involved in biofilm maturation (for review see: [3]).

The third step in fouling formation is the colonization by unicellular eukaryotes; these include photosynthetic taxa such as diatoms, and heterotrophic suspension feeders and predators. The final step is the attachment of propagules of multicellular organisms, invertebrate larvae and algal spores, the predominant organisms differ in temperate zones and the tropics, season and other local conditions. It is now recognized that the nature of biofilms varies widely, and can present a range of positive and negative stimuli to settling larvae. Larvae may respond in the water column to chemical cues emanating from the substratum, and/or upon contact to physicochemical and biological characteristics of the substratum and elect to settle or reject the surface. In this context, is not surprising that the host-specific bacterial communities are maintained by many invertebrates, and may inhibit fouling by chemical deterrence of larvae, or by preventing biofilm formation by inductive strains. Finally, the larval settlement naturally occurs in a turbulent environment, yet the effects of waterborne versus surface-adsorbed chemical inductors and/or defences have not been completely understood [1].

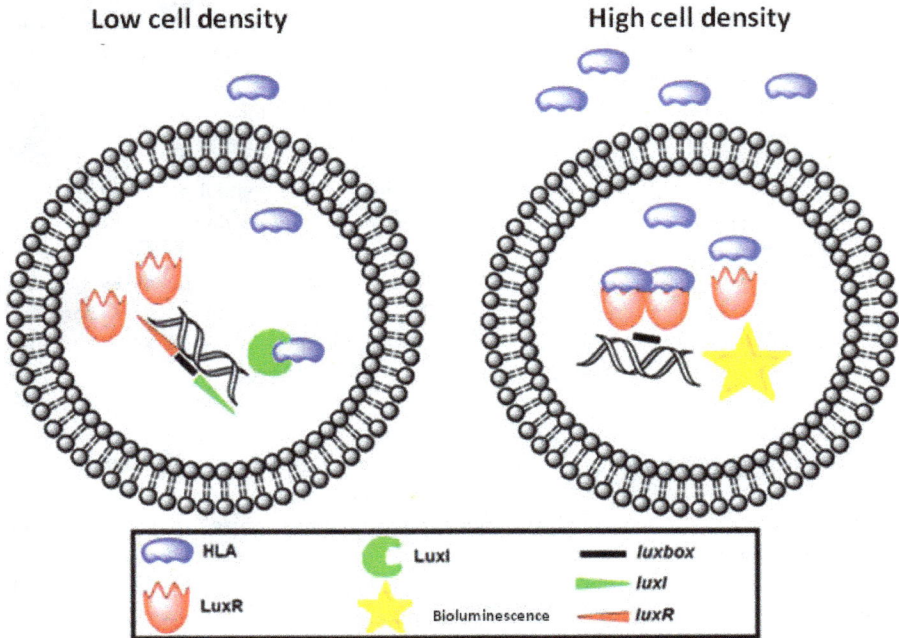

Figure 1. LuxR/I system in *V. fisheri*. QS is defined as the gene expression regulated by the population density. The *luxI* gene encodes the AHL's synthetaze. The AHL's molecules works as an "auto-inductor" that can diffuse freely through the cell membrane, and when it is accumulated up to a specific threshold it binds to the LuxR protein, a transcriptional factor. The complex LuxR-AHL activates the transcription of genes regulated under the *luxbox* promoter, in this case those responsible of bioluminescence.

To sum up, we can distinguish within the fouling structure two levels of organization, the microfouling and the macrofouling. The first one rules the second, and nowadays the efforts to control the fouling phenomena are concentrated in biofilm control, because its inhibition, by the use of quorum sensing inhibitors, could avoid the macrofouling assessment.

The fouling has been identified as a cause of severe problems in different scenarios, e.g., it is estimated that fuel consumption of ships increases 6% for every 100 of hull roughness caused by fouling organisms. Another example is the higher frequency of dry-docking operations required or the invasive species that can be spread inadvertently by fouled ships. Therefore, inadequate protection against fouling is a consequent threat to marine ecosystems with incalculable damage, and just for the shipping industry the cost could be estimated in several billion of dollars [4]. Other industries could be affected by this phenomenon, i.e. marine industries as gas and petroleum exploitation, aquaculture, cooling towers, drinking water distribution systems, building materials, etc; and hull fouling is also a major vector for marine invasive species. The fouling problem is quite important for sensor devices, particularly in marine and riverine sensors [5]; even more, not only the bacterial and marine invertebrates are challenging for antifouling technologies the blood cells are too [6]. In order to deal with this problem the use of antifouling (AF) paints has emerged as the most useful solution; antifouling paints contain biocides that are released during the lifetime of the coating; these biocides are present within a surface micro-layer of water adjacent to the paint surface, avoiding the settlement of juvenile fouling organisms. Due the great number of organisms involved on marine fouling the biocides used in antifouling paints must have a wide spectrum of activity to be able to deter the colonizer organisms on the ship's surface. So, the antifouling products play an important role in the shipping industry and are of significant economic importance [4].

The use of antifoulants to protect the boats is not a new concept; furthermore, the Romans and Greeks coated their boats with lead sheathing. In the discovery and colonization of America the vessels were coated with pitch and tallow, and the British Empire used as antifouling paints grease, sulphur pitch and brimstone, and later copper sheathings were used. Finally, in the mid-1800s the antifouling paints were developed, because with the introduction of iron ships the copper sheathing caused corrosion of the iron, and new formulation were necessaries. These artifouling paints included a lot of toxic compounds as copper oxide, arsenic, and mercury oxide to resin binders. After Second World War, the synthetic copper based paints became most popular until the tributyltin (TBT) proved to be excellent in the prevention of fouling [7]. A lot of advantages were attributed to TBT antifoulant, e.g. it exhibits broad-spectrum biocidal properties, and is effective against most of the colonizing organisms. It could be incorporated into coloured paints because it does not have color, and does not promote galvanic corrosion on iron ships; furthermore, it can also be used on aluminium surfaces. Due it is an organic compound it can be co-polymerised into resin-based paints, and incorporated into self-polishing coatings that remain effective for long periods of time [4]. However, nowadays it is known that organotins, such as tributyltin (TBT) and tributyltin oxide (TBTO), are the most toxic biocides ever introduced to the marine environment [8] because they have important deleterious biological effects over a great number of marine organisms, a lot of them non-target marine organisms, e.g. tissue's analyses of sea mammals, fish and some birds have revealed detectable concentrations of TBT. In this way, the International Maritime Organization (IMO) adopted an international treaty entitled "*The International Convention on the Control of Harmful Anti-Fouling Systems on Ships* in 2001", which banned the presence of organotin compounds in antifouling paints by 1 January 2008 [4].

The prohibition of TBT-based paints forced to develop alternatives that includes booster biocides, like irgarol 1051, sea-nine 211, dichlofluanid, chlorothalonil, zinc pyrithione, diuron, TCMS pyridine, TCMTB, zineb, etc. These biocides are being used in many countries, but they have been found to accumulate in coastal waters and have become a threat to the marine environment as well. These alternatives to TBT are also toxic and their putative impact on non-target organisms is poorly known in some cases, for that reason their contamination in the aquatic environment has been a topic of increasing importance over the last few years, interesting review could be consulted in Antifouling Paint Biocides [7] or Ecotoxicology of

Antifouling Biocides [4]. Nowadays, some countries have signed an agreement to restrict the useof Irgarol 1051 and diuron biocides [7,8].

In this context, the developing of "environmentally friendly" antifoulants is an urgent necessity, which include fouling-release coating and electrical antifouling systems; however, many researchers are trying to employ chemical defense systems from sessile marine organisms for this purpose [8]. These natural compounds incorporated into a painting would mimic the marine organisms, which keep their body surfaces clean due to the natural production of antifouling substances with high anesthetic, repellent, settlement deterrent, and settlement inhibitory properties, but without having biocidal effects. Compounds with different structures have been identified as antifoulants, and include terpenes, nitrogen-containing compounds, phenols, steroids and others. Additionally, mixtures of natural products could be useful taking advantages of the synergistic properties observed; it is expected this mixture to be much better antifoulants than the organotin compounds, see [9]. Dr. Fussetani has presented several reviews about the use of natural products as antifoulants, where different compounds have been identified as the most promising natural product for the antifouling paints. To mention some examples: the sesquiterpene elatol isolated from red alga *Laurencia elata*, the furanones isolated from the red alga *Delisea pulchra* with, some isocyanoterpenes isolated from the sponge *Acanthella cavernosa* and from the nudibranchs *Phyllidia pustulosa, P. ocelata, P. varicosa* y *Phillidiopsis krempfi*, and the 5,6-Dichloro-1-methylgramine (DCMG) inspired in the natural product 2,5,6-tribromo-1-methylgramine.

There is, however, one drawback; the known supply problem for Marine Natural Products (MNP), that has to be overcome in order to apply these products in antifouling technologies. In this context, bacteria and fungi are promising sources, and more efforts towards the development of antifouling compounds from marine microorganisms should be made [1,8,10].

On the other hand, current antifouling paints are not effective against microfouling colonization (bacterial and diatom species) because the microorganisms have the ability to colonize entire surfaces previously treated with common antifouling paints. Therefore, the development of new compounds to regulate the density of microbes on antifouling coatings is urgently needed. In this context, the Quorum Sensing Inhibitors (QSI) are a new alternative, and these compounds can be used too for antimicrobial protection in aquaculture, and even more in the control of medical caterer biofilm development. Dobretsov *et al.* [11]. established the ability to prevent microfouling by the kojic acid in a controlled mesocosm experiment. This acid inhibited formation of microbial communities on glass slides, decreasing the densities of bacteria and diatoms. The study suggested that natural products with QS inhibitory properties can be used for controlling biofouling communities [11-15].

2. Common bioassays used in the evaluation of chemical compounds as candidates to combat the microfouling

As it was mentioned before the fouling is a natural process of colonization of submerged surfaces, involving a wide range of organisms from bacteria until invertebrates. Due to the diverse range of organisms involved, there are not specific assays that may show the anti-fouling potential of a compound, moreover the latest publications agree that no single substance could inhibited the settlement and growth of all the organisms implicated in the marine fouling. However, several bioassays have been developed to determine whether natural products inhibit specific organisms known to be involved in the microfouling process, mainly directed to understand the influence of the initial colonization by bacteria (microfouling) on the subsequent settlement and growth of macrofouling.

Because it is necessary to screen plenty of candidates in order to select the most promising among them, the identification of effective biocides and coatings requires laboratory tools development. The direct evaluation of the candidates in field conditions demands great amounts of each compound, and these assays are affected in an uncontrolled way by numerous factors, including the season, weather etc, so a previous selection is strongly recommended, and the most recommended way is the use of laboratory test. As a consequence, a number of laboratory-based AF assays have been developed in recent years; however few compounds have been tested in field assays or in moving water, which is needed to evaluate the ecological role of a putative antifouling compound. The test could be grouped in three main groups according with the target involved, microfoulers, macrofoulers and enzymes [16].

Because the wide diversity of organisms involved in the fouling process, several different AF targets (organisms) are required for proving the antifouling activity of a particular compound. So, a wide range of test organisms has been used in AF bioassays; however the selection of the target most be done according with the answer to be solved. E.g. For understanding the ecological process must be selected as target organisms those that reflect the potential micro- or macrofoulers of the studied species; for the discovery of new biocides, to be used in AF paints, the AF assays are conducted using the dominant fouling species in the area (marine bacteria, diatoms, algae, mussels and barnacles). However, the difficulty and cost of culturing higher benthic organisms has influenced the final choice of test organism. Some bioassays use a single species of micro- or macrofouler, but for microfouling assays, mixed consortia are sometimes used. In this sense, the most used microorganisms are strains of bacteria, isolated from marine biofilms, especially these pioneer strains since marine bacteria are relatively easy to isolate and maintain. Some diatoms have been also widely used for bioassays, and it is important because many AF coatings fail against microalgal slimes dominated by diatoms. Marine fungi have been used in a few studies, because they are not in the main group of fouler organisms. On the other hand, the most important macroorganisms involved in ship shield colonization are the barnacles; so on the most popular test is involved cypris of the subtropical barnacle *Balanus amphirite*. The second group of macrofouling organisms of importance in terms of the number of publications is the mussels, *Mytilus* spp are the most used for bioassays [16].

Growth inhibition test as microfouling assays could be done both disc diffusion and liquid media; however these assays are not so relevant because the fouling involves biofilm, and these assays do not take account this fact, furthermore, it is well known that the sensitivity of microorganisms growing in a biofilm is lower than those in planktonic culture. On the other hand, these tests could be automated and the compounds evaluated rapidly in multi-well plates by measuring the change in turbidity. E.g. For diatoms, the inhibition of growth can be evaluated in liquid flasks by measuring absorbance or chlorophyll a concentration. The attachment assays (Bacteria and Microalgae) are another test to evaluate microfouling activity; however, most of them generate an *in vitro* biofilm under static water conditions, with a few exceptions using flow chambers, and those conditions don not reflect the real fouling conditions. Additionally, a bacterial multispecies biofilms reflect in the best way the natural conditions because the role of synergy in the resistance to antimicrobial agents. Finally, nowadays it is recognized the fundamental role of Quorum Sensing in the biofilm consolidation and in the fouling process, due this fact the QS inhibition assays are being conducted to determinate the antifouling potential of pure compounds [8,16].

3. Asking for antifouling molecules with the use of bacterial sensors

QS inhibition has been a strategy of algae, animals, plants, bacteria and other microorganism to control its own population and to synchronize the expression of different phenotypes in a community. QS can be inhibited in several points of the communication circuits. The most used inhibition mechanisms are degrading the signaling molecules or competing with the signaling molecule for the binding site in the regulatory protein. Some bacteria have useful phenotypes regulated by QS, for example bioluminescence, or different pigments as violacein, these traits can be exceptional reporters for QS inhibition. Some of these genes have been used in synthetic biosensors to study with a molecular sight the mechanisms of QS inhibition. Those systems are the common tool to evaluate the activity of new molecules using them in simple Petri dishes assays. In this section we introduce our current classification of QS inhibition biosensors and the next generation of biosensors made to order.

3.1. Searching molecules into a complex ecosystem

In a complex social-competitive environment, organisms have developed several mechanisms to control their own populations; one of them is the inhibition of QS, a phenomenon called quorum-quenching (QQ) [30]. Many quorum-quenching molecules have been identified since the 2000 when Rasmussen and Co-workers reported the halogenated furanone produced by the seaweed *Delisea pulchra* (see [17,18]). They show how this molecule and some derivatives could affect the QS and the swarming motility in *Serratia liquefaciens*. Some of the examples are previously mentioned in this chapter, however it is important to mention that moreover that secondary metabolites there are other examples of QS inhibitors including AHL-lactonases, AHL-acylases and paraoxonases (PONs), which degrade AHL signals [19]. *Bacillus sp*. strain 240B1 produces an AHL-lactonase metalloprotein, encoded by the *aiiA* gene, able to attenuate the virulence of *Erwinia carotovora* [20], even many species of *Streptomyces* have been reported to encode AHL-acylases able to degradate AHL signals, decreasing the production of virulence factors such as elastases, proteases and LasA in *P. aeruginosa* [21]. Currently, the use of quorum-quenching molecules could be applied to the control of AHL-mediated pathogenicity and biofilm formation as has been proposed by Park and Co-workers [21].

Because of its own wide-diversity, searching of quorum-quenching molecules in the environment is a complex procedure. Different aspects must be considered, for example the complexity of the holobiont and the conditions that should be carefully selected depending on the organism that may be producing the molecules. In fact, the biggest problem is to be sure of the origin of such a molecule in a complex sample where bacteria, microalgae, fungi, invertebrates among other organism could be included and responsible of the activity. Consequently, the current challenge is to design a wide-ranging method that allows sensing these molecules avoiding the conflicts caused by the origin. The main problem after defining the chemical extraction required and the amount of sample according to the producer organism is it to find a wide detection method for the bioactivity.

Since years, biosensors have been used for the detection of QS activity, induction or inhibition, and recent studies suggest that the biosensor assays in the simple Petri dishes methodology are the best way, up to now, to detect such molecules. Some of the biosensors used as tool for quorum-quenching detection are wild type strains with reporter phenotypes that can be inhibited by the selected substance or could be mutant biosensors with the AHLs synthetase disrupted and a reporter gene induced by synthetic AHL's; this signalization is then impaired by the molecule being tested [22,23]. The main goal now is to find a wide range of detection because usually the biosensor strains have a specific LuxR type protein that can detect a specific range of AHL's.

3.2. Kind of quorum-quenching biosensors:

We define two main approaches to identify quorum-quenching molecules using the biosensors mentioned above: (1) Inhibiting a certain phenotype under the regulation of QS. (2) Disrupting the induction of QS regulated reporter gene in a biosensor strain stimulated with foreign AHL's. Both approaches are illustrated in the Figure 2.

3.2.1. Inhibiting a certain phenotype under the regulation of QS

The best example for the first approach is the inhibition assay using the *Chromobacterium violaceum* wild type as a biosensor strain. This strain is able to produce the pigment Violacein under the regulation of QS when a quorum-quenching molecule is present in the medium C. *violaceum* is not able to produce the pigment (Figure 3).

This approach use wild-type strains able to express phenotypes regulated by QS, some of the most used phenotypes are bioluminescence from *Vibrio* sp, pigments from *Serratia marcescens* or *Cromobacterium violaceum* or antibiotics from *Erwinia carotovora* among others. These phenotypes are multi-factorial traits, which mean there are many genetic and metabolic determinants involved. Consequently special condition should be provided in order to test the quorum-quenching activity of a selected molecule (Figure 3, left).

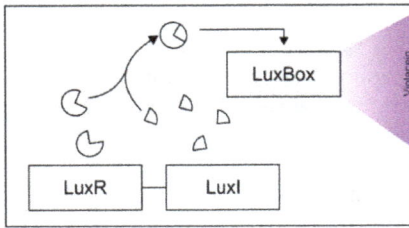

A low number of controled variables   A large number of controled variables

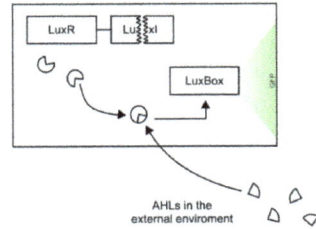

Figure 2. Approaches to identify quorum-quenching molecules. Left: Inhibiting a certain phenotype under the regulation of QS. This approach use wild-type strains able to express phenotypes regulated by QS. Right: Disrupting expression of a reporter gene regulated by QS in a biosensor mutant. This approach use biosensors strains designed by
transformation with synthetic plasmids. The arrows represent potential targets for quorum-quenching molecules:
Left: their own wild-type regulation circuit. Right: the communication between strains, the reporter system and the
stimulatory system.

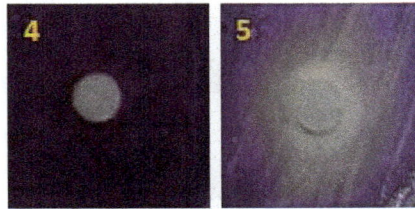

Figure 3. Inhibition of Violaceine in *Chromobacterium violaceum*: Right: control disk with no QS inhibition halo. Left:
Disk impregnated with a QSI compound (results under publication)

3.2.2. Disrupting the induction of QS regulated reporter gene in a biosensor strain stimulated with foreign AHL's

The second approach uses biosensors made with synthetic plasmids; those are more stable
and can be easily manipulated. The biosensors mutants in the Quorum Sensing systems
(mainly from Gram-negative bacteria) have been a helpful tool to analyze the communication system present in a complex environment sample. Therefore, a large number of strains,
genetically transformed with truncated QS circuits, have been developed (for review see
[22]). These mutants require AHLs in the media to induce the reporter phenotype regulated
by QS, consequently the media to test the quorum-quenching activity must be supplement-
ed with the respective AHL's. However, it should be mention that the QS mutated strains
have lots of problems with the laboratory maintenance and the growth conditions.

These transformed biosensors can be classified in two kinds concerning the response level of
the reporter gene:

a. The first one is the plasmid-based biosensors of overall population response, the best
    example are the plasmids pSB403 and pHV200I designed by Winson *et al.* 1998 [23].
    Both used the system LuxR/I able to sense the molecule 3-oxo-C6HL. The strains trans-
    formed with this plasmid are is able to sense the 3-oxo-C6AHL molecule in the growth
    media and induce the expression of the reporter system LuxCDABE This reporter sys-
    tem induces the expression of bioluminescence over all the population and the biolumi-
    nescence is the summatory of the entire bacterial culture.

b. The second type is the plasmid-based biosensors of single cell response, the best exam-
    ple is the plasmid pJBA-132 designed by Andersen *et al.* 2011. This plasmid used also
    the system LuxR/I but with the reporter gen *gfp* (green fluorescent protein). Certain bac-
    terial strain transformed with this plasmid is able to sense, as well the 3-oxo-C6AHL
    (because it has the same LuxR protein), in the growth media and induce the expression
    of GFP. The fluorescence of GFP can be determined in a single cell and therefore it can
    be quantified individually into a population or community.

Additionally, plasmid-based biosensors can be classified regarding the kind of promoter
used. These can be constitutive promoters (for example, the plasmid pSB403) or inducible
promotes (pHV200I). The use of inducible promoters allows the directly manipulation by
activating certain biosensor as desired. The most common used promoter is the PlacO1 promoter. Despite the role of PlacO1 promoter in lactose catabolism has been widely described,
we propose this promoter as a mechanism to switch between an activated or inactive state at
the genetic level of expression of certain reporter system. PlacO1 promoter regulation mechanism works as a response of environmental lactose levels. In an environment without lactose the protein LacI repressor binds to operator region of lac operon inhibiting the
transcription of genes under its control (inactive biosensor). On the other hand, in an environment with lactose or analogous (IPTG), this can bind to LacI to assemble a complex Lac-
tose-LacI and it is possible to induce the expression of genes (active biosensor). We used this
mechanism to switch our biosensor strain whenever we want it. These biosensors are used
in the screening of quorum-quenching molecules adding in the media the respective AHL
molecule to produce the specific stimulus. The use of plasmids with designed circuits improves the knowledge of the variables that should be controlled in the experiment, in con-
trasts with the wild type-strains biosensors mentioned above (Figure 3, right).

Nowadays, it is possible to select the biosensor following the questions we have for the
environmental sample more over a Synthetic Biology approach through successive steps
of in *silico* design, *in vitro* construction and *in vivo* expression is a very useful tool to allow scanning and testing quorum-quenching molecules in an artificial communicational
environment.

3.3. The next generation, biosensors made to order:

As an interesting perspective, we could select our desired biosensor-stimulatory environment from parts assembly by Synthetic Biology approaches to test quorum-quenching
molecules.

Dubrin *et al.* 2007 [24] define Synthetic Biology (SB) as a variety of experimental approaches with the aim to mimic o modify a biological system. This recent field of research has been widely studied [25-27]. A model to test quorum-quenching molecules through SB may be constructed following two main steps: (1) design and (2) manufacturing. Heinemann *et al.* 2006 [28] reported a deep description of each step.

For the particular case of systems of QS, a large number of studies have established a pattern of interactions, levels of regulation, transcription rates and other parameters relating to components of these networks [3,29,30]. This allows us to meet one of the main requirements in SB, in which you have sufficient data to perform a mathematical and computational modelling of various QS systems based on the parameters reported in the literature in order to predict the behaviour of the entire network under certain events that can be simulated (e.g. lactonase activity, AHL acylases concetrations, a mutation in the gene, response to external signals, receptor competition, etc).

The assumptions and standard parameters to be used when performing a computational or mathematical modeling of a QS system have been clearly described in the work of Garcia-Ojalvo, *et al.*, 2004 [31], McMillen *et al.*, 2002 [32] and Dockery & Keener 2001 [33].

The goal of synthetic biology is not only perform a mathematical modelling *in silico*, but to perform an assembly *in vitro* and monitoring *in vivo* of a particular genetic network. In that sense, it is necessary to establish requirements to design a model and that it can be viable at all levels (*in silico, in vitro* and *in vivo*).

Friesen *et al.*, 1993 [34] and Kaznessis 2007 [35] have defined four criteria to design a model in SB, these are: (1) there must be a network topology, where some biomolecules control the concentration of others. (2) Any unit of excitation must activate the system. (3) An oscillating system must include a restoration process that returns the oscillating system to steady state (negative feedback) and (4) there must be a process leading to overcoming the steady state values before the inhibition will be fully effective. A diagram illustrating the basic structure which must have a model designed in SB studies are presented in Figure 4.

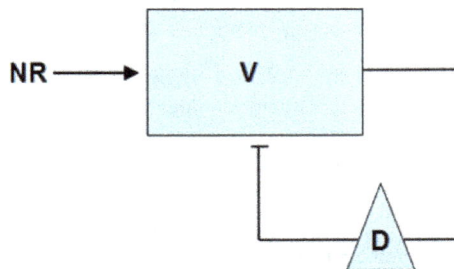

Figure 4. Adapted from Friesen *et al.* (1993) [34]. Basic structure of a model built in Synthetic Biology. NR represents the excitatory input may or may not be rhythmic. V represents the variables and D delays. Arrows indicate induction and perpendicular indicates inhibition.

Many components of QS systems have been employed in the construction of artificial modules and these meet the requirements presented in Figure 4.

Following these criteria it is possible to simulate a QS- communication environment when we can test our samples with the aim to search Quorum-Quenching activity and establishing the presence of quorum quenching molecules.

Finally, we can use bacterial as bio-factories to produce inhibitors molecules in higher amounts than classical approaches under regulation of inducible promoters by using easily manipulated strains.

4. Working model (combined strategy) to measure the potential of inhibitors in marine microfouling process

According to the current knowledge of the fouling process mentioned in the preceding sections, the selection of compounds capable of inhibiting it is not an easy task. For the search of good antifoulants most researchers in this area use the results of bioassays mainly against larvae or spores of macro-organisms. Once selected these compounds, they are applied as additives in antifouling paints. However, there is increasing evidence that microorganisms, in particular bacteria, re-colonize entire surfaces treated with commercial antifoulants [11]. Consequently, there is a real need to find new and more potent antifouling compounds effective against the microorganisms that are deposited in the first stage of the process of colonization.

Therefore, we have developed a strategy using natural compounds or their analogues to interfere with the microfouling considering that the process of biofouling consists of three main stages developed in a sequential manner (Figure 5). The first, usually known as microfouling (attachment of microorganisms mainly bacteria and algal cells), involves the adsorption of dissolved organic molecules; those molecules provide nutrients for attaching the primary colonizers. The subsequent stages involve the recruitment of invertebrate larvae and algal spores (macrofouling) [36,37]. However, it is important to have in mind that these stages can overlap, could be developed in a succession or occurs in parallel [8].

The Figure 5 which illustrates schematically the process of fouling. One can observe several key points within the microfouling step: firstly the reversible attachment of microorganisms (A) (mainly bacteria and algal cells, assay a, blue arrow), at this point they could be detached from the surface just by washing with water; secondly the irreversible attachment of those microorganisms to the solid surface (B) followed by cell division (assay b, red arrow) (C), growth (D), biofilm formation (assay c, yellow arrow) (E) and continuing recruitment (F). At this point Quorum Sensing and biofilm formation of bacteria which are found in all submerged structures in the marine environment frequently in association with algae, protozoa and fungi [38] are key points. Although it is likely that for the majority of organisms a biofilm surface is not a pre-requisite for settlement, in practice colonization by spores and larvae of fouling organisms almost takes place via a biofilmed surface (G). Therefore, we think that in the way to prevent fouling in marine submerged surfaces we could use natural compounds or their analogs selected through their evaluation as potential antifouling agents using antibacterial activity test against marine bacteria associated with heavy fouled surfaces, quorum sensing inhibition QSI and biofilm inhibition assays.

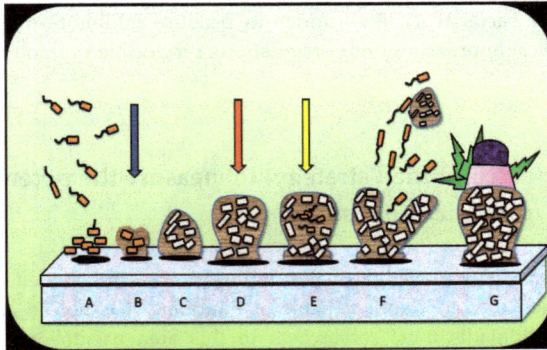

Figure 5. Scheme of the process of fouling. In the Figure 5 which illustrates schematically the process of fouling one can observe several key points within the microfouling step: firstly the reversible attachment of microorganisms (A) (mainly bacteria and algal cells, **assay a,** blue arrow), at this point they could be detached from the surface just by washing with water; secondly the irreversible attachment of those microorganisms to the solid surface (B) followed by cell division (**assay b,** red arrow) (C), growth (D), biofilm formation (**assay c,** yellow arrow) (E) and continuing recruitment (F). At this point Quorum Sensing and biofilm formation of bacteria which are found in all submerged structures in the marine environment frequently in association with algae, protozoa and fungi [38] are key points. Although it is likely that for the majority of organisms a biofilm surface is not a pre-requisite for settlement, in practice colonization by spores and larvae of fouling organisms almost takes place via a biofilmed surface (G). Therefore, we think that in the way to prevent fouling in marine submerged surfaces we could use natural compounds or their analogs selected through their evaluation as potential antifouling agents using antibacterial activity tested against marine bacteria associated with heavy fouled surfaces, quorum sensing inhibition QSI and biofilm inhibition assays.

Although, no laboratory bioassay could replicate the complex process of fouling since it involves a wide range of physical, chemical, and biological interactions [16]; recently, some technical approximations have been reported as indicators of the antifouling potential of chemical compounds. Those strategies use as a model the growth inhibition bioassay and the disruption of biofilm formation in bacteria associated with heavily fouled marine surfaces [8]. As we mention, it has been established that bacteria are the first organisms to colonize immersed surfaces [2] and they are usually associate with soft-bodied organisms or with inert surfaces. The formation of bacterial biofilms is regulated by QS systems, and it has been recognized as fundamental track for the attachment and growth of other organisms such as other bacteria, invertebrate larvae or spores of algae [15]. Thus, the conformation of the bacterial community is involved in the process of fouling by regulating the settlement of organisms and by promoting or inhibiting colonization [15,16,41,47].

Therefore, in our recent research on marine metabolites we have been focused on the search of compounds that exhibit antifouling properties mainly from octocorals that keep their surfaces free of fouling organisms [39-45] using their own metabolites as chemical defenses against the settlement and metamorphosis of invading species. After a quick *in vitro* antifouling test using the bacterial strains associated with fouled surface *Ochrobactrum pseudogringnonense, Alteromonas macleodii; Kocuria* sp., and *Oceanobacillus iheyensis* (described bellow in section a) [41] to evaluate 39 extracts of marine organisms, the *Eunicea knighti* and *Pseudoplexaura flagellosa* extracts showed the strongest antifouling properties [41,43-45]. Consequently, these octocorals *E. knighti* and *P. flagellosa* were collected in Santa Marta bay, Colombian Caribbean Sea by scuba diving. The animals were identified by Prof. Dr. S. Zea, and Prof. Dr. M. Puyana. The fresh coral colonies were immediately frozen after collection and remained frozen until extraction. The organisms *E. knighti* (650 g) and *P. flagellosa* (360 g) were cut in small pieces and separately extracted with a MeOH-CH$_2$Cl$_2$ (1:1 v/v) mixture, concentrated by rotary evaporation and the extracts obtained subjected to reversed-phase HPLC for final purification, to afford pure compounds 1-8 from *E. knighti* and 9-16 from *P. flagellosa* (Figure 6) as was reported by Tello *et al*, 2009 [43], 2011 [44] and 2012 [45].

The structures of compounds 1-16 (Figure 6) were established by the analysis of their spectroscopic features (MS, one- and two- dimensional NMR) and their absolute configurations were determined by a combination of chemical and NMR methods (multiple correlations observed in a ROESY and NOESY experiments, and by the modified Mosher method and the values of their optical rotations). Additionally, the stereostructures of compounds 9-16 (Figure 6) were confirmed by single-crystal X-ray diffraction.

The cembranoid type compounds isolated from *E. knighti* and *P. flagellosa* were subjected to Quorum Sensing and Bacterial Biofilm inhibition assays (Tables 1 and 2) finding interesting values for microfouling activity particularly for compounds 2-5, 7-9, 11, 12, 14, and 15. These results and their structure similarity, indicated an interesting structure-activity relationship, in this specific case of antifouling activity. For this reason, we envisaged a strategy to obtain diverse analogues with a high functional diversity, in order to enhance their activity making possible their use in antifouling technologies. Thus, we selected among the most active natural compounds, six lead compounds (2-5, 9, 11, and 12), based on their antifouling properties, on the easy access to substantial amounts of those compounds and on the presence of highly reactive functional groups in their structure. The latter made them suitable templates for the synthesis of analogues possessing uncommon structural features which enhanced its antifouling properties and leaves the possibility of assessing new interesting biological activities. Thus, we selected a group of regioselective, straightforward, fast, reproducible and high yield reactions to afford the synthetic analogues of cembranoids 17-52 (Figure 7) i.e. epoxide ring opening, oxidation reactions, treatment with iodine, photochemical reactions, methylation and acetylation, and cyclizations [46]. The analogues thus synthesized were subsequently subjected to antimicrofouling assays as explained below to evaluate their properties i.e. if they were capable of inhibiting marine bacteria involve in marine fouling, disrupt Quorum Sensing systems and avoid or inhibit biofilm formation and the subsequent steps of fouling as described in section 1.

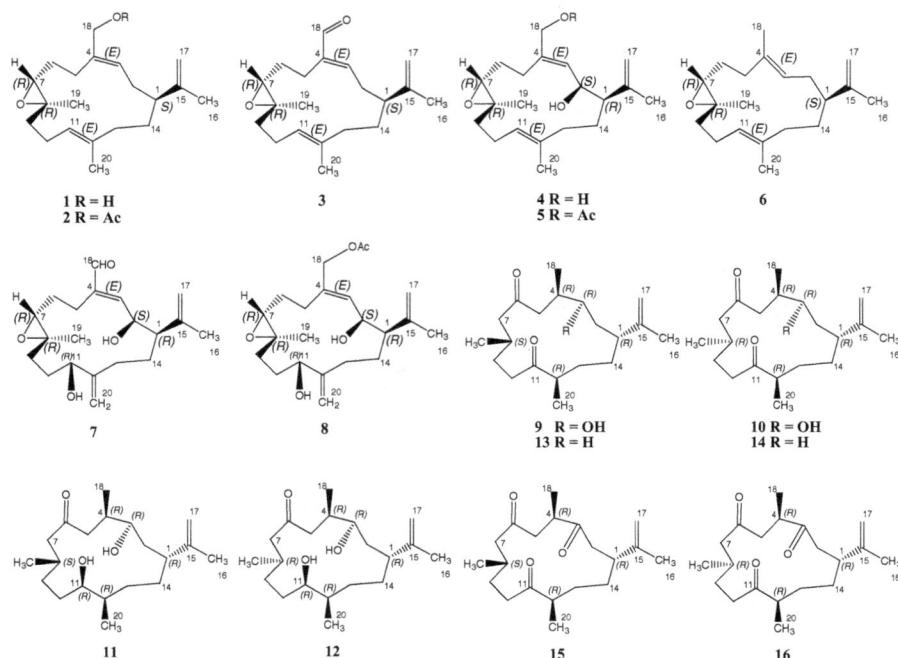

Figure 6. Natural products isolated from *E. knighti* (1-8) and *P. flagellosa* (9-16).

4.1. Marine bacteria involved in fouling process and their use as preliminar sensors to evaluate the antimicrofouling activity of compounds 1-16

As no laboratory bioassay could replicate the complex process of fouling as mentioned before, some technical approximations have been reported as indicators of the antifouling potential of chemical compounds [16]. These approximations use as one of the models, marine bacteria involved in fouling process.

Our experiments used bacterial strains associated with the fouled surface of the sponge *Aplysina lacunosa* and from the calcareous surface of a shell of the bivalve *Donax* sp. Those bacteria were collected at Santa Martha Bay in the Colombian Caribbean, and identified by the 16S rRNA sequencing as *Ochrobactrum pseudogringnonense*, *Alteromonas macleodii*; *Kocuria* sp., and *Oceanobacillus iheyensis* [41]. The natural compounds were tested in a growth inhibition assay against these marine bacterial strains, using the common antimicrobial test (disk–diffusion technique) [41,48,49]. The results showed in general that most of the assayed cembranoids exhibited high activity against most of the isolated bacteria, mainly *O. pseudogringnonense*. Specifically, compound 1 has a wide antibacterial activity and the best potency (halo >10 mm) of all compounds tested, although it was more specific against Gram (+) bacteria. Compounds 4 and 5 exhibited activity against isolates of *O. pseudogringnonense*. The

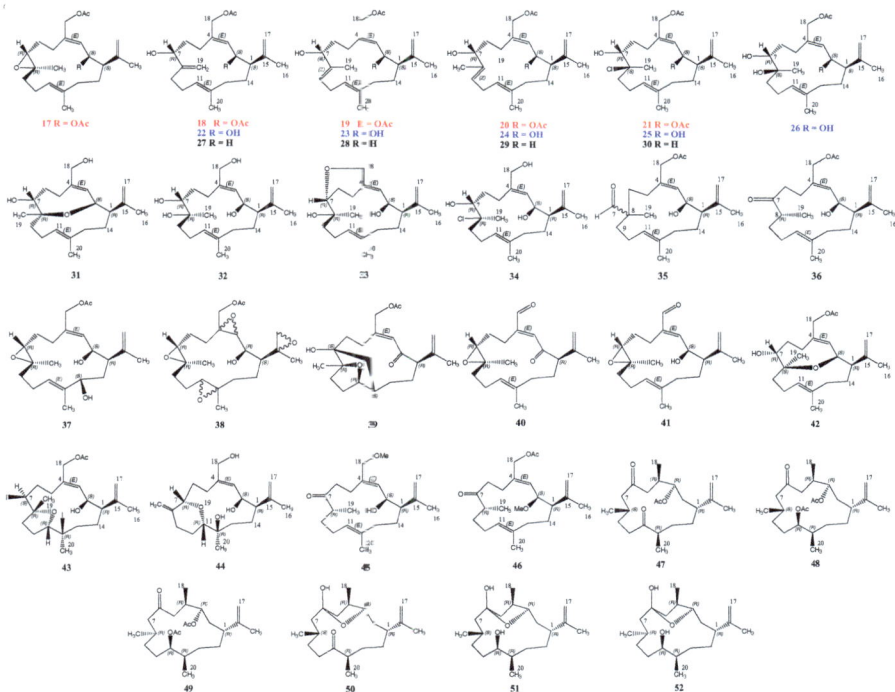

Figure 7. Analogues of cembranoids (17-52).

compound 3 showed a wide activity, but it was more powerful against Gram (+) bacteria as was reported previously by us in Tello *et al.*, 2009 [43].

4.2. Quorum sensing inhibition assay and its role as a tool to decide which of the natural and or the analogues synthesized exhibited the best potency as antimicrofoulants

The Quorum Sensing involves the cell control of bacterial population by communication using chemical signals (molecules) and a complex network of genetic circuits with a positive feedback regulation. Sensing these chemical signals bacteria can respond as groups and detect the "quorum" of a population in order to regulate different phenotypes [11,36], i.e. the biofilm formation and maturation [50] Thus, QS inhibitors can avoid the maturation of bacteria biofilms, and affect the attachment larval indirectly [15]. We decided to evaluate all 52 compounds (naturals and synthetics) in this bioassay, in order to establish whether these cembranoid compounds can interfere with the QS systems signals and at the same time if the chemical transformations to the natural cembranoids led to obtain synthetic analogues with a QS inhibitory activity increased. Thus, to evaluate the QSI activity of the natural and synthetic cembranoid analogues the *Chromobacterium violaceum* ATCC 31532 biosensor strain was used in a standard disk–diffusion assay [49]. The disks were loaded with μg amounts of every compound (1-52) as was previously reported by us in Tello, *et al*, 2011 [44], 2012a [45], and 2012b [46]. The known QS inhibitor kojic acid [11] was used as a positive control together with the common antifouling agent Cu_2O.

Assay	Quorum Sensing Inhibition[a] (μg/disk)[b]		
Compounds	*Chromobacterium violaceum*	Compounds	*Chromobacterium violaceum*
1	-	27	-
2	-	28	5.0
3	7.5	29	-
4	-	30	-
5	-	31	30.0
6	-	32	-
7	7.5	33	-
8	15.0	34	30.0
9	-	35	7.5
10	-	36	30.0
11	7.5	37	30.0
12	-	38	-
13	-	39	15.0
14	7.5	40	-
15	30.0	41	-
16	-	42	-
17	-	43	-
18	7.5	44	-
19	2.5	45	-
20	7.5	46	-
21	-	47	7.5
22	7.5	48	30.0
23	2.5	49	15.0
24	-	50	-
25	7.5	51	7.5
26	7.5	52	-
Kojic acid	90.0	[a]Solvent	-

[a]Activity was measured by the inhibition of violacein pigment. [b]Minimum quantity in μg per disk of compound required to inhibit violacein pigment. – No zone of inhibition was observed, even at 30 μg/disk.

Table 1. Quorum Sensing Inhibition activity of compounds 1-52.

All compounds listed in the Table 1 were tested in the QS inhibition assay and were evaluated. The results showed mainly, that only six of the natural compounds showed QSI activity, being the most active compounds 3, 7, 11, and 14 (all 7.5 µg/disk). The synthetic compounds that showed the best QSI activity were 19 (2.5 µg/disk), 23 (2.5 µg/disk), and 28 (5.0 µg/disk) which present a similar structure and functional groups. All in all, the carried out reaction resulting in an enhanced QS Inhibitors compounds, because, eighteen active analogue compounds were obtained and is noteworthy to say that the most active cembranoid analogues (19, 23, and 28) were most potent than the most active natural products 3, 7, 11, and 14, achieving the main proposal of the work [46].

4.3. Disruption of Bacterial biofilm formation and maturation as indicator of antifouling potency of natural and semisynthetic cembranoid analogues

Bacterial biofilms have long been recognized as fundamental settlement cues for many invertebrate larvae that colonize hard substrata, such as sponges, cnidarians, mollusks, barnacles, bryozoans and ascidians [8] This intrinsically complex process is the result of a network of interactions both in the pioneering biofilm and in the community of colonizers [14]. These interactions in a biofilm determine the composition of different populations and the establishment in a specific environment, all these together create a chemical pattern that gives specific signals for the subsequent colonization of other organisms.

Biofilm inhibition assay[a]			
Compounds	Pseudomonas aeruginosa	Staphylococcus aureus	Vibrio harveyi
1	4,1	2,0	0,3
2	12,8	1,3	9,8
3	9,3	3,2	1,0
4	14,7	0,01	>100,0
5	5,0	0,3	>100,0
6	8,3	0,5	>100,0
7	10,1	0,01	80,2
8	17,8	0,8	>100,0
9	4,0	10,1	17,1
10	4,5	10,0	69,7
11	11.5	"/>100,0	11,0
12	17.2	>100,0	>100,0
13	9,2	20,9	0,3
14	6,4	1,0	>100,0
15	12.2	5,7	9,5
16	6,8	1,4	53,8
Kojic acid	17.2	24.7	>100,0
Solvent	N	NI	NI

[a]Concentration that inhibits 50% of the biofilm and expressed in ppm. NI Non biofilm inhibition.

Table 2. Biofilm inhibition activity of compounds 1-16.

For biofilm inhibition assays, *Pseudomonas aeruginosa* ATCC 27853 (Gram negative), *Staphylococcus aureus* ATCC 25923 (Gram positive) and *Vibrio harveyi* PHY-2A (Gram negative) were used as model bacteria. The pure compounds 1-52 were evaluated in this bioassay to test if they are capable of inhibit the bacterial biofilm formation [44-46]. The bioassay was performed in a 96-well polystyrene microtiter dishes as was previously described in Tello, *et al*, 2011 [44]. For the natural compounds the IC_{50} values were calculated (Table 2).

The results of the disruption of Bacterial biofilm formation showed that approximately 60% of compounds inhibited the biofilm formation in a 50% extension, in at least one strain of the three used for this bioassay. Most of the cembranoid compounds showed outstanding biofilm inhibition activity against all the strains used in this bioassay, mainly against the Gram positive bacterium *S. aureus*. Thereby, fifteen compounds inhibited the bacterium *S. aureus* at lower concentrations than 1.0 ppm, being the compounds 4, 7, 18, 19, 25, 35, and 36 the most active even better than the known QS inhibitor kojic acid (24.7 ppm). Against *P. aeruginosa* nine compounds showed IC_{50} values lower than 10.0 ppm and the most active compounds were 1, 5, 9, and 10. The control kojic acid showed an IC_{50} of 17.2 ppm. Finally, against *V. harveyi* just three natural compounds (1, 3, and 13) presented lower IC_{50} values than 1.0 ppm,, the kojic acid presented an IC_{50} value upper than 100 ppm against this strain (data being published). The kojic acid has been proved by Dobretsov *et al.* [11] to have the ability to prevent microfouling in a controlled mesocosm experiment by inhibition of microbial communities on glass slides, decreasing the densities of bacteria and diatoms. Additionally, the biofilm inhibition in all cases was achieved without interference in the bacterial growth.

Then, based on the results obtained in the above assays for all compounds (1-52), it was possible establish the natural compounds 3 and 7, and the synthetic compounds 18, 19, 25, 35, and 36 as the most active cembranoids in the anti-microfouling bioassays, which could be used as additive in antifouling coating.

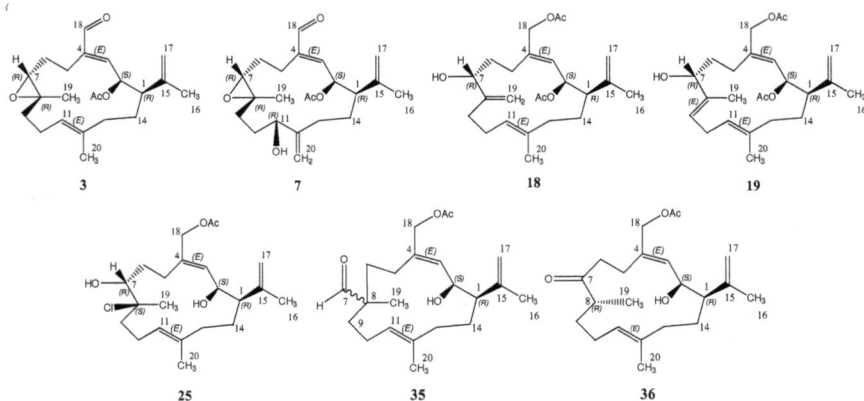

Figure 8. Most active compounds in the anti-microfouling bioassays.

4.4. Field test

Based on the previous results we chose compound (19), one of the most active compounds in the antimicrofouling bioassays to evaluate its activity in a field experiment with natural conditions (the sea) as additive in an industrial coating (Table 3). Thus, white ceramic panels (12 cm × 12 cm × 5 mm) used as surface were polished and then coated with the test paints (code I and II), a copper-based paint (30% of Cu^2O), and a kojic acid-based paint (0.5% of kojic acid).

Code	Paint	Compound 19	epoxymastic (marine paint)	catalyst	Acetone
I	1	2.0%	47.0%	47.0%	4.0%
II	1	0.5%	48.0%	48.0%	3.5%

Table 3. Formulation of the paints used in field tests.

Field experiments were conducted at Rosario Islands, Caribbean coast of Colombia, between 25 April and 9 July of 2012. The test panels were placed at random at a depth of 2 to 6 meters and vertically using SCUBA diving.

The results showed that the copper-based paint was cover with microfouling during the test period, otherwise showed good antifouling activity against macroorganisms after 75 days, the panel control was completely fouled mainly by algae, the test panels treated with the compound 19 completely prevented settlement of macrofouling during the tested period, further, the panels were slightly fouled with microalgae but remained most free of fouling and showed excellent antifouling performance after 75 days of exposure. Finally, barnacles and bivalves were observed alongside the pier wall, but were not found on the test panels. The results of biofilm inhibition activity and the field test will be published in near future.

As summary, QSI and Bacterial biofilm inhibitors compounds were more effective than the known antifoulants kojic acid and Cu_2O, and is noteworthy that the most of the analogues of cembranoids were significantly less toxic against the tested bacteria (*C. violaceum*, *V. harveyi*, *S. aureus* and *P. aureginosa*). Thus, marine cembranoids are recognized to be promising environmentally friendly candidates to be included in industrial coatings as antimicrofouling agents. The results showed that the analogue 19 completely prevented settlement of macrofouling and inhibited most of the microfouling during the period of testing. The above support the use of cembranoids (particularly compound 19) as excellent candidates to be used as antifoulant agents in a commercial antifouling paint, but because of its structural complexity, the synthesis of more simple compounds together with QSAR studies should be the next step in the search for potential non-toxic antimicrofoulant agents.

5. Concluding remarks

Although it is well known that in the marine environment all natural and artificial substrata are quickly colonized by micro- and macroorganisms in a complex physical, chemical and biological process, only recently the studies on microfouling have emerged as a key area of research, in an effort to develop more efficient and environmental friendly antifoulants. Since, the formation of a biofilm is considered an initial step in the development of fouling and taking into consideration that QS controls bacterial biofilm differentiation and matura-tion, the use of chemical compounds (natural or synthetic analogs) as interferences in these processes has been proposed as one potential approach for controlling microfouling. Fur-thermore, since larvae of many marine invertebrates preferentially settle on bacterial bio-film, disruption of bacterial biofilm could lead to the reduction of macrofouling of submerged surfaces as well. In this sense, this chapter, that includes our own working mod-el (combined strategy) to measure using bacterial sensors, the potential of marine isolated compounds and their synthetic analogs as inhibitors in marine microfouling, should surely contribute to the significant expansion of this area of research.

Author details

Carmenza Duque[1], Edisson Tello[1], Leonardo Castellanos[1], Miguel Fernández[2] and Catalina Arévalo-Ferro[1]

1 Departamento de Química, Universidad Nacional de Colombia, Colombia

2 Departamento de Biología, Universidad Nacional de Colombia, Colombia

References

[1] Fusetani N. Biofouling and antifouling. Natural Products Report 2004; 21 94-104.

[2] Krug PJ. Defense of Benthic Invertebrates Against Surface Colonization by Larvae: A Chemical Arms Race in Antifouling Compounds. In: Fusetani, N. Clare A.S. (Eds.) Antifouling Compounds. Berlin-Heidelberg, Springer; 2006. p1-53.

[3] Fuqua WC, Winans SC, Greenberg EP. Quorum Sensing in Bacteria: the LuxR-LuxI Family of Cell Density-Responsive Transcriptional Regulators. Journal of bacteriolo-gy 1994; 176(2) 269-275.

[4] Arai T, Harino H, Ohji M. In: Langston WJ. (Ed) Ecotoxicology of Antifouling Bio-cides. Japan, Springer; 2009, p437.

[5] Whelan A, Regan I. Antifouling strategies for marine and riverine sensors. Journal of Environmental Monitoring. 2006; 8 880-886.

[6] Navarro-Villoslada J, Orellana G, Moreno-Bondi MC, Vick T, Driver M, Hildebrand G, Liefeith K. Fiber-optic luminescent sensors with composite oxygen-sensitive layers and anti-biofouling coatings. Analytical Chemistry2001; 73 5150-5156.

[7] Konstantinou I. Antifouling Paint Biocides. Springer; 2006.

[8] Fusetani N. Antifouling marine natural products. Natural Product Report 2011; 28 400-410.

[9] Omae I. General Aspects of Natural Products Antifoulants in the Environment. In Konstantinou I. (Ed) Antifouling Paint Biocides. Germany, Springer; 2006. pp227-260.

[10] Qian PY, Xu Y, Fusetani N. Natural products as antifouling compounds: recent progress and future perspectivesBiofouling 2010; 26(2) 223–234.

[11] Dobretsov S, Teplitski M, Bayer M, Gunasekera S, Proksch P, Paul, V.J. Inhibition of marine biofouling by bacterial quorum sensing inhibitors. Biofouling 2011; 27(8) 893-905.

[12] Marechal JP, Hellio C. Challenges for the Development of New Non-Toxic Antifouling Solutions. International Journal of Molecular Sciences 2009; 10 4623-4637.

[13] Miller MB, Bassler BL. Quorum sensing in bacteria. Annual Reviews in Microbiology 2001; 55(1) 165-199.

[14] Joint I, Tait K, Wheeler G. Cross-kingdom signalling: exploitation of bacterial quorum sensing molecules by the green seaweed Ulva. Philosophical transactions of the Royal Society of London. Series B, Biological sciences 2007; 362(1483) 1223-1233.

[15] Dobretsov S, Qian PY. The role of epibotic bacteria from the surface of the soft coral Dendronephthya sp. in the inhibition of larval settlement. Journal of Experimental Marine Biology and Ecology 2004; 299(1) 35-50.

[16] Briand JF. Marine antifouling laboratory bioassays: an overview of their diversity. Biofouling 2009; 25 297–311.

[17] Nadell CD, Xavier JB, Levin SA, Foster KR. The Evolution of Quorum Sensing in Bacterial Biofilms. PLoS Biology 2008, 6(1) e14.

[18] Rasmussen TB, Manefield M, Andersen JB, Eberl L, Anthoni U, Christophersen C, Steinberg P, Kjelleberg S, Givskov M. How Delisea pulchra furanones affect quorum sensing and swarming motility in Serratia liquefaciens MG1. Microbiology (Reading, Engl.) 2000; 146 3237–3244.

[19] Dong YH, Wang LH, Zhang LH. Quorum-quenching microbial infections: mechanisms and implications. Philosophical Transactions of the Royal Society B; Biological Science 2007; 362, 1201–1211.

[20] Dong YH, Xu JL, Li XZ, Zhang LH. AiiA, an enzyme that inactivates the acylhomoserine lactone quorum-sensing signal and attenuates the virulence of Erwinia carotovora. Proccedings of the National Academy of Scieces U.S.A. 2000; 97 3526–3531.

[21] Park SY, Kang HO, Jang HS, Lee JK, Koo BT, Yum DY. Identification of extracellular N-acylhomoserine lactone acylase from a Streptomyces sp. and its application to quorum quenching. Applied and Environmental Microbiology 2005; 71 2632–2641.

[22] Steindler L, Venturi V. Detection of quorum-sensing N-acyl homoserine lactone signal molecules by bacterial biosensors. FEMS Microbiology Letters 2007; 266 1-9.

[23] Winson, MK, Swift S, Fish L. Construction and analysis of luxCDABE-based plasmid sensors for investigating N-acyl homoserine lactone-mediated quorum sensing. FEMS Microbiology Letters 1998; 163 185-192.

[24] Drubin DA, Way JC, Silver PA. Designing biological systems. Genes & Development 2007; 21 242-254.

[25] McDaniel R, Weiss R. Advances in synthetic biology: on the path from prototypes to applications. Current Opinion in Biotechnology 2005; 16 476-483.

[26] Andrianantoandro E, Basu S, Karig DK, Weiss R. Synthetic biology: new engineering rules for an emerging discipline. Molecular Systems Biology 2006; 2 2006.0028.

[27] De Lorenzo V, Danchin A. Synthetic biology: discovering new worlds and new words. The new and not so new aspects of this emerging research field. EMBO Reports 2008; 9 822-827.

[28] Heinemann M, Panke S. Synthetic biology--putting engineering into biology. Bioinformatics 2006; 22 2790-2799.

[29] Reading NC, Sperandio V. Quorum sensing: the many languages of bacteria. FEMS Microbiology Letters 2006; 254 1-11.

[30] Zhang HB, Wang LH, Zhang LH- Genetic control of quorum-sensing signal turnover in *Agrobacterium tumefaciens*. Proceedings of the National Academy of Sciences 2002; 99 4638-4643.

[31] Garcia-Ojalvo J, Elowitz MB, Strogatz SH. Modeling a synthetic multicellular clock: Repressilators coupled by quorum sensing. Proceedings of the National Academy of Sciences of the United States of America 2004; 101 10955-10960.

[32] McMillen D, Kopell N, Hasty J, Collins JJ. Synchronizing genetic relaxation oscillators by intercell signaling. Proceedings of the National Academy of Sciences of the United States of America 2002; 99 679-684.

[33] Dockery JD, Keener JP. A mathematical model for quorum sensing in Pseudomonas aeruginosa. Bulletin of Mathematical Biology 2001; 63 95-116.

[34] Friesen WO, Block GD, Hocker CG. Formal Approaches to Understanding Biological Oscillators. Annual Review of Physiology 1993; 55 661-681.

[35] Kaznessis Y. Models for synthetic biology. BMC Systems Biology 2007; 1:47.

[36] Wahl M. Marine epibiosis. I. Fouling and antifouling. Marine Ecology Progress. 1989; 58 175-189.

[37] Aldred N, Clare AS. The adhesive strategies of cyprids and development of barnacle-resistant marine coatings. Biofouling 2008; 24(5) 351-363.

[38] Callow JA, Callow ME. Biofilms. In: Fusetani N. Clare AS. (Eds.) Antifouling Compounds. Berlin-Heidelberg, Springer; 2006; p141.

[39] Cuadrado T, Castellanos L, Osorno O, Ramos FA, Duque C, Puyana M. Estudio Químico y Evaluación de la Actividad Antifouling del Octocoral Caribeño Eunicea laciniata. Quimica Nova 2010; 33 656-661.

[40] Mayorga H, Urrego NF, Castellanos L, Duque C. Cembradienes from the Caribbean Sea whip Eunicea sp. Tetrahedron Letters 2011; 52 2515-2518.

[41] Mora-Cristancho A, Arévalo-Ferro C, Ramos FA, Tello E, Duque C, Lhullier C, Falkenberg M, Schenkel EP. 2011. Antifouling Activities against Colonizer Marine Bacteria of Extracts from Marine Invertebrates Collected in the Colombian Caribbean Sea and on the Brazilian Coast (Santa Catarina). Verlag der Zeitschrift für Naturforschung C 2011; 66 515-526.

[42] Reina E, Puentes C, Rojas J, García J, Castellanos L, Aragón M, Ospina LF, Ramos F. Fuscoside E: A strong anti-inflammatory diterpene from Caribbean octocoral Eunicea fusca. Bioorganic Medicinal Chemistry Letters 2011; 5888-5891.

[43] Tello E, Castellanos L, Arevalo-Ferro C, Duque C. Cembranoid diterpenes from the Caribbean sea whip Eunicea Knighti. Journal of Natural Products 2009; 72 1595-1602.

[44] Tello E, Castellanos L, Arevalo-Ferro C, Rodríguez J, Jimenez C, Duque C. Absolute Stereochemistry of Antifouling Cembranoid Epimers at C-8 from the Caribbean Octocoral Pseudoplexaura flagellosa. Revised structures of plexaurolones. Tetrahedron 2011; 67 9112-9121.

[45] Tello E, Castellanos L, Arevalo-Ferro C, Duque C. Disruption in Quorum sensing Systems and Biofilm Inhibition of Bacteria by Cembranoid Diterpenes Isolated from the Octocoral Eunicea knighti. Journal of Natural Products 2012; 75 1637-1642.

[46] Tello E, Castellanos L, Duque C. Synthesis of Cembranoid Analogues and Evaluation of their Potential as Quorum Sensing Inhibitors. Bioorganic and Medicinal Chemistry 2012. In press. DOI 10.1016/j.bmc.2012.10.022.

[47] Qian P-Y, Lau SCK, Dahms H-U, Dobretsov S, Harder T. Marine biofilms as mediators of colonization by marine macroorganisms: implications for antifouling and aquaculture. Marine Biotechnology 2007; 9 399–410.

[48] Hostettman K, Wolfender L, Rodríguez S. Rapid detection and subsequent isolation of bioactive constituents of crude plant extracts. Planta Medica 1997; 63 2-10.

[49] National Committee for Clinical Laboratory Standards. Performance standards for antimicrobial disk susceptibility test. Fourteenth Informational Supplement. NCCLS document M100-S14. NCCLS, Wayne, PA, 2004.

[50] Dobretsov S, Dahms HU, Qian PY. Inhibition of biofouling by marine microorganisms and their metabolites. Biofouling 2006; 22 43–54.

Chapter 4

Recent Progress in Optical Biosensors for Environmental Applications

Feng Long, Anna Zhu, Chunmei Gu and Hanchang Shi

Additional information is available at the end of the chapter

1. Introduction

The rapid screening and sensitive monitoring of environmental pollutants, such as pesticides, persistent organic pollutants (POPs), endocrine disrupting chemicals (EDCs), explosives, and toxins, is indeed essential to ensure environmental quality, and therefore, human health. Until recently, the quantification of most contaminants has been limited to the traditional chromatographic and spectroscopic technologies. These methods, although accurate with low detection limits, are labor-intensive and require expensive and sophisticated instrumentation, as well as complicated and multistep sample preparation, which prohibits frequent and real-time on-site monitoring of contaminants in environment.[1] Considerable research interests, therefore, have risen for detecting low levels of environmental pollutants in biosensor development because of their simplicity, robustness, sensitivity, specificity and cost-effectiveness. [2]

A biosensor is an analytical device combined a biological sensing element with a physical transducer, in which the binding or reaction between the target and the recognition element is translated into a measurable electrical signal. [3] Among them, optical biosensors are powerful alternative to conventional analytical techniques due to their cost-effective, fast and portable detection, which makes on-site and real-time monitoring possible without extensive sample preparation. Optical biosensors have vast potential applications in environmental monitoring, food safety, drug development, and medical diagnosis.[4] Although the use of optical biosensors in water quality early-warning and pollution control is still in its infancy, research on this topic is an active area and the remarkable technological progress has been made.

The present article gives an overview of the recent advances in optical biosensors and their applications in the environmental field. Functional biorecognition materials (e.g. enzyme, antibody, aptamer, and DNAzyme), a key component of biosensor and specifically binding a broad range of analytes including inorganic, organic, and biomolecules, will be first reviewed. Then, nanomaterials such as quantum dots, graphene, nanogold particles, carbon nanotubes, and magnetic nanoparticles will be introduced, which have been successfully incorporated into optical biosensors to improve the sensibility, sensitivity, and selectivity due to their unique physicochemical properties. In addition, the recent significant improvements in instrumentation will also be discussed, which have allowed a wider variety of pollutants to be analysed in details, and led to the increasing application of optical biosensor technology throughout the environmental detection field. Finally, recent developments of optical biosensors for pollution control and early-warning will be highlighted.

2. Functional biorecognition materials

Functional biorecognition materials are key components of biosensors, and generally have high affinity (low detection limit), high specificity (low interference), wide dynamic range, fast response time, and long shelf life. The antibodies are most frequently used biorecognition molecules in the optical biosensor community. However, enzymes were the first recognition elements used in biosensors. Another frequently used recognition elements are nucleic acids such as aptamer and DNAzyme for the monitoring of environmental pollutants.

2.1. Antibody

Immunosensors, based on specific antigen-antibody interactions, have become the gold-standard technique in clinical diagnostics and environmental monitoring.[4-9] Antibody is a large Y-shaped protein used by the immune system to identify and neutralize a unique part of foreign target, called an antigen, and is produced by white blood cell (a plasma cell). Antibodies are typically made of basic structural units: each with two larger heavy chains and two shorter light chains. The IgG molecule (see Figure 1) is the most used antibody type and is about 150-kDa protein composed of four polypeptide chains.[8] Antibodies are produced as monoclonal and polyclonal varieties, with monoclonal antibodies binding to a single epitope and polyclonal antibodies being capable of binding to multiple epitopes.[9,10] In immunoassay, two antigen binding sites of antibody have a highly specific interaction for one particular target, and this immunochemical reaction can be detected by the transducer (e.g. optical, electronical). [5-9] Therefore, the immunosensor assay provides a highly repeatable and highly specific reaction format, and the capacity for specific recognition of environmental contaminants.

Due to most of the environmental pollutants have the low molecular weight (<1 kDa) and are called haptens which are non-immunogenic, it has to be conjugated to carrier proteins to make them immunogenic.[11] Preparation of antibodies against haptens, such as pesticides, persistent organic pollutants (POPs), and endocrine disrupting chemicals (EDCs), is based on covalent binding of the hapten to a carrier protein and immunisation of animals by the synthesised immunogens. The specificity of antibody is important for immunoassay, while the specificity and quality of antibody is mostly determined by the manner of chemical binding of the hapten to the carrier protein, called complete antigen. In our group, the complete antigen of microcystin-LR (MC-LR-BSA), the most frequent and most toxic hepatotoxin, was synthesized by introducing a primary amino group in the seventh N-methyldehydroalanine residue, and then the product aminoethyl-MC-LR was coupled to bovine serum albumin (BSA) by glutaraldehyde.[11] The residue is located most distantly from both of the variable amino acid residues and Adda, promising active and possibly more specific immunoreactivity. Polyclonal antibodies and a monoclonal antibody (Clone MC8C10) against MC-LR were generated by immunization with MC-LR-BSA, respectively. An indirect competitive enzyme-linked immunosorbent assay (ic-ELISA) with MC8C10 was established to detect the MCs in waters, which showed highly specificity with MC-LR and have a detection limit for MC-LR 0.1 µg L^{-1}.[12]

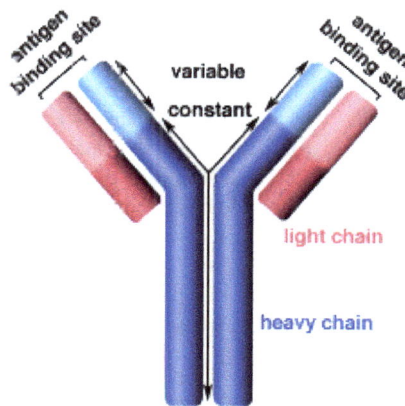

Figure 1. The structure of antibody

A compact, portable, multichannel fiber-optic instrument was reported to detect four targets simultaneously using antibody immobilized fiber-optic probes. [13] This biosensor was simultaneously able to determine 10^5 cfu/mL of *Bacillus globigii*, 10^7 cfu/mL of *Erwinia herbicola*, and 10^9 pfu/mL of MS2 coli phages. A biosensor platform (Analyte 2000) developed by the Naval Research Laboratory (USA) was used to simultaneous determine both the explosives 2,4,6-trinitrotoluene (TNT) and hexahydro-1,3,5-trinitro-1,3,5-triazine (RDX). [14] The limit of detection of TNT and RDX was 5 µg/L and 2.5 µg /L, respectively.

The main advantage of immunosensors over other immnunological methods (e.g., ELISA formats) is the better regeneration and binding properties of the sensing surface, which is critical for the successful reuse of the same sensor surface and the accuracy of detection results.[15] In environmental analysis, targets interest are usually small molecule substances (molecular weight <1kDa), which are greatly difficult to be directly immobilized onto the biorecognition sensing surface. Therefore, antibody immobilisation is always utilized in preparing sensor surface of immunosensors.[5-8,16] However, the control over the number of antibodies and their orientation and position relative to the sensor surface is very difficult. Because of the possibility of inadvertently disrupting the binding site when conjugating antibody with active surface of sensor, the activity loss of antibody is inevitable.[17,18] Most important, due to strong acid being usually used in regeneration process, the recognition ability of antibodies immobilized may be lower after senor surface reuse, which will affect the stability and reliability of the immunosensor. The cycles of regeneration are usually no more than fifteen times and in each cycle, antibody activity decreased, which leads to inaccurate detection results. Therefore, hapten-carrier-protein conjugates as bio-recognition molecules were immobilized onto the surface of immunosensor for obtaining the stable reusable sensor. For example, a reusable immunosurface is provided via the covalent attachment of the 2,4-D-BSA and MC-LR-OVA to a self-assembled monolayer formed onto the fiber optic sensor.[19] The regeneration of the sensor surface allows the performance of more than 100 assay cycles.

2.2. Enzyme

Enzymes are biological molecules that catalyze (i.e., increase the rates of) chemical reactions, and are usually very specific as to which reactions they catalyze and the substrates that are involved in these reactions. Enzymes are historically the first molecular recognition elements included in biosensors and continue to be the basis for a significant number of publications reported for biosensors in general as well as for environmental applications.[4] Enzyme biosensors have several advantages, such as a stable source of material, the possibility of modifying the catalytic properties or substrate specificity by means of genetic engineering, and catalytic amplification of the biosensor response by modulation of the enzyme activity with respect to the target analyte. [20]

Most of enzyme biosensors normally use enzymes as the bioreceptors and achieved pollutants detection based on the enzyme inhibition mechanism.[4][20] Due to ChE enzymes can be inhibited by several toxic chemicals such as organophosphate and pesticides, heavy metals, and toxins, the ChEs biosensors is of particular interest in the area of global toxicity monitoring.[4,21,22]

Due to various pollutants that inhibit the activity of enzymes in a different ways, multi-analytes detection can be achieved by enzyme sensors. For example, simultaneous detection both pesticides and heavy metal ions in a sample solution is possible due to selective inhibition of butyrylcholine esterase by pesticides and urease by heavy metals ions.[4-6,23] Comparing with the inhibition level of urease or butyrylcholine esterase, respectively, the pesticides or heavy metal ions can be determined. The enzyme-inhibition based biosensors array could achieve the simultaneous determination of various pollutants in water samples.

Enzyme biosensors have some limitations for the detection of environmental pollutants, which include the limited number of substrates for which enzymes have been evolved, the limited interaction between environmental pollutants and specific enzymes, and the lack of specificity in differentiating among compounds of similar classes.[6,23] However, artificial or synthetic enzymes could be a useful alternative to natural enzymes for the development of new biosensors, which are more robust, available, chemically malleable and cheap, in comparison with their natural analogues.

No	Target	Aptamer type	Sequence	Reference
1	2,3',5,5'-Tetrachlorobiphenyl; 2,3,3',4,5-Pentachlorobiphenyl	DNA	9.1:CGCTACACCT,CGCCAGCAAA,TTGCCGCCCG,CAGCCCTCTA 9.2:GGGACTCGAG,ACCCGTTCCG,TTCTCCGCTT,GCCCCACAAT	[26]
2	4,4'-methylenedianiline(MDA)	RNA	M1:CUGCGAUCA,GGGGUAAAUU,UCCGCGCAGG,CUCCACGCCG,C M2:CUCGA,GUCCUCUUGA,GCGGUUCCUA,CUUCCCUCUG,CUGUG	[27]
3	Organophosphorus Pesticides:phorate,profenofos, isocarbophos, omethoateas	DNA	1: AAGCTTGCTTTATAGCCTGCAGCGAT TCTTGATCGGAAAAGG CTGAGAGCTACGC 2:AAGCTTTTTTGACTGACTGCAGCGATTCTTGATCGCCACGGTCTGGAAAA AGAG	[28]
4	bisphenol A	DNA	CCGGTGGGTGGTCAGGTGGGATAGCGTTCCGCGTATGGCCCAGCGCATC ACGGGTTCGCACCA	[29]
5	17β-estradiol	DNA	GCTTCCAGCTTATTGAATTACACGCAGAGGGTAGCGGCTCTGCGCATTCA ATGCTGCGCGCTGAAGCGCGGAAGC	[30]
6	Chloramphenicol	DNA	1:ACTTCAGTGA,GTTGTCCCAC,GGTCGGCGAG,TCGGTGGTAG 2:CACCAAGCGC,AGGGAATTAC,ATTGAAGTGT,GGGATTGGCT	[31]
7	Oxytetracycline	DNA	1:CGACCGCAGGTGCACTGGGCGACGTCTCTGGTGTGGTGT 2:CGACGCGCGTTGGTGGTGGATGGTGTGTTACACGTGTTGT	[32]
8	Tetracycline	DNA	T7:GGGCAGCGGTGGTGTGGCGGGATCTGGGGT,TGTGCGGTGT T15:GGAGGAACGGGTTCCAGTGTGGGGTCTATC,GGGGCGTGCG	[33]
9	Kanamycin	DNA	TGGGGGTTGAGGCTAAGCCGA	[34]
10	ricin B chain	DNA	ACACCCACCGCAGGCAGACGCAACGCCTCGGAGACTAGCC	[35]
11	Ochratoxin A	DNA	GATCGGGTGTGGGTGGCGTAAAGGGAGCATCGGACA	[36]
12	E. coli	DNA	ATCCGTCACACCTGCTCTACGGCGCTCCCAACAGGCCTCTCCTTACGGCAT ATTA TGGTGTTGGCTCCCGTAT	[37]
13	Staphylococcus aureus Enterotoxin B	DNA	GGTATTGAGGGTCGCATCCACTGGTCGTTG TTGTCTGTTGTCTGTTATGTTGTTTCGTGATGG CTCTAACTCTCCTCT	[38]
14	Salmonella entericaserovars	DNA	23:CCGCCTTTACTAAATTGACGAACATAGGAATCAATGAAGC 24:GGGAGTCAGAACGCCTGGGCAAGCATAGTACTCGCCGGAA	[39]
17	Ibuprofen	DNA	IBA2:ACAGTAGTGAGGGGTCCGTCGTGGGGTAGTTGGGTCGTGG IBA8:GCGAACGACTTCATAAAATGCTATAAGGTTGCCCTCTGTC	[40]
18	Arsenic	DNA	TTACAGAACAACCAACGTCGCTCCGGGTACTTCTTCATCG	[41]

Table 1. DNA/RNA aptamers

2.3. Aptamer

Aptamer is a single-stranded oligonucleotide that folds into complex three-dimensional structure and bind strongly and selectively to one certain kind of target or one class of targets. [4-6] The aptamer is selected using an *in vitro* process called Systematic Evolution of Ligands by EXponential enrichment (SELEX), which was first put forward by Ellington et al. and Tuerk et al. in 1990.[24,25] For the selection of DNA aptamer, the SELEX starts from the construction of a random pool of DNA sequences ($\sim 10^{15}$), and then the selection procedure could take place, including: (a) binding between the library and the target; (b) separation of the unbound ssDNA and the ssDNA-target complex; (c) elution the bound ssDNA from the ssDNA-target complex; (d) amplification of the bound ssDNA, usually using the PCR method; (e) generation of single strand DNA from the double strand PCR products. Generally, a traditional aptemer selection process need 10-12 cycles. The ssDNA pool from the first selection cycle is used as the starting library for the second selection cycle, then repeating procedure (a), (b),(c),(d) and (e). After the last selection cycle, molecular cloning and DNA sequencing was applied, and the aptamer with high affinity and specifity is obtained.

Aptamers offer a useful alternative to antibodies as sensing molecules and have opened a new era in development of affinity biosensing due to their unique characterizations. In vitro selected aptamers could be produced for any targets such as proteins, peptides, amino acids, nucleotides, drugs, heavy metal ions, and other small organic and inorganic compounds. [26-41] Aptamers could be chemically synthesized without the complicated and expensive purification steps by eliminating the batch-to-batch variation found when using antibodies. Furthermore, modifications in the aptamer through chemical synthesis can be introduced enhancing the stability, affinity and specificity of the molecules. In addition, aptamers are more stable than antibodies and thus are more resistant to denaturation and degradation. Often the affinity parameters of aptamer-target complex can be changed for higher affinity or specificity. In addition, aptamers have the higher temperature stability and can recover their native active conformation after denaturation, whereas antibodies are large, temperature-sensitive proteins that can undergo irreversible denaturation. Recently, several DNA/RNA aptamers, selected for POPs, EDCs, organophosphorus pesticides, antibiotics, biotoxins, and pathogenic microorganisms, are listed in Table 1.

Aptamers have become increasingly important molecular tools for environmental bioassay. EDCs are contaminants of emerging concern and required routine monitoring in water samples, as posed by EPA Unregulated Contaminant Regulation (UCMR3). Gu et al. [42] reported a reusable evanescent wave aptamer-based biosensor for rapid, sensitive and highly selective detection of 17β-estradiol, frequently detected in environmental water samples. In this system, the capture molecular, β-estradiol 6-(O-carboxy-methyl) oxime-BSA, was covalently immobilized onto the optical fiber sensor surface. With an indirect competitive detection mode, the limit of detection of 17β-estradiol was determined as 2.1 nM. Kim et al. [41] developed a high affinity DNA aptamer for arsenic that can bind to arsenate [(As(V)] and arsenite [As(III)] with a dissociation constant of 5 and 7 nM, respectively. Through the "signal-on" mode or the "signal-off" mode, reflecting the extent of the binding process thereby allowing for quantitative measurement of target concentration, several DNA aptamer fluorescence-based sensors have been developed for the detection of Hg^{2+}, Pb^{2+} and other trace pollutants.[43] Although a variety of aptamer has been successfully selected for environmental contaminants, the detection of the real water samples using the right aptamer is still in the cradle.

2.4. DNAzyme

DNAzymes are small single-stranded nucleic acids that fold into a well-defined three-dimensional structure with high specificity to various ligands, such as low-molecular-weight organic or inorganic substrates or macromolecules or metal ions.[43] DNAzymes have a promising capacity to selectively identify charged organic and inorganic compounds at ultratrace levels in environmental samples or biological systems. Furthermore, DNAzymes can perform chemical modifications on nucleic acids, while aptamers can bind a broad range of molecules. A combination of the two has generated a new class of functional nucleic acids known as allosteric DNAzymes or aptazymes. Combining the specificity of nano-biological recognition probes and the sensitivity of laser-based optical detection, DNAzymes are capable of provide unambiguous identification and accurate quantification of environmental pollutants, ranging from small ions to large molecules. RNA-cleaving DNAzymes are the most widely used due to their simple reaction conditions, fast turnover rates and significant modifications of their substrate lengths.

Using the in vitro selection of specifical DNAzymes, several fluorescence biosensors have extensively been developed for the detection of various heavy metal ions, such as Pb^{2+}, Cu^{2+}, Mg^{2+}, Ca^{2+}, Zn^{2+}, Co^{2+}, Mn^{2+}. UO_2^{2+}, Hg^{2+} and Ag^+, etc.[43,44] Moreover, DNAzymes and aptazymes have already found many applications in almost every aspect of DNA nanotechnology, which result to new materials and devices that may penetrate into many other fields for practical applications, including environmental monitoring.

3. Nanomaterials in optical biosensors

Nanomaterials exhibit unique size-tunable and shape-dependent physicochemical properties that are different from those of bulk materials.[45-50] Specially, the interaction of nanomaterials and functional biomaterials opens a new door to develop various novel optical biosensors. NPs such as gold NPs (AuNPs), quantum dots (QDs), magnetic NPs (MNPs), graphene and carbon nanotubes have specific optical, fluorescence and magnetic properties, and interactions between these properties give NPs great potential for environmental screening.[45-51] The extremely high surface-to-volume ratios and exceptional nanoscale properties make NPs useful for next-generation environmental detection.

3.1. Quantum dots

Semiconductor quantum dots (QDs), nanocrystals of inorganic semiconductors, have emerged as promising alternative bioanalytical tools because of their unique optical properties including high quantum yield, photostability, narrow emission spectrum, and broad absorption.[52,53] QDs' band gap depends on the size of the nanocrystal. That is to say, the smaller the nanocrystal, the larger the difference between the energy levels and, therefore, the wider the energy gap and the shorter thewavelength of the fluorescence. [52,53]

The main application of QDs as sensors exploits the Forster resonance energy transfer effect (FRET) due to their narrow, size-tuned, and symmetric emission spectra, which has made them excellent donors for fluorescence resonance energy transfer (FRET) sensors, and greatly reduces the overlap between the emission spectra of donor and acceptor and circumvents the cross-talk in such FRET pairs.[52,53] Meanwhile, QDs have broad excitation spectra as donor, and allow excitation at a single wavelength far removed (>100nm) from their respective emissions, which enables QDs to be used in multiplex assays without the need for multiple excitation sources. In addition, the high photobleaching threshold and good chemical stability of QDs greatly improve the detection sensitivities and detection limits. Therefore, QD-based FRET biosensors have been widely used in environmental monitoring, medical imaging, clinical/diagnostic assays, and biomolecular binding assay. Among available QDs, CdSe/ZnS core-shell quantum dots are most commonly used for biosensing applications. Antibody (or aptamer) bioconjugates of QDs, prepared using covalent or non-covalent linking approaches, are the most developed and widespread detection bioprobes to integrating QDs into bioanalyses. The fluorescent detection of pathogens such as respiratory syncytial virus,[54] E. coli O157:H7,[55] and Bacillus thurinigensis,[56] has been performed based on QDs-FRET.

However, the control over the number of antibodies (or aptamers) per QD and their orientation and position relative to the QD is very difficult. Due to the possibility of inadvertently disrupting the binding site when conjugating QD with antibody, the activity loss of antibody is inevitable.[52,57] Additionally, antibodies usually need to be cryopreserved but QDs cannot be frozen, which makes the storage of QD-antibody a major obstacle for its practical applications. To effectively address these challenges, we have developed carrier-protein-haptens-coupled quantum-dot nanobioprobe protocols to perform rapid and sensitive detection of small targets in real water samples.[58] 2,4-Dicholrophenoxyacetic acid (2,4-D), one of the most widely used pesticides worldwide, was selected as a model target. QD nano-immunoprobe were prepared through conjugating carboxyl quantum dots with 2,4-D-BSA conjugate, which regarded as the immunological recognition of anti-2,4-D antibody as well as for optical transducer. With a competitive detection mode, samples containing different concentrations of 2,4-D were incubated with a given concentration of QD immunoprobe and fluorescence-labeled antibody, and then detected by an all-fiber microfluidic biosensing platform developed by our group. A higher concentration of 2,4-D led to less fluorescence-labeled anti-2,4-D antibody bound to the QD immunoprobe surface, and thus to lower fluorescence signal. The quantification of 2,4-D over concentration ranges from 0.5 nM to 3 μM with a detection limit determined as 0.5 nM. The structure of multiplex-haptens/BSA conjugate coupling to QD greatly improves the FRET efficiency and nanosensor's sensitivity. With the use of different QD immunoprobes modified by the conjugates of other haptens/ carried protein, the methodology presented here has the potential to extend to toward the on-site monitoring other small analytes in a variety of application fields ranged from environmental to biomedical areas.

3.2. Nanogold particles

Gold nanoparticles (AuNPs) typically have dimensions ranging from 1-100 nm and display many interesting electrical and optical properties. Nanogold particles based optical biosensors commonly take use of the fluorescence quenching through fluorescence resonance energy transfer (FRET) or a visible color change due to the aggregation of AuNPs of appropriate sizes.[59] Over the past decades, AuNPs based optical sensors have an important role in the detection of environmental pollutants such as toxins, heavy metals and other pollutants due to their typically high signal-to-noise ratios.[51]

Uzawa et al.[60] developed sugar-coated GNPs for the detection of ricin with visual readout using the naturally occurring infection mechanism and the strong affinity of the toxin ricin to sugar. Many kinds of immunoassays using GNP-antibody conjugates have been developed for detection of ochratoxin A (OTA), zearalenone (ZEA), and aflatoxin B1 (AFB1). [51,61] AuNP-based biosensors have also been used to highly competitive assay technologies for the detection of oligonucleotide targets.[51]

Heavy metal contamination is an ongoing concern worldwide, and it is vital for rapid and simple monitoring technologies of heavy metal ions in environment. Darbha et al.[62] developed a GNP-based sensor for rapid, easy and reliable detection of Hg^{2+} ions in aqueous solutions, which had a detection limit of 5 ng/ml (ppb) through non-linear optical properties. A visual detection methode of Cu^{2+} was reported by Lcysteine-functionalized GNPs in aqueous solution.[63] This colorimetric nanosensor allows rapid, quantitative detection of Cu^{2+} with a sensitivity of 10^{-5} M. Similarly, Xue et al.[64] developed a novel and practical system for room temperature colorimetric detection of mercury based on T-Hg^{2+}-T structure with a sensitivity of 3.0 ppb. Several AuNPs-based optical sensors have been developed on the basis of the principle of FRET. Based on modulating photoluminescent-quenching efficiency between a perylene bisimide chromophore and GNPs in the presence of Cu^{2+}, a homogeneous assay to detect Cu^{2+} was reported.[65] Chen et al.[66] developed a GNP-rhodamine 6G-based fluorescent sensor for detecting Hg^{2+} in aqueous solution with a detection limit of 0.012 ppb. Li et al.[67] used a T-Hg^{2+}-T structure to develop a detection method of aqueous Hg^{2+} with the limit of detection of 50 nM. Freeman et al.[68] showed a multiplex assay for detecting Hg^{2+} and Ag^+ using FRET.

Small molecules, such as hydrogen, carbon dioxide, TNT, and ammonium ions, can also be detected by AuNPs. Dasary et al.[69] developed a cysteine-modified GNP-based label-free surface enhanced Raman spectroscopy probe based on the reaction between TNT and cysteine on the GNP surface. An AuNPs color change is induced in the presence of TNT with a detection limit of 2 pM in water samples.

3.3. Graphene

Graphene, a true two-dimensional material, has received increasing interest due to its unique physicochemical properties such as high surface area, fast electron transportation, high thermal conductivity, high mechanical strength, and excellent biocompatibility.[70] These properties of graphene give it potential applicability in biosensors, especially for electrochemical biosensors. However, the optical properties of graphene have received considerable attention. Wen et al.[71] reported a fluorescence sensor for Ag(I) ions based on the target-induced conformational change of a silver-specific cytosine-rich oligonucleotide (SSO) and the interactions between the fluorogenic SSO probe and graphene oxide. He et al. [72] developed a SERS-based biosensor for DNA detection. The Raman signals of dye were dramatically enhanced by the substrate based on gold nanoparticles-decorated graphene. This platform showed extraordinarily high sensitivity and excellent specificity for DNA detection with a detection limit as low as 10 pM.

Lee et al.[73] reported on a platform based on chemiluminescence resonance energy transfer (CRET) between graphene nanosheets and chemiluminescent donors. In contrast to FRET, CRET occurs via nonradiative dipole-dipole transfer of energy from a chemiluminescent donor to a suitable acceptor molecule without an external excitation source. This graphene-based CRET platform was used for immunoassay of C-reactive protein (CRP) using a luminol/hydrogen peroxide chemiluminescence (CL) reaction catalyzed by horseradish peroxidase with a LOD of 1.6 ng mL^{-1}.

Graphene oxide (GO), a promising precursor for graphene, has great potential for use in biosensors due to its unique characteristics such as facile surface modification, high mechanical strength, good water dispersibility, and photoluminescence.[74,75] The GO has negatively charged functional groups such as carboxylic acids, hydroxy groups, and epoxides, which benefit to the biomolecules bound to the GO sheets. A GO-based immuno-biosensor system has been developed for the detection of rotavirus.[76] The anti-rotavirus antibodies are immobilized on the GO array, and captured the rotavirus cell by specific antigen-antibody interaction. The capture of a target cell was verified by observing the fluorescence quenching of GO by FRET between the GO and AuNPs. AuNP-linked antibodies were bridged with 100-mer single stranded DNA molecules, which provide facile control of distance between Ab and AuNPs. When the Ab-DNA-AuNP complexes were selectively bound to the target cells that were attached to the GO arrays, the fluorescence emission of GO decreased by AuNP quenching, which enabled the identification of pathogenic target cells.

3.4. Carbon nanotubes

Carbon nanotubes (CNTs) are molecular-scale tubes of graphitic carbon with outstanding properties such as high aspect ratios, high mechanical strength, high surface areas, excellent chemical and thermal stability, and rich electronic and optical properties.[77] Carbon nanotubes (CNTs) have been explored for highly sensitive biosensing assay of various types of targets such as cells, proteins, DNA, heavy metal ions, small molecules, and so on.

Compared with biosensors using CNTs' electrochemical or electrical properties, the number of the CNTs' biosensors that exploit the optical properties of CNTs is small. However, several CNTs-based optical biosensing platforms have been developed by the use of the ability of CNTs to quench fluorescence or the near-infrared (NIR) photoluminescence exhibited by semiconducting nanotubes. Due to NIR radiation is not absorbed by biological materials, luminescence of SWNTs is particularly interesting for biosensing, especially within biological samples or organisms. With the ability of CNTs to quench fluorescence, Yang et al.[78] demonstrated a DNA detection system using the preference for single stranded oligonucleotides to wrap around SWNTs compared with the related duplexes. Without the complementary DNA (cDNA), the oligonucleotides labeled with the fluorophore 6-carboxyfluorescein wrap around the SWNTs and the fluorescence will be quenched. With the present of complementary DNA, the fluorescence labeled DNA probes has hybridization with cDNA and forms a rigid duplex, which does not wrap around the nanotubes and hence a fluorescence signal will be observed. The similar methods have also been used to detect the heavy metal ions.[79]

3.5. Magnetic nanoparticles

Magnetic nanoparticles (MNPs), consist of magnetic elements such as iron, nickel and cobalt and their chemical compounds, are a class of nanoparticle which can be manipulated using magnetic field.[80] MNPs are usually prepared in the form of superparamagnetic magnetite (Fe_3O_4), greigite (Fe_3S_4), and various types of ferrites (MeO Fe_2O_3). MNPs provide attractive possibilities in environmental monitoring. On the one hand, MNPs bound to biorecognitive molecules (e.g. DNA or antibody) can be used to enrich the analyte to substantially improve the sensitivity of the biosensors. On the other hand, many of the MNPs are superparamagnetic, which can immediately be magnetised with an external magnetic field and resuspended immediately once the magnet is removed. It is greatly useful for the separation of targets from the complex matrix of environmental samples when developing sensitive and selectively biosensors.

Chemla et al.[81] used MNPs labelled antibodies for detecting biological targets, in which the sensitive superconducting quantum interference device was used to only detect the antigen-antibody magnetic NPs. A good relationship between the luminescence and the mouse IgG concentration was obtained in the $1-10^5$ fg/cm^3 range. Moreover, using magnetic NPs substantially shortened the assay time. Tudorache et al.[82] reported a magnetic-beads-based immunoassay strategy for sensitive atrazine with a limit of detection of 3pg/L.

4. Optical biosensing platform for environmental applications

4.1. Optical waveguide based biosensors

Optical waveguide (e.g. fiber optic and planar waveguide) transmit light on the basis of the principle of total internal reflect (TIR). When the incident light is totally reflected, the evanescent wave that penetrates essentially into the surrounding cladding of lower refractive index, decays exponentially with distance[83] (Figure 3):

$$E(z) = E_0 \exp\left(- \, d/ \, d_p\right) \qquad (1)$$

Where δ is the distance from the interface. For multimode waveguides, the penetration depth d_p, is a function of the two refractive indices, the angle of incidence of the light, and the wavelength, is given by:

$$d_p = \frac{\lambda_{ex}}{2\pi}\left[(n_2) \, \sin^2 \alpha - (n_1)\right]^{2-1/2} \qquad (2)$$

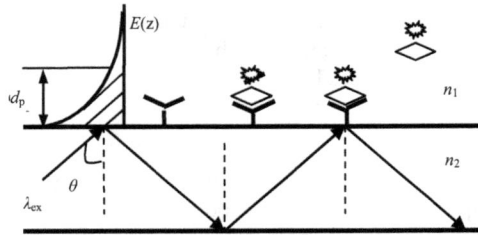

Figure 2. The principle of evanescent wave fluorescence biosensor

In evanescent wave fluorescence biosensor, the evanescent wave can excite fluorescence primarily from the fluorescently labelled analyte complexes that have been bound to the surface through affinity recognition interactions. Not only does this decreases the need for washing or separation procedures to avoid optical interference or contribution from free components, but the signal obtained is directly related to the binding kinetics of the detection interaction.[84]

4.1.1. Evanescent wave fiber-optic biosensor

Evanescent wave fiber optic biosensors, one of the most promising detection technologies to achieve the rapid, specific, sensitive, cost-effective, and real-time on-site detection of the environmental pollutants. They have been applied to detect a wide variety of analytes such as TNT, 2,4-D, atrazine, Escherichia coli O157:H7, and Staphylococcal enterotoxin B, etc.[19] Despite the technological leaps made in the past decades, Evanescent wave fiber optic immunosensor, based on the selective interaction between antigen and antibody, has few actual applications to routine analysis. The following problems should be responsible for this situation. The conventional evanescent wave fiber optic biosensor always have the large size, number of optic components, such as chopper, off-axis parabolic reflector, biconvex silica lens and so on. Such a conventional bulk optics arrangement is costly and requires crucial optical alignment. Once the direction of any element appears inaccurate, the whole system will be destroyed and be difficult to reconvert.

Recently, we reported a portable evanescent wave all-fiber biosensor (EWAB) (Figure 3),[85] whose configuration is simple, compact and portable, has been developed.With a single–multi-fiber optic coupler, both the transmission of the excitation light and the collection and transmission of the fluorescence are achieved by fiber optic in this system, which reduces optical components required and does rarely need optical alignment. Meanwhile, the efficiency of light transmission is higher, light loss lower, and the S/N ratio improved.

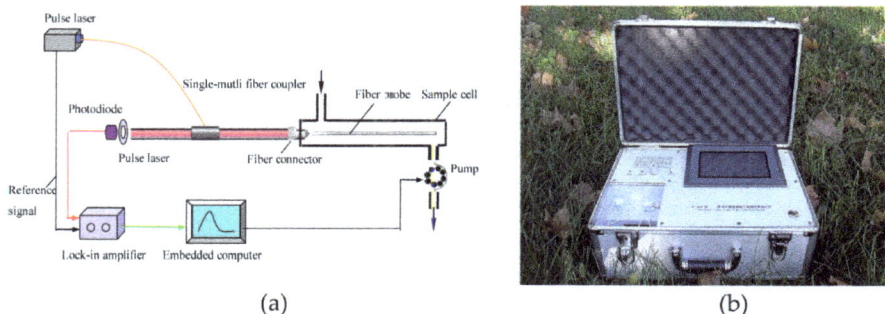

Figure 3. (a) Schematic of EWAB; (b) Photograph of EWAB.

Fast and sensitive detection of microcystin-LR (MC-LR) was conducted with this portable evanescent wave biosensor based on the principle of immunoassay and total internal reflection fluorescence. The reusable biosensing surface was produced by covalently immobilizing a MC-LR-ovalbumin (MC-LR-OVA) conjugate onto a self-assembled thiol-silane monolayer of fiber optic sensor through a heterobifunctional reagent. The MC-LR-OVA immobilized fiber optic probe is highly resistive to non-specific binding of proteins, and can be reused more than 150 times. The limit of detection (LOD) of MC-LR is 0.03 µg/L. The developed immunosensor method was applied to the monitoring of MC-LR in various types of water with the recovery ratio ranged from 80 to 110%. The sensitive and rapid detection of the herbicide 2,4-D has also been achieved with the EWAB. Under optimum conditions, calibration curve obtained for 2,4-D had detection limits of 0.07µg/L. The portable biosensor is commercially obtained from the company JQ-environ Co. Ltd. (China).

Ultrasensitive DNA detection was achieved by the EWAB based on quantum dots (QDs) and total internal reflection fluorescence (TIRF), which featured an exceptional detection limit of 3.2 amol of bound target DNA.[86] The ssDNA coated fiber probe was evaluated as a nucleic acid biosensor through a DNA-DNA hybridization assay for a 30-mer ssDNA, the segments of the *uidA* gene of *Escherichia coli*, labeled by QDs using avidin/biotin interaction. Based on our proposed theory, the quantitative measurement of binding kinetics can be achieved with high accuracy, indicating 1.38×10^6 $M^{-1}s^{-1}$ for association rate and 4.67×10^{-3} s^{-1} for dissociation rate.

Moreover, based on a direct structure-competitive detection mode, we report a rapid and highly sensitive Hg^{2+} detection method using the EWAB.[87] In this system, a DNA probe covalently immobilized onto a fiber optic sensor contains a short common oligonucleotide sequences that can hybidize with a fluorescently labeled complementary DNA. The DNA probe also comprises a sequence of T-T mismatch pairs that binds with Hg^{2+} to form a T-Hg^{2+}-T complex by folding of the DNA segments into a hairpin structure. With a structure-competitive mode, higher concentration of Hg^{2+} lead to less fluorescence-labelled cDNA bound to the sensor surface and thus in lower fluorescence signal. The total analysis time for a single sample, including the measurement and surface regeneration, was <6 min with a detection limit of 2.1 nM.

4.1.2. Surface Plasmon Resonance (SPR) biosensors

SPR is a surface sensitive optical technique for monitoring biomolecular interactions exploiting special electromagnetic waves due to fluctuations in the electron density at the boundary of two materials.[88] SPR has given it a great potential for the real-time and label-free study of the binding interactions between a biorecognition molecules immobilized on sensor surface with its special receptors (analyte). The SPR biosensors have been used to investigate protein binding, association/ dissociation kinetics, and affinity constants, and have wide applications such as clinical diagnosis, drug discovery, food analysis, environmental monitoring. [89]

The principle of SPR biosensor was shown in Figure 4. Using a Kretschmann configuration, SPR detects a small refractive index change at the metal/analyte interface, and the information of the molecular interactions can be obtained by measuring the optical intensity (or phase/polarization) of light reflected from the optical instrument.[88,89] In SPR sensors, changes in the plasmonic resonance signals at a thin metal film are strongly dependent on the refractive index (RI) of the medium. SPR biosensors containing a biorecognition molecule layer can detect minute changes in RI on binding of the special receptors. The sensitivity of the SPR biosensor is limited by the magnitude of the refractive index change at the metal surface, and the minimum SPR shift is detectable by the instrument as a result of recognition events occurring between a surface-bound receptor and analyte of interest.

Recently, there is a growing interest to use indirect competitive SPR immunoassays for detection of environmental contaminants including atrazine, dichloro diphenyl trichloroethane (DDT), 2,4-D, Benzo(a)pyrene (BaP), biphenyl derivatives, carbaryl, 2,3,7,8-tetrachlorodibenzop-dioxin (TCDD), TNT, and so on.[89] BaP, a potential marker of environmental pollution, is a carcinogenic endocrine disrupting chemical and its content well correlates with the total amount of polycyclic aromatic hydrocarbons (PAHs) in the environment. An SPR immunosensor for BaP was reported using the indirect competitive immunoreaction principle with a detection limit of 10 ppt.[90] Svitel et al.[91] showed the sensitive detection of 2,4-D by exploring the binding interaction of dextran matrix with D-glucose and concanavalin A. Shimomura et al.[92] developed an immunosensor for the detection of TCDD, polychlorobiphenyl (PCB) and atrazine, and found a higher sensitivity with the indirect competitive assay than the direct assay. A possibility of ultra-highly sensitive detection of

TNT has been shown by an indirect competitive SPR immunoassay using commercial and home-made antibodies. Mauriz et al.[93] showed the detection of carbaryl, DDT and chlorpyrifos using a portable SPR instrument, where the immunosensor fabricated by a self-assembly method is highly stable and regenerable for more than 250 cycles. Despite the progress has been made, the complex matrix of environmental water samples will still be a great challenge for the practical applications of SPR biosensor.

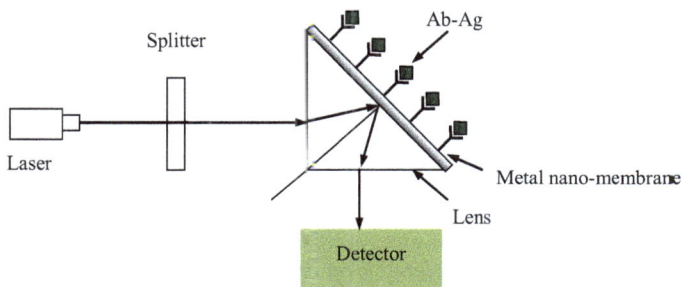

Figure 4. The schematic of SPR biosensor

4.2. Optical biosensor arrays

Analytical microarrays have emerged as powerful tools for high-throughput and rapid analysis of multiple analytes.[94] Antibody and hapten arrays are specific quantitative analytical techniques using antibodies/antigens as highly specific biological recognition elements. They possess the capability to simultaneously detect numerous analytes in low sample volumes. Because antibodies have been generated which specifically bind to individual compounds or groups of structurally related compounds with a wide range of affinities, immunosensors are inherently more versatile than enzyme-based biosensors. Recent advances reported for immunosensor arrays for environmental applications have primarily been focused on using analyte derivatives as immobilized recognition molecules For example, Jin et al.[95] have developed a fluorescent immunosensor system for the detection of bioterrorism agents with high sensitivity, specificity, and reproducibility.

We developed a proof-of-concept development of a novel optic fiber-based immunoarray biosensor for the detection of multiple small analytes.[96] This was developed through immobilization of two kinds of hapten conjugates, MC-LR-OVA and NB-OVA, onto the same fiber optic probe. The technique is significantly different from conventional immunoarray sensors. Microcystin-LR and trinitrotoluene (TNT) could be detected simultaneously and specifically within an analysis time of about 10 min for each assay cycle. The limits of detection for MC-LR and TNT were 0.04 μg/L and 0.09mg/L, respectively. Good regeneration performance, binding properties, and robustness of the sensor surface of the proposed immunoarray biosensor ensure the cost-effective and accurate measurement of small analytes. This compact and portable quantitative immunoarray provides an excellent multiple-assay platform for clinical and environmental samples.

There are, however, several limitations in the use of immunosensors for environmental monitoring applications. For example, the complexity of assay formats; and the number of specialized reagents (e.g., antibodies, antigens, tracers, etc.) that must be developed and characterized for each compound; and the limited number of compounds typically determined in an individual assay as compared to the multiple compounds that contaminate environmental samples.[4-8]

4.3. Emerging optical biosensors

Label-free optical biosensing is a rapidly emerging research area with potential applications ranging from medical and clinical diagnostics to food safety and environmental detection, especially for portable, easy-to-use devices.[88,89] Without the use of radioactive or fluorescent labels, the complexity in the detection and screening process significantly reduces and the intrinsic properties of the target molecules have been few influenced.[88,89] To dated, the most well established technique for label-free optical biosensing is surface plasmon resonance (SPR) based biosensor. Recently, several novel optical biosenors such as optical ring resonator based biosensor, photonic crystal biosensors, and optical nano-biosensors, have been developed in very small dimensions and allowed to fabricate with standard CMOS techniques. Although they have little applications in the monitoring of environmental pollutants, these biosensors have great potential posibilities for on-site real-time detection of micro-environment.

4.3.1. Optical ring resonator based biosensors

Optical ring resonator is an emerging sensing technology, in which at least one is a closed loop coupled some sort of light input and output (see Figure 5).[97] In a ring resonator, the light propagates in the form of whispering gallery modes (WGMs) or circulating waveguide modes. When light of the resonant wavelength transports through the loop from input waveguide, it builds up in intensity over multiple round-trips due to constructive interference and is output to the output bus waveguide which serves as a detector waveguide. [97]

The WGM spectral position is related to the refractive index (RI) through the resonant condition:$\theta = 2\pi r n_{eff}/m$, where r is the ring outer radius, n_{eff} the effective RI experienced by the WGM, and m is an integer. n_{eff} changes when the RI near the ring resonator surface is modified due to the capture of target molecules on the surface, which in turn leads to a shift in the WGM spectral position.[98] Thus, by directly or indirectly detecting the WGM spectral shift, the quantitative detection of targets will be achieved.

4.3.2. Photonic crystal biosensors

Photonic crystal fibres have wavelength-scale morphological microstructures that run along the entire fiber length by corralling it within a periodic array of microscopic air holes.[99] Overcoming the limitations of conventional fiber optics, photonic crystal fibers are proving to have a multitude of important technological and scientific applications including biosensors. Due to their well-defined physical properties such as reflectance/ transmittance, photonic crystal biosensors are enabled superior levels of sensitivity resulting in precise detection limits. Photonic crystal biosensors are very small and are possible through coupling the incident and reflected/transmitted light to optical fibers and analyzing them in remote locations.

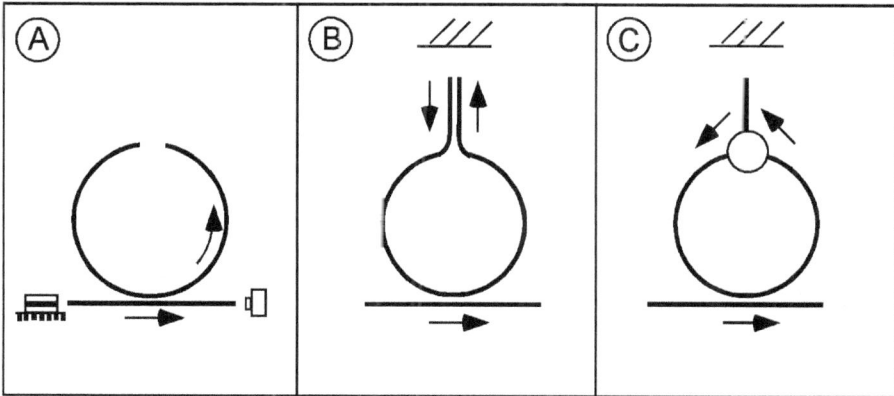

Figure 5. Various ring resonator biosensors

To design a photonic crystal biosensor, some portion of the resonant electric field must be in contact with liquid media that contains the analyte, providing a surface on which biorecognition molecules may be adsorbed. Label-free photonic crystal biosensors generally detect shifts in resonant wavelength or coupling angle caused by the interaction between the target molecule and the evanescent wave.[99] The narrow spectral linewidth achieved by using high Q factor passive optical resonators enables sensor systems to resolve smaller wavelength shifts associated with the detection of analytes at low concentration such as environmental pollutants. Photonic crystals biosensors have been applied for sensing the pH and ionic strength of solutions, metal ions and trace orgnic pollutants. [100]

4.3.3. Optical nano-biosensors

Recent developments have greatly improved the sensitivity of optical sensors based on nano-structures and nanoparticles.[101] Optical biosensors have been used to provide a reliable method of monitoring various chemicals in microscopic environments and to detect different entities within single cells.

Vo-Dinh et al.[102] have designed fiber-optic nanosensors for environmental and biochemical monitoring. The nanosensors were fabricated with tapered optical fibers with distal ends with a 20~500m diameter. Biorecognition molecules, such as antibody, peptides, and nucleic acids, are immobilized on the fiber tips and designed to be selective to bind target molecules (analyte) of interest. This fiber-optic nanosensor has become a powerful tool for measurements in submicron environments and for probing individual chemical species in specific locations throughout a living cell due to their small nanoscale sizes. In their previous work, [103] the nanosensors have been developed for in situ measurements of the carcinogen BaP inside single cells using the antibody probe. In another study,[102] nanosensors have been used for the measurement of intracellular concentrations of benzopyrene tetrol (BPT) in the cytoplasm of human mammary carcinoma cells and rat liver epithelial cells. They performed calibration measurements of solutions containing different BPT concentrations ranging from 1.56×10^{-10} to 1.56×10^{-8} M. Fiber-optic nanosensors for monitoring single cells have opened up new applications due to their small sizes, which provide important tools for minimal invasive analysis at single cellular or subcellular level.

5. Optical biosensors for pollution control and early-warning

The new technologies for environment pollutants, which are rapid, specific, sensitive, cost-effective, and suitable for real-time on-site detection, have a strong demand due to a large number of pollutants and their derivates present in surface and ground waters and stricter regulations for the detection of these pollutants set out by the legislative bodies.[104] Existing analysis methods, such as HPLC or GC/MS, are very sensitive at detecting these toxic targets, however, the analytical procedure are rather complicated and therefore labour-intensive and time-consuming. Moreover, contaminant concentrations in water courses are dynamic, changing both as a result of inputs and changes in water flow. With monthly sampling and analysis, it is extremely unlikely that the maximum concentration for a period of time can be detected. The need for cheap and general network system (multiple autonomous analytical stations that extensively control the sites of interest in rivers and lakes) for pollution control and early warning has generated great interest.

The EWS is an integrated system for monitoring, analyzing, interpreting, and communicating monitoring data, which identify low probability/high-impact contamination events in sufficient time to be able to safeguard the public health. The ideal integrated EWS should demonstrate a number of characteristics as following: [104]

- provide a rapid response and warning in sufficient time for action

- covers all potential threats

- exhibit a significant degree of automation, including automatic sampling

- allows acquisition, maintenance, and upgrades at an affordable cost

- require low skill and training

- identify the source of the contaminant

- demonstrate sufficient sensitivity

- give minimal false-positives/false-negatives

- exhibit robustness and ruggedness in long-term monitoring
- reproducible and verifiable
- allow remote operation

Single device alone may not satisfy all of these requirements. Therefore, an early warning system network (EWSN) including various detection technologies will be usesful for homogeneous environments such as for rivers and coastal areas. A new generation of monitoring tools based on sensor technology has emerged in the last decades. Optical biosensors have proven advantages over other types of sensors for multitarget sensing and continuous real-time on-site monitoring. Optical biosensors have been integrated into many early warning systems (EWS) that can provide easy, rapid and on-site measurements. These EWS are useful for mapping of contamination such as after accidental spills or pollution events.

A number of early warning systems have been developed. For example, J-Mar Biosentry™ can perform low density microbial suspension detection in drinking water using eight on-line sensors and instruments.[105] This system was able to indicate significant visual responses to the introduction of E. coli and B. globigii down to concentrations of 600 cfu/mL. The YSI Sonde™ system can simultaneously achieve measurement of conductivity, salinity, temp, depth, pH, dissolved oxygen, turbidity, chlorophyll and blue-green algae, which is ideal as early warning of algae blooms with good sensitivity at natural levels. [106]

TOXcontrol™ uses freshly cultivated light emitting bacteria (Vibrio fischeri) as a biological sensor,[107] which combines the advantages of whole organism toxicity testing and instrumental precision. The luminescence is measured before and after exposition to calculate the inhibition in percent. The more toxic the sample, the greater the percent light loss from the test suspension of luminescent bacteria.

The DaphTox II[108] is a new sensitive system to detect hazardous compounds in water from rivers (source-water protection) based on the Extended Dynamic Daphnia Test. Sample water (0.5-2 L/h) continuously runs through the measuring chamber containing the daphnia. The live images obtained using a CCD-camera are evaluated online with an integrated PC to analyse changes in the behaviour of the daphnia. If the change is statistically significant, an alarm is triggered. The method of image analysis enables a series of measurement methods and plausibility tests to assess the daphnia's behaviour using different criteria.

Supported by the Water Framework Directive (WFD), the automated water analyser computer-supported system (AWACSS)[109], based on an optical immunosensor, was the establishment of an early-warning system by means of a network of measurement and control stations. The AWACSS system included four major components: the AWACSS instrument with fluidics control and optical transducer chip, the HTC PAL autosampler for sample preparation, the personal computer at the sampling site and the server with database and web site. The sampling site software allows for bi-directional autosampler control. Using fluorescence-based immunoassay technology, this system can measure several organic pollutants at low ng/L in a single few-minutes analysis without any prior sample neither pre-concentration nor pre-treatment steps. A web-based AWACSS system allows for the internet-based networking between the measurement and control stations, global management, trend analysis, and early-warning applications.

6. Key trends and perspectives

There is no doubt that the progress of biosensor technology in recent years makes an important contribution to protect human health and local ecosystems.[2-5] However, biosensors are not as successful as was expected initially, and there is a challenge to creating improved, cost-effective, and more reliable instruments. There are many reasons for this, and a few of them are mentioned here. First, most of the biosensor systems commercially available today are either prohibitively costly or highly inflexible. Second, the content of the environmental samples is complex and vary, which is not like that of clinical samples. It is essential to reduce the effect of matrix of environmental samples on the bioassay. Third, the storage of bio-reagents is one of the key issues to be resolved in long-term monitoring. Finally, the stability and reliability of biosensors should be improved to satisfy the practical applications.

Optical biosensors have proven advantages over other types of sensors for multi-target sensing and continuous monitoring.[4,6-8] Development of new functional materials allows the optical biosensor to have more practical applications. The unique properties of nano-materials offer excellent prospects for interfacing biological recognition events with optical transductor and for designing next-generation of biosensors exhibiting novel functions. Recent technological developments in miniaturizing optical biorecognition elements and wireless-communication technology have led to the emergence of environmental sensor networks, which will greatly enhance on-site long-term monitoring ability of the natural environment and provide more effective way to deal with the pollution incidents with less effort and cost. The trend toward multianalyte sensing and toward biosensor arrays allow optical biosensors become more compact, robust, smaller and adaptable for rapid toxicity screening, multianalyte testing, and continuous on-site monitoring of environmental pollutants. Moreover, biosensor will offer strong potential for researchers to more effectively investigate and understand diverse environmental phenomena, including the fate and transport of contaminants, which provide novel insights into the mechanisms of remediation. The number of opportunities to incorporate new science and technology into optical biosensor systems is almost overwhelming. In the near future, we believe that optical biosensor will provide the most productive paths to solve real problems in everyday life.

Acknowledgement

This research was financially supported by the National Natural Science Foundation of China (21077063), the 863 National High Science and Technology Development Programs of China (2009AA06A417-07), and the Supervisor's Project of Outstanding Doctoral Dissertation Award of Beijing (YB20091000302).

Author details

Feng Long[1], Anna Zhu[2], Chunmei Gu[3] and Hanchang Shi[3]

1 School of Environment and Natural Resources, People's University of China; State Key Joint Laboratory of ESPC, School of Environment, Tsinghua University, Beijing, China

2 Research Institute of Chemical Defence, Beijing, China

3 State Key Joint Laboratory of ESPC, School of Environment, Tsinghua University, Beijing, China

References

[1] A. Jang, Z. Zou, K.K. Lee, C.H. Ahn, P.L. Bishop, Meas. Sci. Technol. 22(2011)1-18.

[2] K.R. Rogers, Anal. Chim. Acat. 568(2006)222-231.

[3] D.R.Thevenot, K.Toth, R.A. Durst, G.S.Wilson, Biosens. Bioelectron. 16(2001) 121-131.

[4] S.M. Borisov and O.S. Wolfbeis. Chem. Rev. 108(2008)423-461.

[5] N.J. Ronkainen, H.B. Halsall, W.R. Heineman. Chem. Soc. Rev. 39(2010)1747-1763.

[6] M.M.F. Choi, Microchim. Acta, 148(2004)107-132.

[7] E.M.D. Barcelo, C.B.G. Gauglitz, R.Abuknesha, Trends. Anal. Chem. 20 (2001)124-132.

[8] B. Byrne, E. Stack, N. Gilmartin, R. O'Kennedy, Sensors, 9(2009) 4407-4445.

[9] S. Tokonami, H. Shiigi, T. Nagaoka, Anal. Chim. Acta, 641(2009)7-13.

[10] J. Woof, D. Burton. Nat. Rev. Immunol.4 (2004) 89–99.

[11] J.W. Sheng, M. He, H.C. Shi, Y. Qian, Anal. Chim. Acta, 572(2006)309-315.

[12] J.W. Sheng, M. He, H.C. Shi, Anal. Chim. Acta, 603(2007)111-118.

[13] F.P. Anderson, K.D. King, K.L. Gaffney, L.H. Johnson, Biosens.Bioelectron. 14(2000)771-777.

[14] I.B. Bakaltcheva, F.S. Ligler, C.H. Patterson, L.C. Shriver-Lake, Anal. Chim. Acta, 399 (1999) 13-30.

[15] O. Hofstetter, H. Hofstetter, M. Wilchek, V. Schurig, B.S. Green, Nat. Biotechnol. 17(1999) 371-374.

[16] L. Tedeschi, C. Domenici, A. Ahluwalia, F. Baldini, A. Mencaglia, Biosens. Bioelectron. 19(2003) 85-93.

[17] W.R. Algar, A.J. Tavares, U.J. Krull, Anal. Chimi. Acta, 673(2010)1-25.

[18] S.J. Rosenthal, J.C. Chang, O. Kovtun, J.R. McBride, I.D. Tomlinson, Chemistry & Biology, 18(2011)10-24.

[19] F. Long, M. He, H.C. Shi, A.N. Zhu, Biosens. Bioelectron. 23(2008)952-958.

[20] L. Vial, P. Dumy, New J. Chem. 33(2009)939-946.

[21] R.E. Luckham, J. D. Brennan. Analyst 135(2010)2028-2035.

[22] C.R. Ispas, G. Crivat, S. Andreescu. Analytical Letters, 45(2012)168-186.

[23] C. Malitest, M. R. Guascito, Biosens. Bioelectron. 20(2005)1643-1647.

[24] C. Tuerk, L. Gold, Science, 249 (1990) 505-510.

[25] A.D. Ellington, J.W. Szostak, Nature, 346 (1990) 818-822.

[26] J. Mehta, E. Rouah-Martin. Anal. Chem. 84(2011)1669-1676.

[27] U. Brockstedt, A. Uzarowska. A. Montpetit, W. Pfau, D. Lauda. Biochem. Biophy. Res. Commun. 313(2004) 1004-1008.

[28] L. Wang, X. Liu, Q. Zhang, C. Zhang, Y. Liu, K. Tu, J. Tu. Biotechnol. Lett. 34(2012)869-874.

[29] M. Jo, J.Y.Ahn. Oligonuleotide, 21(2011)85-92.

[30] Y.S. Kima, H.S. Jung, T. Matsuura, H.Y. Lee, T. Kawai, M.B. Gu. Biosens. Bioelectron. 22(2007)2525-2531.

[31] J. Mehta, B. Van Dorst, E. Rouah-Martin, W. Herrebout, M.L. Scippo, R. Blust, J. Robbens. J. Biotechnol. 155(2011)361-369.

[32] J.H. Niazi, S.J. Lee. Bioorgnic, 16 (2008).1254-1261.

[33] J.H. Niazi, S.J. Lee, M.B. Gu. Bioorgan. Med. Chem. 16(2008)7245-7253.

[34] K.M. Song, M. Cho, H. Jo, K. Min. Anal. Biochem. 415(2011)175-181.

[35] E.A. Lamont, L. He, K. Warriner, T.p. Labuza, S.A. Sreevatsan. Analyst, 136(2011) 3884-3895.

[36] J.A. Cruz-Aguado, G. Penner. Agric. Food Chem. 56 (2008)10456-10461.

[37] J.G. Bruno, M.P. Carrillo, T. Phillips, C.J. Andrews. J. Fluoresc 20(2010)1211-1223.

[38] J.A. DeGrasse. PLoS ONE 7(2012): e33410.

[39] R. Joshi, H. Janagama, et al. Molecular and Cellular Probes 23(2009)20-28.

[40] Y.S. Kim, C.J. Hyun, et al. Bioorgan. Med. Chem. 18(2010) 3467-3473.

[41] M. Kim, H. Um, et al. Environ. Sci. Technol. 43(2009) 9335-9340.

[42] N. Yildirim, F. Long, C. Gao, M. He, H.C. Shi, A.Z. Gu, Environ. Sci. Technol. 46(2012)3288-3294.

[43] X.B. Zhang, R.M. Kong, Y. Lu, Annu. Rev. Anal. Chem, 4(2011)105-128.

[44] M. Hollenstein, C. Hipolito, C. Lam, D. Dietrich, D.M. Perrin, Angew. Chem. Int. Ed. 47(2008)4346-4350

[45] A. Gómez-Hens, J.M. Fernández-Romero, M.P. Aguilar-CaballosL. Trends in Anal. Chem. 27(2008)394-406.

[46] C.Y. Zhang, H.C. Yeh, M.T. Kuroki, T.H. Wang, Nat. Mat. 4(2005)826-831.

[47] S.S. Agasti, S. Rana, M. Park, C.K. Kim, C. You, V.M. Rotello. Adv. Drug Deliv. Rev. 62(2010)316-328.

[48] W.R. Algar, A.J. Tavares, U.J. Krull. Anal. Chimi. Acta, 673(2010)1-25.

[49] F. Amaro, A.P. Turkewitz , A. Martín-González , J.C. Gutiérrez. Microb. Biotechnol. 4(2011)513-522.

[50] Y. Cui, Q. Wei, H. Park, C.M. Lieber, Science, 293(2001)1289-1292.

[51] L. Wang, W. Ma, L. Xu, W. Chen, Y. Zhu, C. Xu, N.A. Kotov, Mater. Sci. Eng. R, 70(2010)265-274.

[52] I.L. Medintz, A.R. Clapp, H. Mattoussi, E.R. Goldman, B. Fisher, J.M. Nat. Mat. 2(2003)630-638.

[53] W.R. Algar, A.J. Tavares, U.J. Krull, Anal. Chimi. Acta, 673(2010)1-25.

[54] E.L. Bentzen, F. House, T.J. Utley, J.E. Crowe, D.W. Wright, Nano Lett. 5(2005) 591-595.

[55] M.A. Hahn, P.C. Keng, T.D. Krauss, Anal. Chem. 80(2008)864-872.

[56] M. Ikanovic, W.E. Rudzinski, J.G. Bruno, A. Allman, M.P. Carrillo, S. Dwarakanath, S. Bhahdigadi, P. Rao, J.L. Kiel, C.J. Andrews, J. Fluoresc. 17(2007)193-199.

[57] S.J. Rosenthal, J.C. Chang, C. Kovtun, J.R. McBride, I.D. Tomlinson, Chemistry & Biology 18 (2011)10-24.

[58] F. Long, C.M. Gu, A.Z. Gu, H.C. Shi, Anal.Chem. 84(2012)3646-3653.

[59] K. Saha, S.S. Agasti, C. Kim, X. Li, V.M. Rotello, Chem. Rev. 112(2012)2739-2779.

[60] H. Uzawa, K. Ohga, Y. Shinozaki, I. Ohsaw, T. Nagatsukac, Y. Setob, Y. Nishidad,,Biosens. Bioelectron. 24 (2008) 923-927.

[61] W.B. Shim, K.Y. Kim, D.H. Chung, J. Agric. Food Chem. 57 (2009) 4035-4041.

[62] G.K. Darbha, A.K. Singh, U.S. Rai, E. Yu, H.T. Yu, P.C. Ray, J. Am. Chem. Soc. 130 (2008) 8038-8043.

[63] W.R. Yang, J.J. Gooding, Z.C. He, Q. Li, G.N. Chen, J. Nanosci. Nanotechnol. 7 (2007)712-716.

[64] X.J. Xue, F. Wang, X.G. Liu, J. Am. Chem. Soc. 130 (2008) 3244-3245.

[65] X.R. He, H.B. Liu, Y.L. Li, S. Wang, Y.J. Li, N. Wang, J.C. Xiao, X.H. Xu, D.B. Zhu, Adv.

[66] Mater. 17 (2005) 2811-2815.

[67] J.L. Chen, A.F. Zheng, A.H. Chen, Y.C. Gao, C.Y. He, X.M. Kai, G.H. Wu, Y.C. Chen, Anal. Chim. Acta 599 (2007) 134-142.

[68] T. Li, S.J. Dong, E. Wang, Anal. Chem. 81 (2009) 2144-2149.

[69] R. Freeman, T.L. Finder, I. Willner, Angew. Chem. Int. Ed. 48 (2009) 7818-7821

[70] S.S.R. Dasary, A.K. Singh, D. Senapati, H.T. Yu, P.C. Ray, J. Am. Chem. Soc. 131(2009) 13806-13812.

[71] C.N.R. Rao, A.K. Sood, K.S. Subrahmanyam, A. Govindaraj, Angew. Chem. Int. Ed. 48(2009), 7752-7777.

[72] Y.Q. Wen, F.F. Xing, S.J. He, S.P. Song, L.H. Wang, Y.T. Long, D. Li, C.H. Fan, Chem. Commun. 46(2010)2596-2598.

[73] S.J. He, B. Song, D.Li, C.F. Zhu, W.P. Qi, Y.Q. Wen, L.H. Wang, S.P. Song, H.P. Fang, C.H. Fan, Adv. Funct. Mater. 20(2010)453-459.

[74] J.S. Lee, H. Joung, M. Kim, C.B. Park, ACS Nano, 6(2012)2978-2983.

[75] Y. Liu, X. Dong, P. Chen, Chem. Soc. Rev, 41(2012)2283-2307.

[76] G. Eda, Y.Y. Lin, C. Mattevi, H. Yamaguchi, H. A. Chen, I.S. Chen, C.W. Chen, M. Chhowalla, Adv. Mater. 22(2010)505-509.

[77] J.H. Jung, D.S. Cheon, F. Liu, K.B. Lee, T.S. Seo, Angew. Chem. Int. Ed. 49(2010)5708-5711.

[78] W. Yang, K.R. Ratinac, S.P. Ringer, P. Thordarson, J.J. Gooding, F. Braet, Angew. Chem. Int. Ed. 49(2010)2114-2138.

[79] R.H. Yang, Z.W. Tang, J.L. Yan, H.Z. Kang, Y.M. Kim, Z. Zhu, W.H. Tan, Anal. Chem. 80(2008)7408.

[80] D.A. Heller, E.S. Jeng, T.K. Yeung, B.M. Martinez, A.E. Moll, J.B. Gastala, M.S. Strano, Science 311(2006)508-511.

[81] S. Andreescu, J. Njagi, C. Ispas, M.T. Ravalli, J. Environ. Monit. 11(2009)27-40.

[82] Y.R. Chemla, H.L. Grossman, Y. Poon, R. McDermott, R. Stevens, M.D. Alper, J. Clarke. Proc. Natl. Acad. Sci. 97(2000)14268-14272.

[83] M. Tudorache, A. Tencaliec, C. Bala, Talanta, 77(2008)839-843.

[84] J.D. Andrade, R.A. Vanwagenen, D.E. Gregonis, IEEE Trans. Electron Devices, 32(1985)1175-1179

[85] J.P. Golden, E.W. Saaski, L.C. Shriver-Lake, G.P. Anderson, F.S. Ligle, Opt. Eng. 36(1997)1008-1013.

[86] F. Long, M. He, A.N. Zhu, H.C. Shi, Biosens. Bioelectron. 24(2009)2346-2351.

[87] F. Long, S. Wu, M. He, T. Tong, H. Shi, Biosens. Bioelectron. 26(2011)2390-2395.

[88] F. Long, C. Gao, H.C. Shi, M. He, A.N. Zhu, A.M. Klibanov, A.Z. Gu, Biosens. Bioelectron. 26(2011)4018-4023

[89] M.A. Cooper, Optical biosensors in drug discovery, Nat. Rev. 1 (2002)515-528.

[90] D.R. Shankaran, K.V. Gobi, N. Miura, Sensors and Actuators B, 121(2007)158-177.

[91] N. Miura, M. Sasaki, K.V. Gobi, C. Kataoka, Y. Shoyama, Biosens. Bioelectron. 18 (2003) 953-959.

[92] J. Svitel, A. Dzgoev, K. Ramanathan, B. Danielsson, Biosens. Bioelectron. 15 (2000) 411–415.

[93] E. Mauriz, A. Calle, A. Abad, A. Montoya, A. Hildebrandt, D. Barcelo, L.M. Lechuga, Biosens. Bioelectron. 21 (2006) 2129-2136.

[94] T.M. Blicharz, W.L. Siqueira, E.J. Helmerhorst, F.G. Oppenheim, P.J. Wexler, F.F. Little, D.R. Walt, Anal. Chem. 81(2009)2106-2114.

[95] W. Lian, D. Wu, D.V. Lim, S. Jin, Anal Biochem, 401 (2010) 271-279.

[96] F. Long, M. He, A. Zhu, B. Song, J. Sheng, H. Shi, Biosens. Bioelectron. 26(2010)16-22.

[97] I.M. White, X. Fan, Opt. Express 16 (2008) 1020-1028.

[98] X. Fan, I.M. White, S.I. Shopova, H. Zhu, J.D. Suter, Y. Sun, Anal. Chim. Acta, 620(2008)8-26.

[99] J.C. Knight, Nature, 424(2003)847-851

[100] R.V. Nair, R. Vijaya, Progress in Quantum Electronics, 34(2010)89-134.

[101] B. Lee, S. Roh, J. Park, Optical Fiber Technology, 15(2009)209-221.

[102] T. Vo-Dinh, Spectrochim Acta Part B 63(2008)95-103.

[103] T. Vo-Dinh, J.P. Alarie, B.M. Cullum, G.D. Griffin. Nat Biotechnol. 18(2000)764-767.

[104] USEPA, 2005. Technologies and Techniques for Early Warning Systems to Monitor and Evaluate Drinking Water Quality: A State-of-the-art Review, EPA-600-R-05e156. U.S. Environmental Protection Agency, Office of Research and Development, National Homeland Security Research Center, Cincinnati, OH.

[105] USEPA, 2010. Detection of Biological Suspensions Using Online Detectors in a Drinking Water Distribution System Simulator. EPA/600/R-10/005, Washington, DC.

[106] S.F. Atkinson, J.A. Mabe, Environ. Monit. Assess. 120(2006)449-460.

[107] J.L. Zurita, A. Jos, A.M. Cameán, M. Salguero, M. López-Artíguez, G.Repetto, Chemosphere, 67(2007)1-12.

[108] C.J. de Hoogh, A.J. Wagenvoort, F. Jonker, J.A. Van Leerdam, A.C. Hogenboom, , Environ. Sci. Technol. 40(2006)2678-2685.

[109] J. Tschmelak, G. Proll, J. Riedt, J. Kaiser, P. Kraemmer, J.S. Wilkinson, Biosens. Bioelectron. 20(2005)1499-1508.

Chapter 5

Impedimetric Immunosensor for Pesticide Detection

Saloua Helali

Additional information is available at the end of the chapter

1. Introduction

Pollution of surface water by chemicals can disturb aquatic eco-systems and cause loss of habitats and reduce biodiversity. Pollutants may accumulate in the food chain, and harm predators consuming contaminated fish. Humans are exposed to pollutants through the aquatic environment by fish or seafood consumption, drinking water and possibly recreational activities. Pollutants may be found in the environment many years after being banned; some may be transported over long distances and can be found in remote areas. Pollutants may be released in the environment from various sources, e.g., agriculture, industry, incineration, as products or as unintended by-products, they may have been released in the past or continue to be released from consumer products used in everyday life.

Recently, the European Commission adopted a proposal for a new Directive to protect surface water from pollution by chemicals (COM(2006)397 final) [1]. The proposed Directive will set limits on concentrations in surface waters of 41 dangerous chemical substances including 33 priority substances and 8 other pollutants that pose a particular risk to animal and plant life in the aquatic environment and to human health. Pesticides are important pollutants and are hazardous to human health and life. During the past 50 years, pesticides have been used in increasing amounts throughout the word. Among pesticides, atrazine (2-chloro-4-ethylamino-6-isopropylamino-1,3,5-triazine) is the most extensively applied herbicide to control broad-leaf plants and grassy weeds, because of its high relative mobility in the soil in the world [2-7]. Atrazine is a putative endocrine disruptor and may cause serious health risks even at very low levels (parts-per-billion concentration). At high concentrations (100 μgl^{-1}), it causes dramatic effects on the photosynthesis, growth, chlorophyll content and biomass of most aquatic producers [5]. Long-term exposure of humans and animals to atrazine at low concentrations may induce subacute injury and potential hazards to the body. Although studies on the toxicity of atrazine on humans have not been completely conclusive, atrazine exposure in rodent models has identified reproductive and developmental abnormalities. The maximum level for atrazine contamination is 3.0 ppb in drinking water, as established by the US Environmental Protection Agency [8].

The standard procedure for pesticide determination in water is extraction (liquid-liquid, liquid-solid), followed by chromatographical separation and specific detection (UV-visible spectroscopy)[9]. Such techniques are reliable and currently used but they require purification of samples prior to assay, thus limiting the number of samples that can be analyzed [10]. In addition, these conventional approaches are expensive, time consuming, frequently generate considerable waste, and require highly trained personnel. Moreover, conventional environmental monitoring typically involves several steps such as sampling, sample handling, and sample transportation to a specialized laboratory that prevents real-time on-site detection of the sample toxicity. These disadvantages of traditional analytical methods have paved the way for the development of atrazine biosensors as simple, fast, sensitive, selective, cost-effective, real-time, on-site, and field portable monitoring technologies with negligible waste generation[11-13].

Due to the highly sensitive and selective nature of the recognition between antigen (Ag) and antibody (Ab), immunoassays are very useful in widespread applications such as medical detection, processing quality control, and environmental monitoring. Traditional methods used in immunoassays involve radioimmunoassay (RIA) and enzyme-linked immunosorbent assay (ELISA). Although they are sensitive, RIA exposes laboratory workers to a significant safety hazard, and ELISA is tedious and time-consuming. New techniques, such as electrochemistry, chemiluminescence, piezoelectricity and surface Plasmon resonance have attracted extensive interest in immunoassays due to their simple and specific characteristics. Among these techniques, electrochemical immunoassay has received much attention for its high sensitivity and low cost. As most antibodies and antigens are electrochemically inert, the label-free technique of electrochemical impedance spectroscopy (EIS) is developed to provide a direct detection of immunospecies by measuring the change of impedance. In addition to its convenience, EIS provides a nondestructive means for the characterization of the electrical properties in biological interfaces [14-15].

2. Biosensors overview

A biosensor can be described as a transducer that incorporates a biological recognition component as the key functional element. It consists of three main components as illustrated in Fig. 1: the biorecognition element, the transducer and the signal display or readout [16]. The interaction of the analyte with the biorecognition element is converted to a measurable signal by the transduction system. The signal is then converted into a readout or display. Analytical immunosensors are a subset of biosensors which utilize either antigen or antibody as the biospecific sensing element.

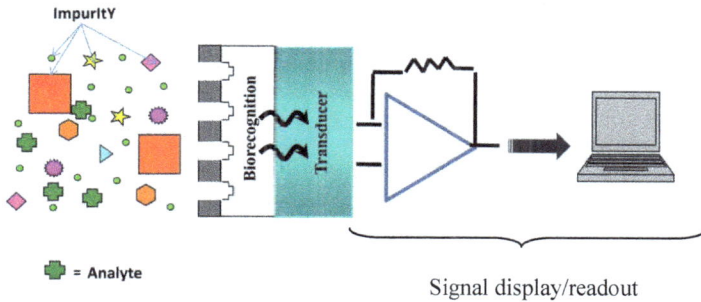

Figure 1. Composition of a biosensors

When antibodies or antibody fragments are used as molecular recognition element for specific analytes (antigens) to form a stable complex, the device is called immunosensor [17].

Many kinds of nanomaterials, including metal nanoparticles (gold, magnetic beads), polymer (polypyrrole), and carbon nanotubes (CNTs) have been widely used in immunosensor (Fig.2). The common characteristics of these nanomaterial in immunosensor are providing signal amplifications comparing to the traditional metal ion labels, enzyme labels and redox probe labels[18].

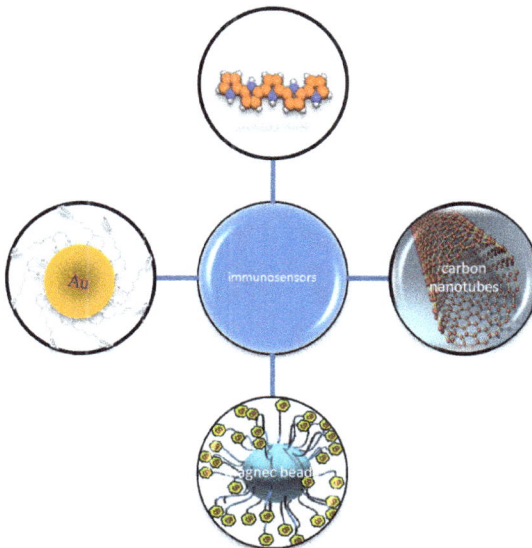

Figure 2. Nanomaterial used in immunosensor

Electrodes are commonly made of inert metals such as platinium, gold or carbon, either in the of graphite, glassy carbon as a solid or as a paste. These electrodes are commonly used to detect chemical compounds produced or consumed by binding. Therefore, for the construction of impedimetric immunosensors are needed.

2.1. Biosensors for monitoring pesticides

The first synthetic pesticides became available during the 1940s, generating large benefits in increased food production. Each year an estimated 2.5 million tons of pesticides are applied to agricultural crops worldwide. Pesticides can be classified into three different groups: insecticides, herbicides and fungicides. Insecticides are usually organophosphorous compounds (e.g. parathion), organochlorine compounds (e.g. DDT) or carbamates (e.g. carbofuran). Fungicides are either sulphur, copper or organic based compounds, while herbicides can be either organic or inorganic compounds. Pesticides can be applied as dust or granules, as a vapour or more commonly applied as, or in the presence of a liquid (water or oil usually). Pesticides, depending on their water solubility can either remain in the soil to be broken down by the action of certain organisms [19], or washed off, eventually washing into rivers and sometimes water supplies. The persistency of pesticides and of their degradation products in the geosphere causes environmental problems. The transfer of pesticides from treated soil to surface and ground water leads to contamination of drinking water resources and to subsequent intake of pesticides by man.

Pesticides can be carcinogenic, citogenic, they can produce bone marrow diseases, infertility, nerve disorders and immunological and respiratory diseases [20]. For these reasons, there is a growing demand for fast and reliable pesticide monitoring in agriculture and food production. Recently there has been a large increase in the number of publications concerning biosensors for monitoring pesticides.

A biosensor for the detection of carbamate insecticides was based on the action of enzymes, acetylcholinesterase [21]. Carbamate pesticides inhibited acetylcholinesterase and the decrease in the enzyme activity was used to determine these pesticides. Acetylcholinesterase was immobilized on silica gel by covalent binding. pH and conductivity electrodes were used to detect the ionic change of the sample solution due to hydrolysis of acetylcholine. The biosensors were used to analyze carbaryl in water.

The optical transducer of CdTe semiconductor quantum dots (QDs) has been integrated with acetylcholinesterase enzyme (AChE) by the layer-by-layer assembly technique, resulting in a highly sensitive biosensor for detection of organophosphorus pesticides (OPs) in vegetables and fruits based on enzyme inhibition mechanism[22]. The detection limits of the proposed biosensors are as low as 1.05×10^{-11} M for paraoxon and 4.47×10^{-12} M for parathion.

An amperometric biosensor which used the enzyme an acetylthiocholine (ATCh) has been used for detection and quantification of three organophosphorus pesticides – paraoxon ethyl, monocrotophos and dichlorvos. The inhibition curves for each pesticide was plotted and the linear intervals were determined along with the corresponding equations and detection limits – 0.87×10^{-11} M for paraoxon, 1.08×10^{-11} M for monocrotophos and 1.22×10^{-10} M for dichlorvos [23].

Clemens Steegborn and Petr Skliidal described a piezoelectric immunosensor for determination of the herbicide atrazine. The Interaction of the anti-atrazine monoclonal antibody (MAb, clone D6F3) with the immobilized atrazine was characterized using both crude ascetic fluid and Protein A-purified MAb preparates. As expected, the higher dilutions of MAb provided improved sensitivity for the analyte. For the 1000X diluted ascetic fluid, 0.1 and 1 μg/l atrazine caused 5 and 30% decreases of the relative binding of MAb, respectively [24].

3. Electrochemical impedance spectroscopy (EIS)

Impedimetric immunosensors have recently received particular attention since they possess a number of attractive characteristics associated with the use of electrochemical transducers, namely, low cost of electrode mass production, cost effective instrumentation, the ability to be miniaturized and to be integrated into multi-array or microprocessor-controlled diagnostic tools, remote control of implanted sensors, etc. Indeed, due to the above-mentioned characterics, electrochemical impedance spectroscopy (EIS)-based sensors are considered as promising candidates for use at on-site applications [25,26].

3.1. Fundamentals [27-29]

Electrochemical Impedance Spectroscopy (EIS) is the method in which the impedance of an electrochemical system is studied as a function of the frequency of an applied a.c. wave.

When the system is perturbed (by applied a.c. voltage) it relaxes to a new steady state. The time taken for this relaxation is known as the time constant, τ, and given by:

$$\tau = RC$$

where R is the resistance and C the capacitance of the system. The analysis of this relaxation process would provide information about the system. The ratio of the response to the perturbation is the transfer function. When the applied perturbation is an a.c. potential and the response an ac. current, the transfer function is the impedance. To simplify calculations further, the perturbation and response are transformed from a function of time into the frequency domain via a Laplace transformation. The applied potential is given by

$$\underline{E} = E_0 \exp(j\omega t) \qquad (1)$$

where E0 is the amplitude of the signal and $\omega = 2\pi f$ is the radial frequency and f is frequency

The output current of the system is also a sinusoidal, is shifted in phase (φ) and has a different amplitude, I_0 and it is given by,

$$\underline{I} = I_0 \exp(j\omega t + \varphi) \tag{2}$$

According to Ohm's law, impedance (Z) of the circuit at any frequency (ω) can be represented by:

$$\underline{Z} = \frac{E}{\underline{I}} = \left(\frac{E_0}{I_0}\right)\exp(-j\varphi) = Z_0 \exp(-j\varphi) \tag{3}$$

The impedance is therefore expressed in terms of a magnitude, Z_0, and a phase shift, φ. It is possible to express the impedance as:

$$\underline{Z} = Z_0 \cos(\varphi) - jZ_0 \sin(\varphi) = Z' - jZ'' \tag{4}$$

The expression for $\underline{Z}$ is composed of a real and an imaginary part.

Usually a low a.c. voltage of about 10 mV is applied to keep the system linear.

The most popular formats for evaluating electrochemical impedance data are the Nyquist and Bode plots. In the former format, the imaginary impedance component (Z'') is plotted against the real impedance component (Z') at each excitation frequency, whereas in the latter format, both the logarithm of the absolute impedance, $|Z|$ and the phase shift, φ, are plotted against the logarithm of the excitation frequency.

3.2. Equivalent circuit

Interpretation of EIS measurements is usually done by fitting the impedance data to an equivalent electrical circuit which is representative of the physical processes taking place in the system under investigation. In fact, one of advantages of EIS is that impedance functions frequently display many of the features exhibited by passive electrical circuit.

The most important elements that can be used in equivalent circuits are summarized in table 1. The resistor, R, represents the resistor that charge carriers encounter in a specific process or material. The capacitor, C, represents the accumulation of charged species. The inductance, L, is used to represent the deposition of surface layers such as the passive layer. The Warburg element, W, is used to model linear semi-infinite diffusion which occurs when the diffusion layer has infinite thickness. The constant phase element, CPE, is a general element which can represent a variety of elements such as inductance (n=-1), resistance (n=0), Warburg (n=0.5), capacitance (n=1).

Element	Symbol	Impedance expression
Resistance	R	R
Capacitance	C	$1/(jc\omega)$
Inductance	L	$jl\omega$
Warburg	W	$1/[Y(j\omega)^{1/2}]$
CPE	Q	$1/[Y(j\omega)^{n}]$

Table 1. Electrical impedance elements

Capacitive immunosensors exploit the change in dielectric properties and/or thickness of the dielectric layer at the electrolyte–electrode interfaces, due to the antibody–antigen interaction, for monitoring this process. An electrolytic capacitor (working electrode/ electrolyte) allows the detection of an analyte specific to the receptor that has been immobilized on the surface of the working electrode. Ideally, this configuration resembles a capacitor in its ability to store charge and thus, the electric capacitance between the working electrode and the electrolyte is given by equation 5:

$$C = \frac{\epsilon_0 \epsilon S}{d} \tag{5}$$

Where ε, is the dielectric constant of the medium between the plates, εo, is the permittivity of free space (8.85419pF/m), S, is the surface area of the plates, and, d, is the thickness of the insulating layer (m).

A decrease of the total capacitance, due to the increase of the distance between the plates is thus expected upon the binding of the analyte to its specific receptor.

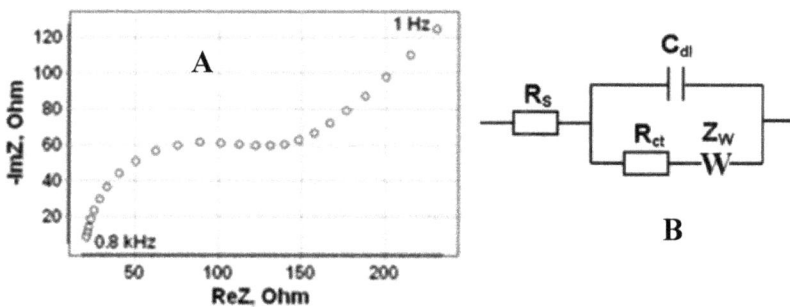

Figure 3. Nyquist plot arising from a Randles circuit showing in the side panel

While no equivalent model can be guaranteed to be unique, simulation of the recorded impedimetric data to an equivalent electric circuit is a common strategy for understanding the physical origin of the observed response. The simplest, and in fact the most frequently used equivalent circuit for modelling of EIS experimental data is the so called Randles circuit (Fig. 3.B), which comprises the uncompensated resistance of the electrolyte (Rs), in series with the capacitance of the dielectric layer (Cdl) and the charge-transfer resistance (Rct), if a redox probe is present in the electrochemical cell. The latter two components are connected in parallel. An additional component, connected in series with Rct, the Warburg impedance (Zw) accounts for the diffusion of ions from bulk electrolyte to the electrode interface. A typical shape of the impedance spectrum of this circuit presented in a Nyquist plot (Fig. 3.A) includes a semicircle region lying on the real axis followed by a straight line. The linear part ($\phi=\pi/4$), observed at the low frequency range, implies a mass-transfer limited process, whereas the semicircle portion, observed at high frequency range, implies a charge-transfer limited process.

4. Bioreceptors

The sensitivity of immunosensor is strongly dependant on the amount of immobilized antibodies and their remaining antigen binding properties. A variety of immobilization methods for proteins has been reported in the literature [30]. The choice of the optimal immobilisation method not only depends on the surface linking layer and its specific functional end groups but also on the free functional groups in the antibodies or their respective fragments. Numerous bi-functional cross linking reagents have been developed using reactive chemical groups such as succinimide esters and aldehyde groups. These linkage strategies selectively form covalent bonds with the lysine residues randomly present in the antibodies, giving rise to a random orientation of the receptor molecules immobilized on the sensor surface.

4.1. Mixed biotinylated self-assembled monolayer

Self-assembled monolayers consisting of long alkyl-thiols chains on gold have been shown to be stable in air, water and organic solvents at room temperature [31]. On the other hand, the biotin/neutravidin couple has a quite high binding affinity and can act as a bridge to anchor bioreceptor species. Therefore, a stable self-assembling system combined with a biotin/neutravidin couple has potential application for construction of biosensors. In this study, a mixed monolayer is chosen, which is composed of 1,2 dipalmitoyl-sn-glycero-3-phosphoethanolamine N-(biotinyl) (biotinyl-PE) and 16-mercaptohexadecanoic acid (MHDA). It possess a thiol group allowing its immobilization on gold surfaces, and an hydrophilic terminal carboxyl group. The final stability of the mixed SAM layer is obtained through hydrophobic interaction between long alkyl (C16) chains. Fig.4 shows the assembly of the mixed SAM.

After the mixed MHDA/biotinyl-PE self-assembled monolayer was formed on gold electrode and in order to reduce non-specific adsorption, an anti-goat IgG was used to block the free space between biotinyl-PE molecules in the mixed SAM. Then, neutravidin was bound on the biotinyl-PE. Like streptavidin, neutravidin can bind four biotinyl groups, so it can act as a cross-linking agent between different molecular layers. Neutravidin is used as an alternative to streptavidin as it is carbohydrate free and has a neutral isoelectric point, which provides exceptionally low nonspecific binding properties. In the following step, biotinyl-Fab fragment K47 antibody was anchored onto neutravidin to allow the specific affinity immobilization of atrazine.

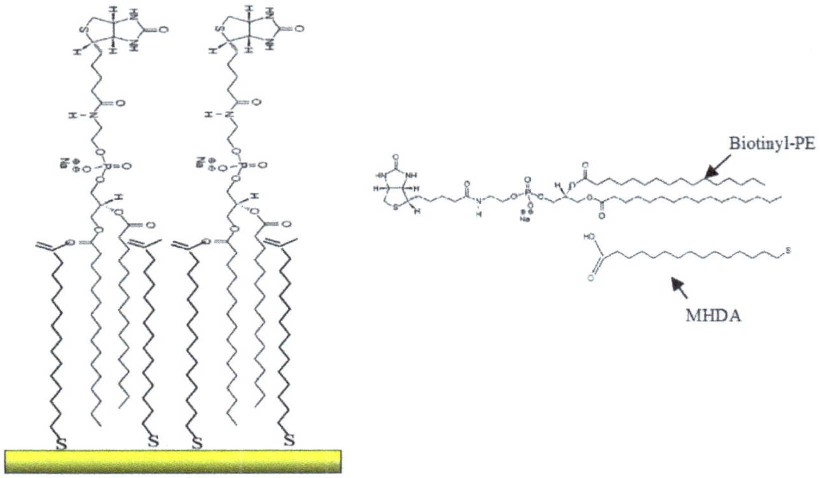

Figure 4. Schematic showing the assembly of a m xed self-assembled monolayer.

4.1.1. Electrochemical characteristics of mixed self-assembled monolayer [32]

The properties of mixed monolayer were characterized by cyclic voltammetry and impedance spectroscopy techniques.

Complex impedance plots of bare electrode (a) and mixed self-assembled monolayers (b) are shown in Fig. 6.A. The impedance spectra of bare gold electrode fit the theoretical profile and include a semi-circle region in the frequency range from 0.5 to 1×10^5 Hz. For the mixed self-assembled monolayers deposited on the gold electrode, the impedance spectra include a semi-circle region observed at high frequencies, corresponding to a change in the SAM structure, followed by a linear region characteristic of lower frequencies attributed to diffusion phenomenon. The respective compressed semi-circle diameters correspond to the membrane resistance at the electrode surface, and increase on the addition of self-assembled monolayers on the electrode surface.

The experimental non-faradic impedance spectra were fitted with computer simulated spectra using an electronic circuit shown in Fig. 5. This equivalent circuit includes the ohmic resistance of the electrolyte solution, Rs, the Warburg impedance, Zw, from the diffusion, the constant phase element, CPE, and membrane resistance, Rm. The latter three components,

Zw, CPE and Rm, represent interfacial properties of the electrode, and they are affected by the surface modification.

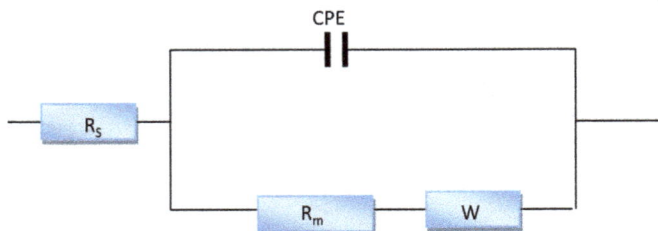

Figure 5. Equivalent circuit used to model impedance data in PBS solution.

An excellent fitting between the simulated and experimental spectra was obtained for the bare Au-electrode and the mixed monolayer-modified Au-electrode Fig. 6.A. It can be seen that the diameter of semi-circle at high frequency increases upon the stepwise formation of modifier on the electrode surface. The membrane resistance values, Rm, were extracted from the computer simulated spectra which are 2180 and 13967 Ωcm^2 for bare Au-electrode and mixed modified Au-electrode, respectively.

The values of the fractional coverage area of the mixed monolayer (θ) can be calculated from the impedance diagrams using equation 6 [33]:

$$\theta = 1 - \frac{R_m}{R_m^*} \tag{6}$$

where R_m and R^*_m are the values of the membrane resistance derived from the impedance diagram of the bare gold electrode and mixed self-assembled monolayer, respectively. In our system the fractional coverage area was equal to 0.84. The high value of the Warburg impedance of the gold electrode with the covered SAM shows that the layer is not acting as a blocking layer but as a diffusion layer because of the low percentage of area coverage.

The experimental variations of impedance versus time show a stability in the range of 8%, which proves the mixed SAMs is quite stable. Such stability of the mixed monolayer offers us a very good basis for further construction.

Cyclic voltammetry experiments further confirmed that the mixed SAM was successfully formed on the gold surface. When the electrode surface was modified by addition of material, the electron transfer kinetics of $Fe(CN)_6^{-4/-3}$ were perturbed. As shown in Fig. 6.B, the stepwise assembly of bare gold and mixed SAMs is accompanied by a decrease in the peak to peak separation between the cathodic and anodic waves of redox probe. This shows the formation of the mixed monolayer.

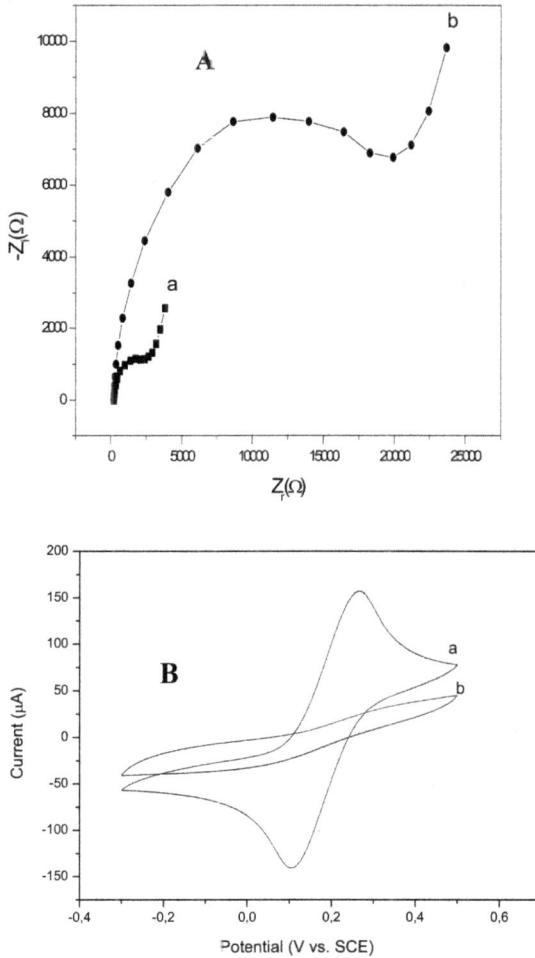

Figure 6. (A) Nyquist diagram (Zr vs.Zi) for the ncn faradic impedance measurements corresponding to (a) bare Au-electrode and (b) mixed self-assembled monolayer functionalized Au electrode. Symbols show the experimental data in PBS solution. Solid curves show the computer fitting of the data using the equivalent circuits shown in Fig. 5. (B) Cyclic voltammetry for bare gold and mixed SAM electrodes in 5 mM K3[Fe(CN)6]/K4[Fe(CN)6] in PBS pH 7.0.

4.1.2. Recognition properties of the self-assembled multiplayer

In order to evaluate the recognition properties of the system, in terms of sensitivity and se-lectivity, we exposed the gold electrode with self-assembled multilayers to various concen-trations of atrazine after immobilization of biotinyl-Fab fragment K47 antibody. The corresponding Nyquist plots of impedance spectra are shown in Fig.7 the semicircle diame-ter in Nyquist plot seems to decrease with the antigene concentration, implying that more amount of antigen was linked to the interface and the mixed SAMs change its structure with different concentration adding of antigen as the antigen was not immobilized on the entire surface and thus do not act as a blocking layer.

When the concentration of antigen was increased over 300 ng/ml, the change of impedance spectroscopy become gradually weak, showing that immobilization of the antibody on a gold electrode trends to saturation situation. So we can conclude that the self-assembled multilayer system that we developed allows distinguishing between specific immobilization of a given receptor and parasitic adsorption of the other proteins present, and thus repre-sents an efficient means to constitute a biosensor for detection of atrazine. Taking into ac-count the blank and the signal fluctuation (noise), the detection limit for the binding of atrazine on the self-assembled multilayer is 20ng/ml.

4.2. Functionalized magnetic beads

Recently, many improvements including enhanced sensitivity and reduced detection time have been made to immunosensors. These improvements are considerably attributed to the use of magnetic beads (MBs) in immunosensors. MBs allows (i) easy separation and localiza-tion of target proteins by an external magnet, (ii) fast immunoreactions between antigen and antibody, and (iii) low nonspecific binding by surface modification[34]. Recently, MB-based immunoassay systems are widely used in clinical laboratories, and MB-based immunoassay becomes one of standard formats in high-throughput assay. The lowest detection limit of these immunoassay system is generally in the pg/mL range. Accordingly, a more sensitive detection method based on MB is required. Recently, many electrochemical immunosensors using MB have been developed to achieve low detection limits. The high interest in electro-chemical immunosensors is due to their easy miniaturization and operation [35-37].

EIS correlates parameters of the system that are purely electrochemical, i.e. current/potential [38]. Nevertheless, electrochemical systems have also been studied by extending the impe-dance concept to measurement obtained by application of perturbations of a non-electrical character, such as temperature, magnetic field, illumination, etc., and in the measurement of the responses of the responses of non-electrical character, such as optical transmittance, mass determination via a quartz balance etc., after an electrochemical perturbation

Here, we present an ultrasensitive and promising analytical method employing an electro-chemical immunosensing strategy based on magnetic monolayer of magnetic particles coat-ed with streptavidin. This novel strategy takes advantage of easy magnetic separation and immunoreaction by MB and a high binding affinity between the biotin/ streptavidin couple.

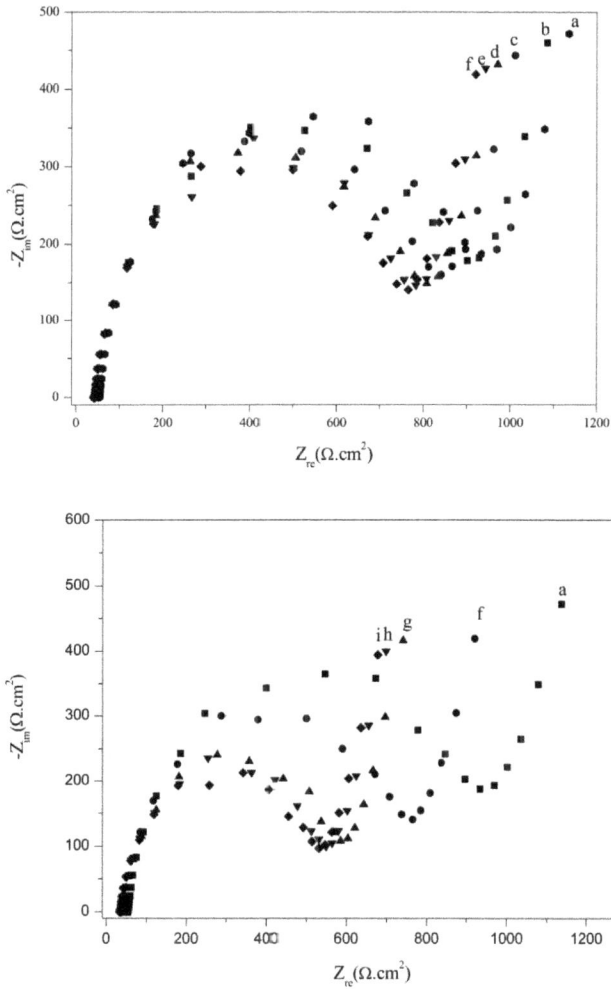

Figure 7. Complex impedance plots of antigen–antibody/neutravidin/blocking with IgG/mixed SAM/gold electrode under various concentrations of atrazine. The concentrations of atrazine (ng/ml): (a) 0; (b) 10; (c) 30; (d) 50; (e) 80; (f) 120; (g) 200; (h) 600; (i) 1100.

4.2.1. Characterization of magnetic monolayer [39]

The magnetic coated streptavidin nanoparticles that display a diameter of 200 nm and an iron oxide content of about 70%. A schematic illustration of the nanoparticle is presented in Fig. 8. After application of a 300 mT magnetic field, a layer of magnetic particles coated with streptavidin was formed on the gold electrode.

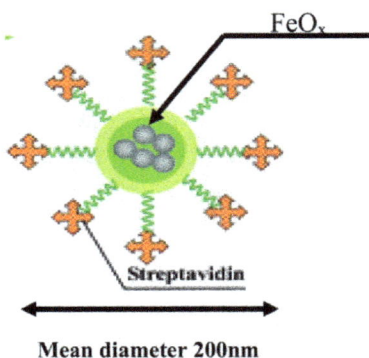

Figure 8. Nanoparticles structure.

The magnetic monolayer was characterized using faradaic impedance spectroscopy, cyclic voltammetry and atomic force microscopy (AFM) techniques.

Cyclic voltammetry experiments further confirmed that a magnetic monolayer was successfully formed on the gold surface. When the electrode surface was modified by addition of material, the electron transfer kinetics of $[Fe(CN)_6]^{4-/3-}$ were perturbed.

The stepwise assembly of bare gold and magnetic beads is accompanied by a decrease in the peak to peak separation between the cathodic and anodic waves of the redox probe. This result shows that the magnetic monolayer formed on gold electrode is not a real insulating layer but presents a conductivity near the gold surface.

In order to characterize the formation of the magnetic monolayer and obtain information on its architecture, AFM measurement were taken in tapping mode. Fig.9 shows an AFM image of the bare gold and of the magnetic beads layer. The image shows the formation of magnetic monolayer homogeneous and dense. The line profile measurement of the magnetic beads give diameter distribution 200 nm witch fit will with real beads dimension.

The faradic impedance spectra for bare gold electrode and the magnetic monolayer show a strong decrease in the constant phase element. It is evident such a constant phase element decrease can simply be attributed to a change in the thickness. Further, we have a decrease in electron transfer resistance. This decrease could be due to changes in surface conductivity.

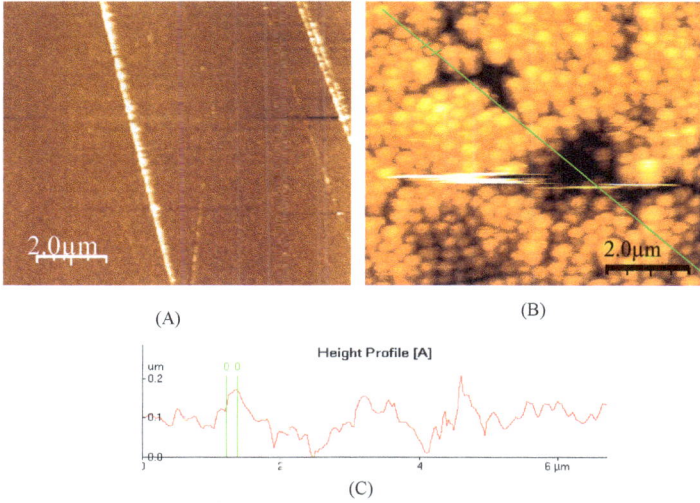

(A)   (B)

(C)

Figure 9. AFM image of (a) bare gold, (b) deposited magnetic beads on gold electrode and (c) profile measurement of magnetic beads.

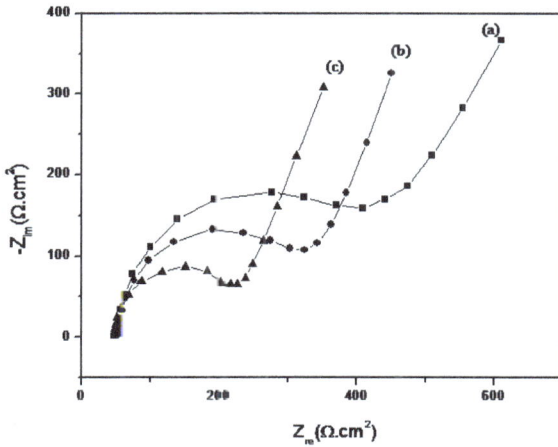

Figure 10. Nyquist diagram (Zr vs. Zi) for the non-Faradaic impedance measurements corresponding to: (a) magnetic beads/Au-electrode; (b) Fab fragment K47 antibody/magnetic beads/Au-electrode; (c) 600ng/ml Atrazine/Fab fragment K47 antibody/magnetic beads/Au-electrode. Solid curves show the computer fitting of the data. Symbols show the experimental data.

4.2.2. Atrazine detection

The antibody, biotinyl–Fab fragment K47, forms a quite stable layer onto the magnetic monolayer due to the high affinity of the biotin/streptavidin interaction. After the antibody layer formation an antigen, atrazine was injected to react with the antibody.

Complex impedance plots of the successive buildind-up of the sensing layer are shown in fig. 10. To analyze the complex impedance spectra, data were fitted with the commercially available software Zplot/Zview (Scibner Associates Inc.) to equivalent circuit:

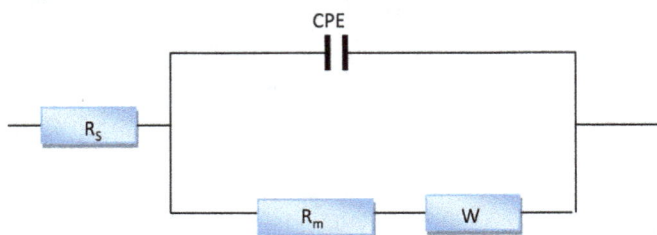

CPE

R_s

R_m W

The electron transfer resistance values were 304.6 Ωcm^2, 204.5 Ωcm^2 and 188.5 Ωcm^2 for the magnetic monolayer, the antibody layer and after injection of 600 ng/ml of atrazine, respectively. The decreases of electron transfer resistance could be attributed to a reorganization of the beads as the constant phase element decreases too. The constant phase elements, Q, extracted from the computer fitting for the same steps were 17$\mu F/cm^2$, 15$\mu F/cm^2$ and 14.29$\mu F/cm^2$, respectively. This decrease is due to an increase in thickness.

In order to test the specific binding of the magnetic monolayer, we exposed the substrate to various concentrations of atrazine, and Nyquist plots were recorded using EIS. A significant difference in the impedance spectra was observed with increasing receptor concentration. A linear relationship between the ΔR_{et} values and the concentration of atrazine was established in the range from 50ng/ml to 500ng/ml.

4.3. Polypyrrole–neutravidin layer

A variety of methods capable of immobilizing biologically active material onto or in close proximity of the transducer surface have been reported. The conducting polymers can be considered as effective material for immobilization of biomolecules and for transducing/ amplification of electrical signal in design of immunosensing devices [40,41]. The conjugated double bonds in the backbone of the conducting polymers allow free movement of electrons within the conjugating length, which makes them electrically conductive [42]. Till now, polypyrrole (PPy) has mostly been applied because of its high conductivity, high storage ability, good thermal and environmental stability, high redox and capacitive current and biocompatibility. Polypyrrole can be synthesized by chemical polymerization [43], photoinduced synthesis [44] and electrochemical activation by anodic current[45]. Electrochemical polymerization is the most commonly used procedure to deposit conductive polymers due to its simplicity and rapidity. During the electrochemical oxidation process, biomolecules can be added into the monomer solution for subsequent entrapment. Typically, this polymer/biomolecule layer is developed at fixed potentials or by cyclic voltammetry. The method allows the polymeric layer to be controlled, but could also lead to the denaturation of the biologically active element during the immobilization process.

Moreover, by using this technique, thickness and morphology of deposited layer might be controlled by application of well-defined potential and known current passing through the electrochemical cell

In this approach, we present another way to immobilize biomaterials based on neutravidin entrapment during electrochemical deposition of polypyrrole. The attachment of biotinyl–Fab fragment K47 antibody through the specific biotin–neutravidin interaction was therefore done on the total volume of the polypyrrole layer.

4.3.1. Preparation of the PPy/neutravidin layer [46]

In order to obtain a semi-transparent and thin polypyrrole film, we minimize the duration of the cyclic voltammetry. The polypyrrole film with yellow color can be obtained with cyclic voltammetry between 600 and 900mV. A reproducible PPy/neutravidin thickness layer can be realized with the same method. However, the majority of proteins are not highly charged in neutral pH. We have therefore used an anionic surfactant (SDS) as a co-dopant in order to add an ionic behavior to the proteins which is very useful for the entrapment process within the biofilm [47]. The confirmation the incorporation of neutravidin inside the Polypyrrole film was studied by impedance spectroscopy. The impedance spectroscopy measurement give big change in charge transfer resistance. If large size dopant molecules such as neutravidin were incorporated into PPy film during electropolymerization, the polypyrrole layer might to have a porous structure. Therefore, we can confirm that neutravidin was inside the PPy film since the high resistance deduced from the impedance plot was generally expected to be related to the porous structure.

4.3.2. Immunoassay

After blockage of nonspecific sites close to the biofilm with Bovine serum albumin (BSA), the subsequent grafting of Fab fragment will therefore be based only on biotin–neutravidin bonding. This interaction is very rapid and the strongest known non-covalent binding with a dissociation constant of 10^{-15} M [33]. The Nyquist plot after antibody attachment had a semicircle shape which indicating a film charge-transfer resistance. It is clear that Fab fragment loading affect the dynamics of charge transfer near the electrode interface due to differences in the electroconductivity between PPy/neutravidin film and Fab immobilized surface. A non-homogenous insulating protein layer was therefore added on the biofilm after fixation of the immunoreceptor.

In order to assess the immunosensor sensitivity, different concentrations of atrazine in the range of 0.1–200ng/ml were injected on the Fab fragment modified electrode show an increase in the semicircle diameter with antigen concentration. This was related to the positive change in the film charge transfer resistance after atrazine injection. An insulating organic layer was progressively added to the biofilm since an immunoreactions near the electrode surface was established between the Fab fragment and the antigen. This interaction might affect the thickness of the double couche at the electrode interface. Fig.11 illustrates very well the good sensitivity of our impedimetric immunosensor even with 0.1ng/ml of atrazine. However, this excellent antigen detection limit could have an important non-specific part. Thus, different concentrations of rabbit IgG (non-specific antigen) in the range of 0.1–200 ng/ml were added to the biosensor. The impedance measurements show a very little increase in the Nyquist plot diameter with respect to the atrazine detection.

Based on all these results and in order to illustrate the sensitivity and the selectivity of the immunosensor, two curves corresponding to the variation of ΔRt (ΔRt is the change of charge transfer resistance obtained by subtracting the resistance of the immobilized biotinylated Fab fragment from the resistance of the immune complex) with atrazine and rabbit IgG concentrations were plotted (Fig. 11).

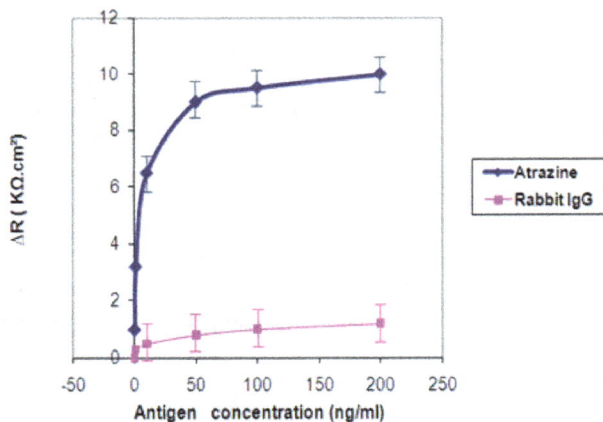

Figure 11. The variation of ΔRt with atrazine and rabbit IgG concentrations

As can be seen in Fig. 11, the plot for the atrazine detection was almost linear and tend to reach saturation next to 200ng/ml. Whereas the response of the immunosensor to different concentrations of non-specific antigen was clearly non-significant. These results have been obtained with a good reproducibility. We are therefore sure that the above-observed impedance changes after atrazine injection were generated from the result of specific Fab fragment-antigen interaction.

4.4. Labeled magnetic nanoparticles assembly on polypyrrole film

In recent years, conducting polymers combined with metallic nanoparticles have been paid more attention due to their potential applications in microelectronics, microsystems, optical sensors and photoelectronic chemistry. In many recent works, PPy films are found to be associated with metallic nanoparticles (NPs) [48]. The development of such nanocomposites is essentially motivated by their high analytical sensitivity in sensing applications. Different properties emerging from the nanostructuration with NPs are at the origin of the increased sensing sensitivity. The NPs size and high surface area have first the ability to facilitate direct and fast electron transfer between the nanocomposite and the transducer. Second, when compared to homogeneous bulk matrices, the high surface area of the NPs assembly also leads to nanoporosity for signal amplifications and increased sensitivity toward surface adsorption or surface reactions. Because of the same geometric properties the NPs assembly also allows minimum diffusion of the target molecule, and in the same time, miniaturization of the device. Finally it was shown that the selectivity of the sensor could been hanced by tuning the molecular interactions between the NPs and linker molecules. The improvement of the sensing properties resulting from the nanostructuration is such that various routes were proposed in order to incorporate NPs either in the PPy film or by synthesis of the metallic NPs directly on the PPy film.

We addressed this study to the preparation and characterization of a nanocomposite composed by a thin polypyrrole (PPy) film covered with an assembly of magnetic nanoparticles (NPs). The magnetic particles were immobilized on PPy films under appropriate magnetic field in order to control their organization on the PPy film and finally to improve the sensitivity of the system in potential sensing applications (Fig.12).

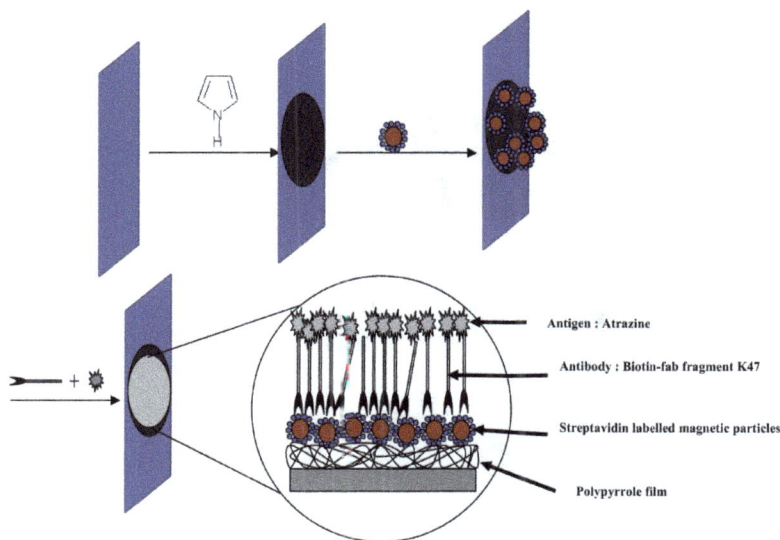

Figure 12. A schematic diagram of the immunosensor showing the stepwise immunosensor fabrication process.

4.4.1. The polypyrrole film/magnetic nanoparticles composite [49]

The atomic force microscopy images of the PPy film shown in Fig.13.A. These micrographs reveal the cauliflower morphology usually observed for electrochemically deposited PPy films which is attributed to the nodular fractal-type growth of these polymers. However, the high roughness of this surface does not allow atomic force microscopy acquisitions on large area. The NPs assembly is clearly observable with atomic force microscopy as shown in Fig. 13.B. Cracks in the NPs assembly are again present at microscopic scale. AFM observations allow us to assess for the distribution of both the PPy and the NPs on the electrode, which should provide enhanced adsorption and sensitivity in sensing application.

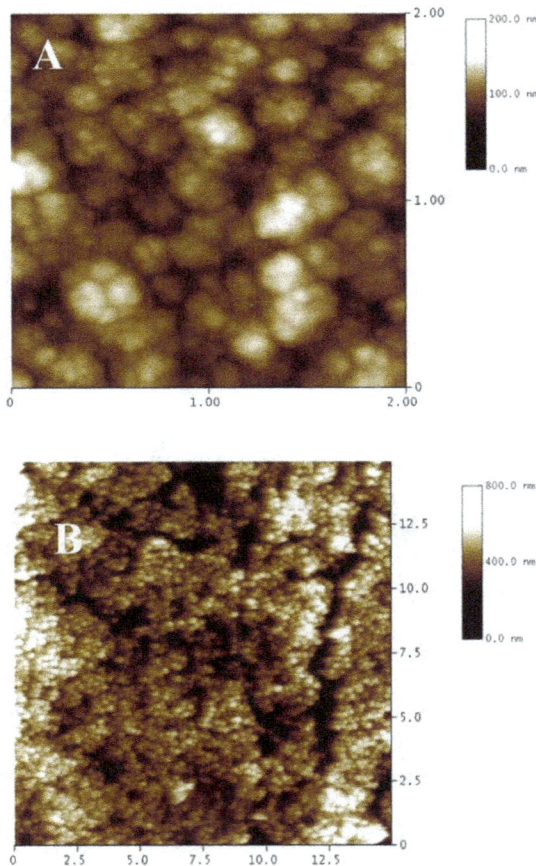

Figure 13. Top AFM images of: (A) PPy film showing the cauliflower structure of this polymer, (B) PPy film covered with an array of nanoparticles

Cyclic voltammograms of the gold electrode present a reversible phenomenon, which is the typical behaviour of gold surface with redox couple (Fig.14). The two peaks of the cathodic and anodic waves of redox probe have been obtained. After modification of the gold surface with PPy, the dc-current increases due to the conducting properties of the PPy film. After immobilization of the magnetic particles on the PPy film, the direct current decreases due to the insulating properties of the functionalized film (streptavidin) covering the particles. The same explanation applies for the decrease of the direct current after immobilization of the antibody and after the BSA blocking step.

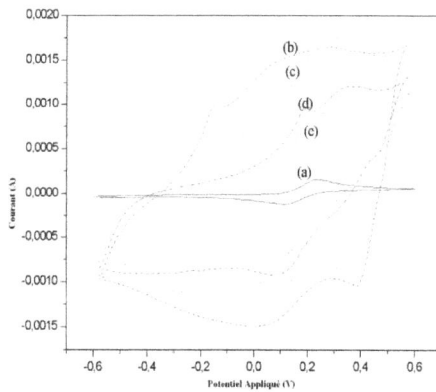

Figure 14. Cyclic voltammograms after different steps of modification: (a) bare gold electrode, (b) PPy film modified gold electrode, (c) PPy covered with streptavidin labelled magnetic particles,(d) immobilization of the antibody biotin-Fab fragment K47 and (e) BSA blocking layer.

Fig. 15 shows the impedance spectrum of a PPy film coated gold electrode (curve a) compared with streptavidin labeled magnetic particles surface (curve b), with an antibody-immobilized surface (curve c) and with BSA blocking layer (curve d).

Note that all the spectra are almost similar, containing a distorted semicircle. The diameter of the semicircle provides an estimate of the film charge transfer resistance. The resistance of the studied interface increases after the immobilization of each step. This increase is due to the decrease of the conductivity due to the insulating properties of grafted layers. This confirms the results obtained with cyclic voltammetry.

4.4.2. Biosensor application

Biotinylated Fab fragment K47 antibody was covalently bound to the particles through streptavidin/biotin linkage. This allows specific bounding of atrazine on the NPs and variation of the electrode impedance response. Atrazine–antibody interactions were monitored by impedance spectroscopy at −1200mV. The semicircle diameter in the impedance spectroscopy measurement is increasing with the atrazine concentration, implying that more amount of atrazine was linked to the interface. These results revealed that the presence of the PPy film under the NPs assembly increases over four orders of magnitude the sensitivity of the sensor as compared to only the NPs assembly, with an excellent detection limit of 5ng/ml [38].

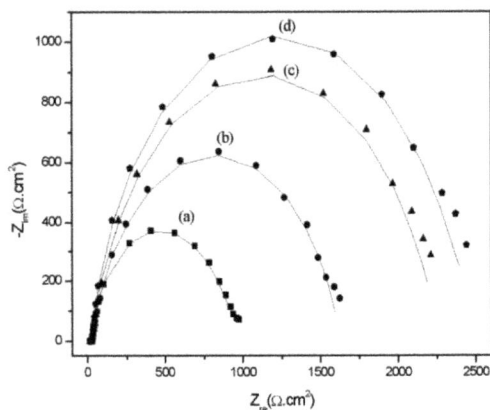

Figure 15. Nyquist diagram for the faradic impedance measurements corresponding to (a) PPy film/Au-electrode, (b) streptavidin labeled magnetic particles/PPy film/Au-electrode, (c) biotin-Fab fragment K47 antibody/streptavidin labeled magnetic particles/PPy film/Au-electrode and (d) blocked layer with BSA/biotin-Fab fragment K47 antibody/streptavidin labeled magnetic particles/PPy film/Au-electrode. Solid curves show the computer fitting. Symbols show the experimental data.

5. Conclusion

Over the last few years, there have been an increasing number of publications concerning the application of biosensors to environmental analysis. Biosensors have been reported for the commonly used pesticides and industrial chemicals. In some cases there is a need for sensitivity and lifetime improvements as conventional techniques can outperform biosensors in these respects at present. The health and safety of workers applying pesticides or indeed any other chemical could be protected by providing them with biosensors for monitoring the levels of pesticides or chemicals in the air around them. Biosensors could contribute towards monitoring the progress of clean-up operations after environmental spillages of certain chemicals.

Author details

Saloua Helali

Address all correspondence to: salwaHeli@yahoo.fr

Research and Technology Centre of Energy, Hammam Lif, Tunisia

References

[1] COM(2006) 398 final: Proposal for a Directive of the European Parliament and of the Council of 17 june, 2006 on environmental quality standards in the field of water policy and amending Directive 200C/60/EC.

[2] Nanoporous impedemetric biosensor for detection of trace atrazine from water samples: Pie Pichetsurnthorn, Krishna Vattipalli , Shalini Prasad. Biosensors and Bioelectronics, Volume 32, 2012, Pages 155-162.

[3] A direct optical immunosensors for atrazine detection: Andreas Brecht, Jacob Piehler, Gerd Lang, Günter Gauglitz: Analytica Chimica Acta, Volume 311, 1995, Pages 289-299.

[4] Separation-free electrochemical immunosensor for rapid determination of atrazine: R.W. Keay, C.J. McNeil. Biosensors and Bioelectronics, Volume 13, 1998, Pages 963-970.

[5] Effects of atrazine on periphyton under grazing pressure: Isabel Muñoz, Montserrat Real, Helena Guasch, Enrique Navarro, Sergi Sabater: Aquatic Toxicology, Volume 55, 2001, Pages 239-249.

[6] Pesticide determination in tap water and juice samples using disposable amperometric biosensors made using thick-film technology: Miquel Albareda-Sirvent, Arben Merkoçi, Salvador Alegret: Analytica Chimica Acta, Volume 442, 2001, Pages 35-44.

[7] Plant tissue electrode for the determination of atrazine: Franco Mazzei, Francesco Botrè, Giampiero Lorenti, Giovanra Simonetti, fernando Porcelli, Giancarlo Scibona, Claudio Botrè: Analytica Chimica Acta, Volume 316, 1995, Pages 79-82.

[8] Council Directive of 15 July 1980 relating to the quality of water intended for human consumption (80/778/EEC).

[9] Construction and characterization of the direct piezoelectric immunosensor for atrazine operating in solution: Clemens Steegborn & Petr Skliidalt. Biosensors & Bioelectronics Vol. 12. pp. 19-27, 1997.

[10] Immunochemical methods for environmental monitoring: Gianfranco Giraudi, Claudio Baggiani, Nuclear Medicine and Biology, Volume 21, 1994, Pages 557-572.

[11] Applications of electrochemical immunosensors to environmental monitoring: Omo-wunmi A. Sadik, Jeanette M. Van Emon: Biosensors and Bioelectronics, Volume 11, 1996.

[12] Biosensors for pesticide detection basedon alkaline phosphate-catalyzed chemilumi-nescence: Madhu S.Ayyagari, Sanjay Kamtekar Rajiv Pande, Kenneth A. Marx, Jay-ant Kumar, Sukant K. Tripathy, David L. Kaplan: Materials Science and Engineering: C, Volume 2, 1995, Pages 191-196

[13] Conductimetric immunosensor for atrazine detection based on antibodies labelled with gold nanoparticles: Enrique Valera, Javier Ramón-Azcón, F.-J. Sanchez, M.-P. Marco, Ángel Rodríguez: Sensors and Actuators B: Chemical, Volume 134, 2008, Pa-ges 95-103.

[14]] Label-free impedimetric immunosensor for sensitive detection of atrazine: Rodica E. Ionescu, Chantal Gondran, Laurent Bouffier, Nicole Jaffrezic-Renault, Claude Marte-let, Serge Cosnier: Electrochimica Acta, Volume 55, 2010, Pages 6228-6232.

[15] Impedimetric immunosensor for atrazine detection using interdigitated μ-electrodes (IDμE's): Enrique Valeraa, Javier Ramon-Azconb, Angel Rodrigueza, Luis M. Casta-nera, F.-J. Sanchezb, M.-P. Marco: Sensors and Actuators B 125 (2007) 526–537

[16] Antibodies for immunosensors: A review, Bet-told Hock, Analytica Chimica Acta 347 (1997) 177-186

[17] Immunosensors for detection of pesticide residues: Xuesong Jianga, Dongyang Li, Xia Xu, Yibin Ying, Yanbin Li, Zunzhong Ye, Jianping Wang, Biosensors and Bioelec-tronics 23 (2008) 1577–1587.

[18] Nanomaterial labels in electrochemical immunosensors and immunoassays: Guo-dong Liu, Yuehe Lin, Talanta 74 (2007) 308–317.

[19] Lewis, Lord of Newnham (1992) (chairman) commission on environmental pollu-tion,16th report, HMSO, London.

[20] Needs for reliable analytical methods for monitoring chemical pollutants in surface water under the European Water Framework Directive. Peter Lepom, Bruce Brown, Georg Hanke, Robert Loos, Philippe Quevauviller, Jan Wollgast, Journal of chroma-tography A, 1216 (2009) 302-315;

[21] Semi disposable reactor biosensors for detecting carbamate pesticides in water: Siri-wan Suwansa-ard, Proespichaya Kanatharana, Punnee Asawatreratanakul, Chusak Limsakul, Booncharoen Wongkittisuksa, Panote Thavarungkul, Biosensors and Bioe-lectronics, Volume 21, 2005, Pages 445-454.

[22] highly sensitive organophosphorous pesticide biosensors based on nanostructured films of acetylcholinesterase and CdTe quantum dots: Zhaozhu Zheng, Yunlong Zhou, Xinyu Li, Shaoqin Liu, Zhiyong Tang,Biosensors and Bioelectronics, Volume 26, 2011, Pages 3081-3085.

[23] Amperometric inhibition-based detection of organophosphorus pesticides in unary and binary mixtures employing flow-injection analysis: Ivaylo Marinov, Yavor Ivanov, Nastya Vassileva, Tzonka Godjevargova,Sensors and Actuators B: Chemical, Volume 160, 2011, Pages 1098-1105

[24] Construction and characterization of the direct piezoelectric immunosensor for atrazine operating in solution: Clemens Steegborn, Petr Skládal, Biosensors and Bioelectronics, Volume 12, 1997, Pages 19-27.

[25] Local electrochemical impedance spectroscopy: A review and some recent developments, Vicky MeiWen Huang, Shao-Ling Wu, Mark E. Orazem, Nadine Pébère, Bernard Tribollet, Vincent Vivier, Electrochimica Acta

[26] Application of electrochemical impedance spectroscopy to study the degradation of polymer-coated metals: A. Amirudin, D. Thierry, Progress in Organic Coatings 26 (1995) 1-28.

[27] Impedance spectroscopy: Over 35 years of electrochemical sensor optimization: Bobby Pejcic, Roland De Marco, Electrochimica Acta 51 (2006) 6217–6229

[28] Impedimetric immunosensors-A review: Mamas I. Prodromidis, Electrochimica Acta 55 (2010) 4227-4233

[29] Impedance spectroscopy: J.Ross Macdonal, Annals of biomedical Engineering, volume 20, 1992, pages 289-305.

[30] Comparison of Different Protein Immobilization Methods on Quartz Crystal Microbalance Surface in Flow Injection Immunoassay: Yung-Chuan Liu, Chih-Ming Wang, and Kuang-Pin Hsiung. Analytical Biochemistry 299, 130–135 (2001)

[31] Micocontact printing of proteins on mixed self-assembled monolayer J.L. Tan, J. Tien, C.S. Chen, Langmuir 18 (2002) 519.

[32] Atrazine analysis using an impedimetric immunosensor based on mixed biotinylated self-assembled monolayer: S. Hlelia,, C. Martelet, A. Abdelghani, N. Burais, N. Jaffrezic-Renault. Sensors and Actuators B 113 (2006) 711–717

[33] A correlation study between the conformation of the 1,4-dithiane SAM on gold and its performance to assess the heterogeneous electron-transfer reactions: J.R. Sousa, M.M.V. Parente, I.C.N. Diogenes, L.G.F. Lopes, P.L. Neto, M.L.A. Temperini, Al.A. Batista, I.S. Moreira, J. Electroanal. Chem. 566 (2004) 443.

[34] Ultrasensitive electrochemical immunosensing using magnetic beads and gold nanocatalysts: Thangavelu Selvaraju, Jagotamoy Das, Sang Woo Han, Haesik Yang, Biosensors and Bioelectronics 23 (2008) 932–938.

[35] Magnetic bead-based DNA detection with multi-layers quantum dots labeling for rapid detection of Escherichia coli O157:H7, Yi-Ju Liua, Da-Jeng Yao, Hwan-You Chang, Chien-Ming Liu, Chih Chen, Biosensors and Bioelectronics Volume 24, 2008, Pages 558-565.

[36] Controlled torque on superparamagnetic beads for functional biosensors: X.J.A. Janssen, A.J. Schellekens, K. van Ommering, L.J. van IJzendoorn, M.W.J. Prins, Biosensors and Bioelectronics, Volume 24, Issue 7, 15 March 2009, Pages 1937-1941.

[37] Monitoring the growth and drug susceptibility of individual bacteria using asynchronous magnetic bead rotation sensors: Paivo Kinnunen, Irene Sinn, Brandon H. McNaughton, Duane W. Newton, Mark A. Burns, Raoul Kopelman, Biosensors and Bioelectronics, Volume 26, 2011, Pages 2751-2755.

[38] An impedimetric DNA sensor based on functionalized magnetic nanoparticlesfor HIV and HBV detection: Walid Mohamed Hassen, Carole Chaix, Adnane Abdelghani, François Bessueille, Didier Leonard, Nicole Jaffrezic-Renault, Sensors and Actuators B: Chemical, Volume 134, 2008, Pages 755-760.

[39] A disposable immunomagnetic electrochemical sensor based on functionalised magnetic beads on gold surface for the detection of atrazine: Saloua Helali, Claude Marteleta, Adnane Abdelghani, Mhamed Ali Maaref, Nicole Jaffrezic-Renault. Electrochimica Acta 51 (2006) 5182–5186.

[40] Stable enzyme biosensors based on chemically synthesized Au–polypyrrole nanocomposites: John Njagi, Silvana Andreescu Biosensors and Bioelectronics 23 (2007) 168–175

[41] Polypyrrole based amperometric glucose biosensors: Minni Singh, Pavan Kumar Kathuroju, Nagaraju Jampana, Sensors and Actuators B: Chemical, Volume 143, 2009, Pages 430-443.

[42] Molecular electronics of conducting polymers: S. Roth, G. Mahler, Y. Shen, F. Coter, Synth. Met. 28 (1989) 815–822.

[43] Chemical synthesis and characterization of polypyrrole coated on porous membranes and its electrochemical stability: H. S. Lee, J. Hong, Synth. Met. 113 (2000) 115–119.

[44] Conducting polymer films by UV photo processing : Q.Fang, D.G.Chetwynd, J.W.Garden, Sens. Actuators A 99 (2002) 74–77.

[45] Polypyrrole, a new possibility for covalent binding of oxido reductases to electrode surfaces as a base for stable biosensors: W. Schuhmann, R. Lammert, B. Uhe, H.L. Schmid, Sens. Actuators B 1 (1990) 537–541.

[46] Polypyrrole–neutravidin layer for impedimetric biosensor: Chiheb Esseghaier, Saloua Helali, Heikel Ben Fredj, Asma Tlili, Adnane Abdelghani, Sensors and Actuators B 131 (2008) 584–589

[47] Synthesis and characterization of high molecular weight, highly soluble polypyrrole in organic solvents, E.J.Oh, K.S.Jang, Synth. Met. 119 (2001) 109–110

[48] Conducting Polymer Nanocomposites: R. Gangopadhyay, A. De, Chem. Mater. 12 (2000) 608.

[49] Labeled magnetic nanoparticles assembly on polypyrrole film for biosensor applications: H. Ben Fredj, S Helali, C. Esseghaier, L. Vonna, L. Vidal, A. Abdelghani. Talanta 75 (2008) 740–747

Inhibitive Determination of Metal Ions Using a Horseradish Peroxidase Amperometric Biosensor

B. Silwana, C. van der Horst, E. Iwuoha and
V. Somerset

Additional information is available at the end of the chapter

1. Introduction

The development of electro-analytical methods for the determination of mercury, lead, cadmium and various other trace metals in acidic media or at different pH values are not new and for that reason, the investigation of alternative techniques have been ongoing and especially to find mercury-free electrodes (Ugo et al., 1995). Stripping voltammetry has been widely used for trace metal analysis with mercury as the working electrode due to its remarkable analytical properties. However, due to the toxicity of mercury and the human health risk that it poses (bioaccumulation in the food chain), there have been in-sistent efforts to remove the use of mercury completely. Electroanalysis has therefore seen the use of mercury-free sensors, while much attention has been dedicated to the develop-ment of such sensors over the last decade (Hwang et al., 2008; Sonthalia et al., 2004). Sev-eral heavy metals create environmental and human health concerns when elevated concentrations of these metals are present in the environment. In this regard, lead (Pb) and mercury (Hg) and more increasingly cadmium (Cd) heavy metals are of prime envi-ronmental concern, since they are significant for environmental surveillance, food control, occupational medicine, toxicology and hygiene (Ensafi and Zarei, 2000). Lead (Pb) is fur-thermore constantly monitored in natural and drinking water due to the harmful effects that are often manifested in young children (Zen et al., 2002). It is also known that several trace metals are regarded as essential micro-nutrients and play an integral role in the life processes of living organisms. In contrast, metals such as aluminium, silver, cadmium, gold, lead and mercury play no biological role in living organisms and lead to toxicity and adverse human health effects when present (Somerset et al., 2010a; Estevez-Hernandez et al., 2007; Honeychurch et al., 2002).

The simultaneous analysis of metal ions is typically performed with inductively coupled plasma atomic emission spectrometry (ICP-AES), inductively coupled plasma mass spectrometry (ICP-MS), X-ray fluorescence spectrometry (XRF), or atomic absorption spectrometry (AAS). These are well established methods that are characterised by low detection limits, but these methods require expensive instrumentation and trained personnel and cannot be used for field and on-site measurements. On the other hand, anodic stripping voltammetry (ASV) is one of the most favourable techniques for the determination of heavy metal ions due to its low cost, high sensitivity, easy operation and the ability of analysing element speciation (Li et al., 2010; Somerset et al., 2010b).

In this chapter the use of a conducting polymer modified platinum surface on which horseradish perioxidase (HRP) has been immobilised was investigated as an alternative transducer platform for the amperometric analysis of Hg^{2+}, Pb^{2+} and Cd^{2+} ions in aqueous solutions. The results obtained for the quantitative analysis of the metal ions included the detection limit, linear range, sensitivity and R.S.D. for individual metal ions and are discussed in this chapter. Results for the biosensor storage stability and response reproducibility were also investigated and reported.

2. Materials and methods

2.1. Chemicals

The reagents aniline (99%), 2,2´-dithiodianiline (98%), potassium dihydrogen phosphate (99%), hydrogen peroxide (30%), disodium hydrogen phosphate (98%) and diethyl ether (99.9%) were obtained from Sigma-Aldrich, Germany. The enzyme peroxidase (EC 1.11.1.7 type IV from horseradish) was also purchased from Sigma-Aldrich, Germany. The potassium chloride, sulphuric acid (95%), ethanol (98%) and hydrochloric acid (32%) were obtained from Merck, South Africa. The standards for cadmium (Cd), lead (Pd) and mercury (Hg) were purchased as atomic adsorption standard solutions (1000 mg/l) and purchased from Sigma-Aldrich, Germany.

2.2. Apparatus

Electrochemical protocols were performed with a PalmSens portable potentiostat / galvanostat, with the PSTrace program and accessories (PalmSens® Instruments BV, 3992 BZ Houten, the Netherlands). The portable potentiostat was interfaced with a microcomputer controlled by PS 2.1 software for data acquisition and experimental control.

A conventional three-electrode system was employed consisting of a platinum (Pt) working electrode, a BAS 3 M NaCl-type Ag/AgCl reference electrode, and a platinum wire auxiliary electrode (Somerset et al., 2010a).

2.3. Construction of Pt/PANI/HRP biosensor

2.3.1. Electrosynthesis of polymer film on platinum electrode

During electropolymerisation a three-electrode cell with a 10 ml capacity was utilised. Electropolymerisation from a 0.2 M aniline solution dissolved in a 1 M hydrochloric acid (HCl) solution onto a thoroughly cleaned and polished Pt electrode was performed. The aniline / HCl solution was first degassed by passing argon (Ar) through the solution, followed by electropolymerisation to obtain a smooth polymer coated electrode surface. During electropolymerisation the potential was repeatedly scanned from – 200 mV to + 1200 mV, at a scan rate of 40 mV/s. This was done for 20 voltammetric cycles, to ensure a relatively thick polymer film was obtained for enzyme immobilisation (Mathebe et al., 2004; Somerset et al., 2009; Somerset et al., 2010a; Nomngongo et al., 2011).

2.3.2. Enzyme immobilisation and biosensor preparation

Electropolymerisation of a fresh PANI polymer film on a Pt electrode was followed by activation of the polymer film for enzyme attachment. The Pt/PANI electrode was transferred to a batch cell, containing 1 ml of a 0.1 M phosphate buffer (pH 6.8) solution, degassed with argon. The fresh PANI polymer film was first reduced at a potential of - 500 mV (vs. Ag/AgCl) until a steady current was achieved. The Pt/PANI bioelectrode was next transferred to a 0.1 M phosphate buffer (pH 6.8) solution, containing HRP (2 mg/ml in 1.0 ml fresh buffer solution). The enzyme solution was then argon degassed, after which enzyme immobilisation onto the Pt/PANI bioelectrode was achieved by oxidation of the PANI film in the presence of HRP at a potential of + 400 mV (vs. Ag/AgCl), until steady-state current was achieved. The resulting biosensor will be referred to as Pt/PANI/HRP biosensor (Morrin et al., 2003; Mathebe et al., 2004; Somerset et al., 2009; Nomngongo et al., 2011).

2.4. Biosensor response evaluation

2.4.1. Voltammetric measurements

Cyclic and differential pulse voltammetry measurements were used to monitor the responses of the Pt/PANI/HRP biosensor towards hydrogen peroxide (H_2O_2) as substrate (Morrin et al., 2005; Mathebe et al., 2004; Nomngongo et al., 2011). Cyclic voltammetry (CV) was performed at a slow scan rate of 10 mV/s in order to study the catalytic oxidation of H_2O_2 by applying a potential scan between + 400 mV and - 1200 mV (vs. Ag/AgCl). Furthermore, to differentiate between the voltammetric responses, the cyclic voltammogram was first recorded in the absence of the substrate, followed by analysis in the presence of H_2O_2 as substrate. This was achieved by sequential addition of 1 mM of H_2O_2 solution to the 1 ml of 0.1 M phosphate buffer (PB) solution, degassed with argon that was repeated after each addition of the substrate (Chen and Gu, 2008; Nomngongo et al., 2011).

Differential pulse voltammetry (DPV) immediately followed the CV analysis in the same solution mentioned in the previous paragraph. The cathodic difference differential pulse voltammogram (DPV) was collected in the reduction direction only by scanning the potential between

+400 mV and - 1200 mV (vs. Ag/AgCl), at a potential step of 20 mV and a pulse amplitude of 20 mV. As for the CV analysis, the DPV measurements were first obtained in the absence of the substrate, followed by analysis in the presence of H_2O_2 as substrate (Nomngongo et al., 2011).

2.4.2. Inhibition response measurements

The electrochemical cell prepared for biosensor inhibition measurements consisted of the Pt/PANI/HRP bioelectrode, a platinum wire and Ag/AgCl as the working, counter and reference electrode, respectively. Inhibition measurements were performed in a 1 ml test solution containing 0.1 M PB solution that was degassed with argon before any substrate was added and after each addition of small aliquots of 1 mM H_2O_2 solution.

Inhibition plots for each of the heavy metals studied (e.g. Cd^{2+}, Pb^{2+}, and Hg^{2+}) were obtained using the percentage inhibition method. This procedure involved the study of the Pt/PANI/HRP biosensor in the presence of H_2O_2 solution first, followed by exposure to sequential additions of the heavy metal solutions. The heavy metal concentrations evaluated during sequential addition were 0.001 ppb, 0.005 ppb and 0.01 ppb for each of Cd^{2+}, Pb^{2+}, and Hg^{2+}.

For the inhibition studies, the Pt/PANI/HRP biosensor was first placed in a stirred 1 ml of 0.1 M PB solution (anaerobic conditions) and multiple additions of a standard peroxide substrate solution was added until a stable current and a maximum concentration of 6 mM were obtained. This steady state current was related to the activity of the biosensor with no inhibitor present. In the second phase of the inhibition studies, the biosensor was transferred to a fresh 1 ml of 0.1 M PB solution (anaerobic conditions) and multiple additions of a standard heavy metal solution (e.g. Cd^{2+}, Pb^{2+}, and Hg^{2+}) was again added, until a stable current was obtained (Nomngongo et al., 2011).

The percentage inhibition was then calculated using the formula (Somerset et al., 2007; Guascito et al., 2008; Nomngongo et al., 2011):

$$I\% = \frac{I_1 - I_2}{I_1} \times 100\% \tag{1}$$

where $I\%$ is the degree of inhibition, I_1 is the steady-state current obtained in buffer solution with no heavy metal ion present, while I_2 is the steady-state current obtained after the biosensor was exposed to sequential additions of the separate heavy metal ions of Cd^{2+}, Pb^{2+}, and Hg^{2+} respectively.

3. Results and discussion

3.1. Cyclic voltammetric characterisation of PANI electropolymerisation

In Figure 1, the cyclic voltammogram (CV) for the electropolymerisation of polyaniline (PANI) on a Pt electrode is shown, which was obtained by cycling the potential between − 200 and + 1100 mV at a scan rate of 40 mV/s.

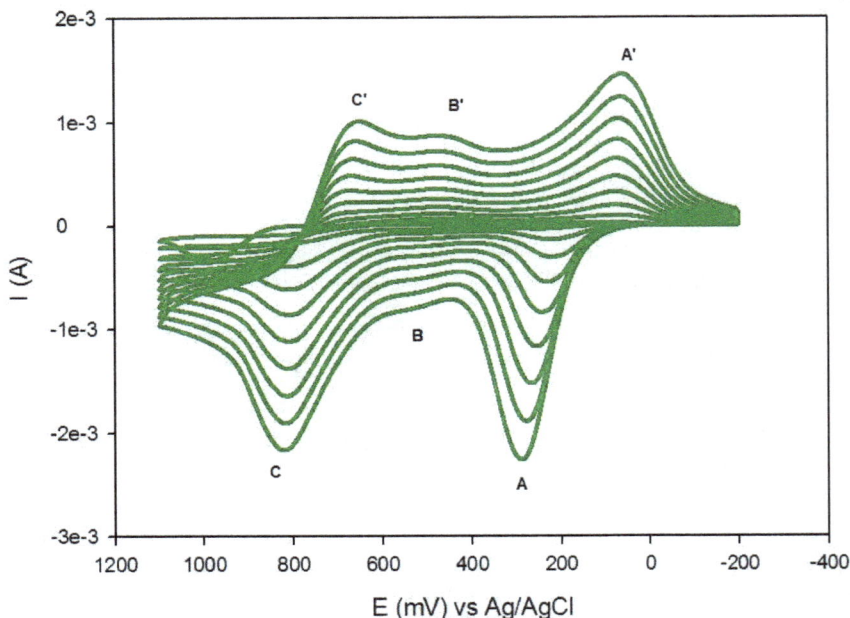

Figure 1. Results for the electropolymerisation of PANI in a 1 M HCl solution on a Pt electrode with the potential scanned from − 200 to +1100 mV at a scan rate of 40 mV/s.

The results obtained for the electropolymerisation of PANI in Figure 1, shows that PANI displays very good redox activity at acidic pH. The CV obtained in Figure 1 further shows two main anodic peaks in A and C, which corresponds to the transformation of leucoemeraldine base to emeraldine salt and the emeraldine salt to pernigraniline salt forms. The reverse scan in the cathodic direction shows the main peaks C′ and A′ that corresponds to the conversion of pernigraniline salt to emeraldine salt and emeraldine salt to leucoemeraldine base. The small redox couple of (B/B′) in the centre of the centre of the CV can be attributed to impurities such as benzoquinone and hydroquinone. With repetitive cycling of the potential an increase in the redox peaks was observed, which indicated the formation of a conducting polymer on the electrode surface (Mathebe et al., 2004; Morrin et al., 2005; Somerset et al., 2007; Somerset et al., 2010a).

Further characterisation of the electrosynthesised PANI polymer was done to compare the results obtained to that of other researchers. To determine the surface concentration of the PANI film, Γ_{PANI}, Brown-Anson analysis (Bard and Faulkner, 2001) was performed. Similarly, Randles-Sevcik analysis (Bard and Faulkner, 2001) of peak current (Ip) versus square root of scan rate ($v^{1/2}$) was performed, to estimate the electron transport diffusion coefficient, D_e, for electrons within the polymer backbone. A summary of the results are shown in Table 1.

Polymer parameter	Mathebe et al., 2004	Nomngongo et al., 2011	This study
Γ_{PANI}	1.85×10^{-7} mol/cm^2	7.8×10^{-7} mol/cm^2	6.19×10^{-8} mol/cm^2
D_e	8.68×10^{-9} cm^2/s	4.07×10^{-8} cm^2/s	4.94×10^{-10} cm^2/s

Table 1. Comparison of the results obtained for the surface concentration and electron transport diffusion coefficient of the electrosynthesised polymers.

Analysis of the results in Table 1 shows that both the results for the surface concentration of the PANI polymer and the electron transport diffusion coefficient obtained in this study, compares well to that of similar studies performed previously.

3.2. Optimisation of solution pH for Pt/PANI/HRP biosensor

After construction of the Pt/PANI/HRP biosensor, evaluation of the biosensor was performed over the pH range from 4.5 to 7.2, to confirm the optimum current response for the constructed biosensor. A fresh biosensor was constructed and evaluated at each of the pH values evaluated from 4.5 to 7.2. The results obtained are displayed in Figure 2.

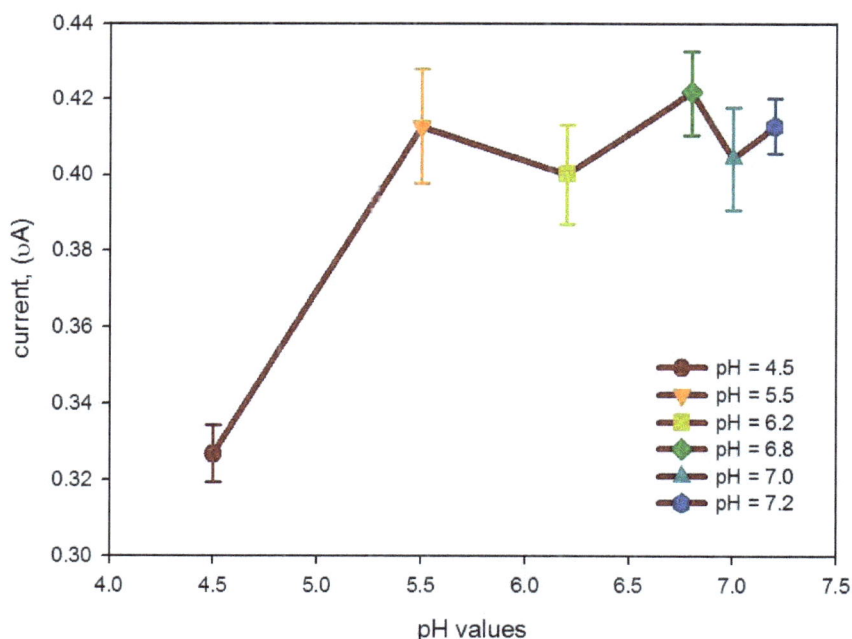

Figure 2. Results obtained for the optimisation of pH for the PtPANI/HRP enzyme electrode in 0.1 M phosphate buffer solution.

Analysis of the results in Figure 2 indicates indicate that a maximum amperometric response and sensitivity was obtained for the biosensor at a pH of 6.8. Similar studies performed on a similar biosensor construction have shown it to operate at an optimum pH range from 6.8 to 7.2 in buffered electrolyte solution. All further biosensor studies was performed at a pH = 6.8 (Mathebe et al., 2004; Nomngongo et al., 2011).

3.3. Differential pulse voltammetric characterisation of Pt/PANI/HRP biosensor

After construction, the amperometric behaviour of the Pt/PANI/HRP biosensor was evaluated in the presence and absence of H_2O_2 as substrate. Both the cyclic and differential pulse behaviour of the biosensor in the presence and absence of the substrate were evaluated, although only the differential pulse voltammetric (DPV) results (Figure 3) will be discussed in this section.

The results obtained for the cyclic voltammetric (CV) behaviour of the Pt/PANI/HRP biosensor (not shown here), have shown that with sequential addition of H_2O_2 to the phosphate buffer (pH = 6.8) solution, the reduction peak current shifted and increased with addition of the substrate (Mathebe et al., 2004; Nomngongo et al., 2011).

In Figure 3 the DPV results obtained for the evaluation of the Pt/PANI/HRP biosensor are shown.

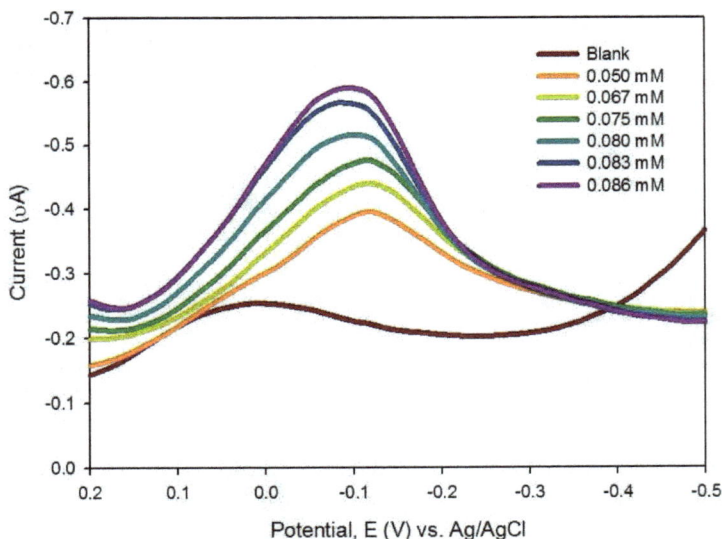

Figure 3. Net cathodic differential pulse voltammograms (DPVs) of the Pt/PANI/HRP biosensor in the presence of increasing concentrations of H_2O_2 as substrate. The experimental conditions were: frequency, 20 Hz; amplitude, 20 mV; and potential step, 20 mV.

The evaluation of the DPV results was done by cycling the potential in the cathodic direction from + 0.4 to – 1.2 V (vs. Ag/AgCl). The results obtained in Figure 3 have shown that the in the absence of the substrate, no electrocatalytic reduction of H_2O_2 was observed. However, as soon as H_2O_2 was added, an increase in the current of the peak observed at approximately – 0.2 V (vs. Ag/AgCl) occurred, thereby demonstrating effective electrocatalytic reduction of H_2O_2. With subsequent additions of substrate to the 0.1 M PBS, both a positive shift in the reduction peak and reduction peak current was evident, proportional to increased H_2O_2 concentration. Similar observations for the same biosensor were made by Mathebe et al. (2004) and Nomngongo et al. (2011).

In order to explain the DPV results obtained, the mechanism involved for the Pt/PANI/HRP biosensor is shown in Figure 4. This scheme shows that when the Pt/PANI/HRP bioelectrode is charged at a constant potential of – 0.2 V (vs. Ag/AgCl), the H_2O_2 substrate reduction charge is propagated along the PANI polymer chain to the Pt electrode surface, by fast electron transfer reactions. It is further shown that the PANI redox species (PANI^0 $\leftrightarrow$ PANI $^+$) is involved in the reduction charge propagation.

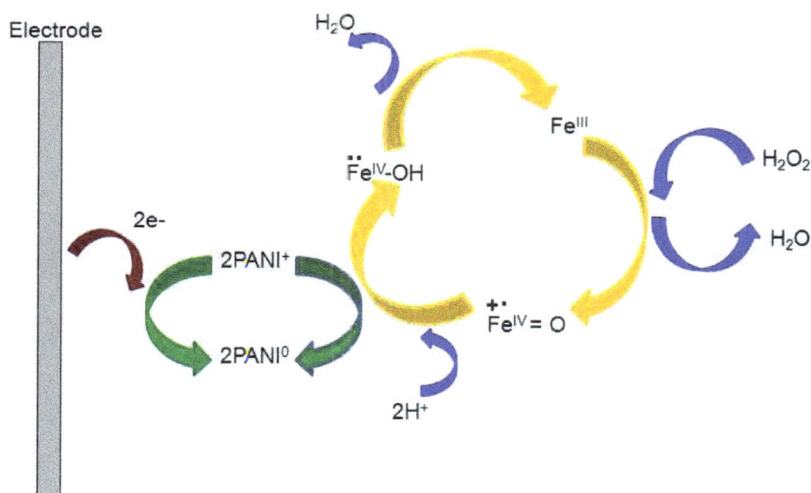

Figure 4. The Pt/PANI/HRP biosensor mechanism indicating the redox species that are either electron donors or hydrogen donors in the reaction mechanism (Iwuoha et al., 1997).

Figure 4 also shows that the substrate H_2O_2 is reduced by HRP (in the ferric (Fe^{III}) resting state) to form water, which results in the oxidation of HRP to form the oxyferryl HRP-I (Fe^{IV} = 0) compound. In turn this compound undergoes a two-electron reduction step to form an intermediate compound called hydroxyferryl HRP-II (Fe^{IV}-OH). With continued charge propagation taking place, the hydroxyferryl HRP-II compound goes back to the ferric HRP resting state and the process is repeated (Iwuoha et al., 1997; Nomngongo et al., 2011).

3.4. Cadmium (II), lead(II) and mercury (II) inhibition studies

The results obtained for the inhibition studies for each of the Cd^{2+}, Pb^{2+} and Hg^{2+} metal ions determined with the Pt/PANI/HRP biosensor are discussed in this section. Other studies (Zhao et al., 1996; Shyuan et al., 2008; Nomngongo et al., 2011) have shown that enzymes (e.g. HRP, alkaline phosphatase) are known to be inhibited by metals such as Cd^{2+}, Co^{2+}, Cu^{2+}, Fe^{3+}, Ni^{2+}, Pb^{2+} and Hg^{2+}. Three different heavy metal concentrations were evaluated, ranging from a relatively low, intermediary to higher concentrations. The inhibition results obtained for each of these concentrations (added sequentially) are evaluated and discussed in the following paragraphs.

3.4.1. Inhibition results for lowest metal concentration investigated

The percentage inhibition plots obtained for the inhibition of HRP when aliquots of 0.001 ppb of Cd^{2+}, Pb^{2+} and Hg^{2+} was sequentially added to the 0.1 M PB (pH = 6.8) solution, are shown in Figure 5.

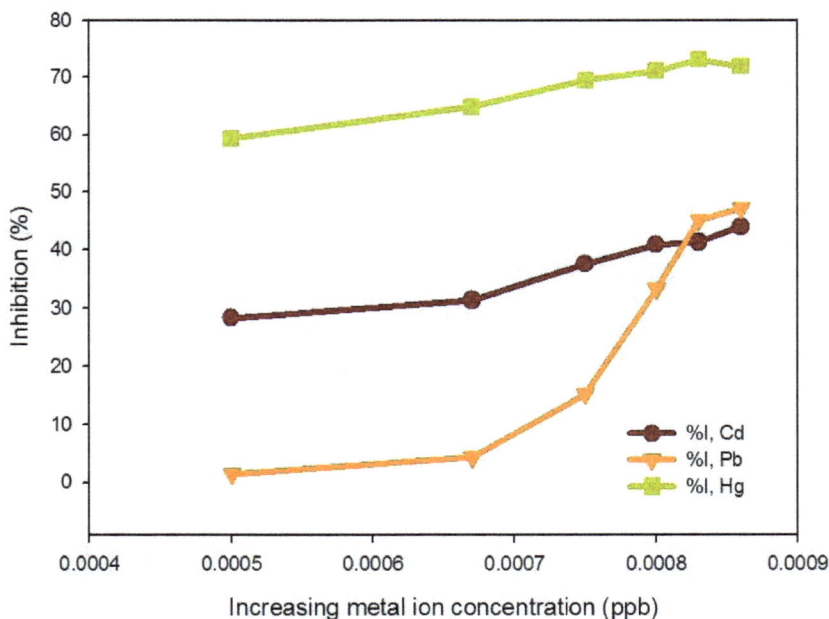

Figure 5. Results obtained for inhibition of the Pt/PANI/HRP biosensor in the presence of 0.001 ppb of Cd^{2+}, Pb^{2+} and Hg^{2+}, respectively.

For the results shown in Figure 5 it was observed that three distinctive patterns of inhibition was obtained for each of the Cd^{2+}, Pb^{2+} and Hg^{2+} metal ions investigated. For metal concen-

trations ranging from 0.001 – 0.004 ppb, it was observed that the decreasing trend of inhibition was $Hg^{2+} > Cd^{2+} > Pb^{2+}$. For higher concentrations the inhibition trend was $Hg^{2+} > Pb^{2+} >$
Cd^{2+}. It was further observed that after the initial concentration of 0.001 ppb for each metal
was added to the Pt/PANI/HRP biosensor, the first inhibition results were 1.5% (Pb^{2+}), 28.4%
(Cd^{2+}) and 59.4% (Hg^{2+}), respectively. This was a clear indication of the initial toxicity of the
respective metal ions to HRP as enzyme. The inhibition plots have also shown that for Pb^{2+} a
gradual increase in the inhibition was observed as the metal ion concentration was increased, with the final percentage inhibition at 47.2%. On the other hand, in the case of Hg^{2+}
and Cd^{2+} an initial high percentage inhibition was obtained that gradually increased slightly
with increased metal ion concentration. The highest percentage inhibition obtained for Cd^{2+}
and Hg^{2+} were 44.1% and 71.9%, respectively.

3.4.2. Inhibition results for intermediary metal concentration investigated

Figure 6 displays the results obtained for the percentage inhibition plots of HRP when aliquots
of 0.005 ppb of Cd^{2+}, Pb^{2+} and Hg^{2+} was sequentially added to the 0.1 M PB (pH = 6.8) solution.

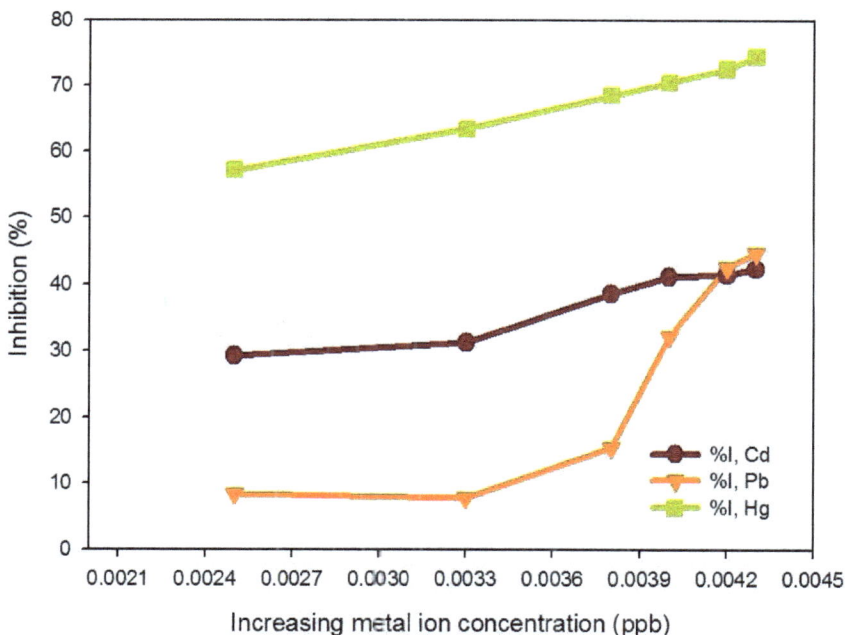

Figure 6. Results obtained for inhibition of the Pt/PANI/HRP biosensor in the presence of 0.005 ppb of Cd^{2+}, Pb^{2+} and Hg^{2+}, respectively.

Figure 6 also displayed the same characteristic trend for the percentage inhibition results, compared to that in Figure 5. For metal concentrations ranging from 0.005 – 0.02 ppb, it was observed that the decreasing trend of inhibition was $Hg^{2+} > Cd^{2+} > Pb^{2+}$. For higher concentrations the inhibition trend was $Hg^{2+} > Pb^{2+} > Cd^{2+}$. The results in Figure 6 are characteristically similar to that obtained in Figure 5 and differences were only observed when the individual percentages were compared. For Pb^{2+} a gradual increase was again observed as the concentration was increased, with the final percentage inhibition obtained at 44.7%. In the case of Hg^{2+} and Cd^{2+} an initial high percentage inhibition was obtained that gradually increased slightly with increased metal ion concentration. The highest percentage inhibition obtained for Cd^{2+} and Hg^{2+} were 42.4% and 74.4%, respectively.

3.4.3. Inhibition results for highest metal concentration investigated

The percentage inhibition plots obtained for the inhibition of HRP when aliquots of 0.01 ppb of Cd^{2+}, Pb^{2+} and Hg^{2+} was sequentially added to the 0.1 M PB (pH = 6.8) solution, are shown in Figure 7.

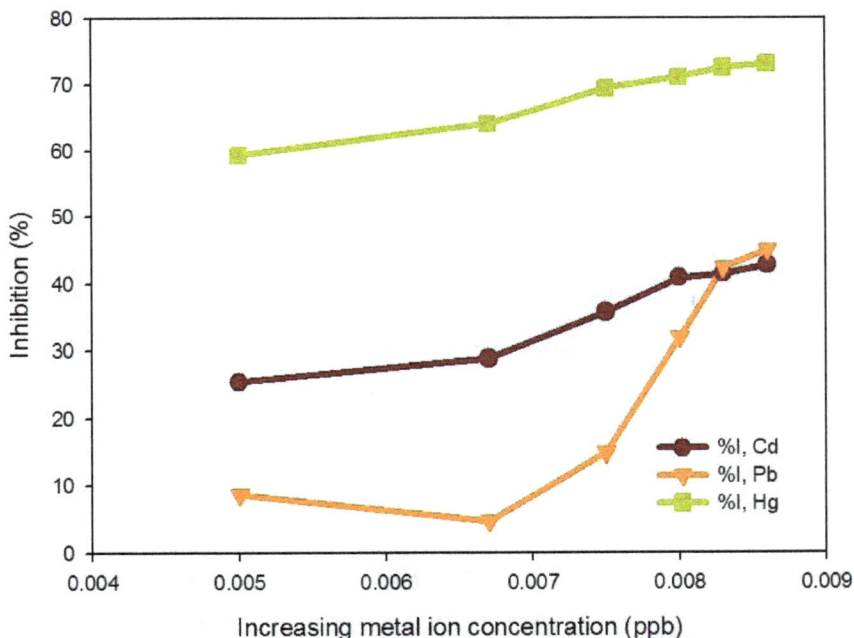

Figure 7. Results obtained for inhibition of the Pt/PANI/HRP biosensor in the presence of 0.01 ppb of Cd^{2+}, Pb^{2+} and Hg^{2+}, respectively.

Similarly, the same percentage inhibition trends as observed for the two previous investigations were observed for the results shown in Figure 7. The decreasing trend of inhibition was $Hg^{2+} > Cd^{2+} > Pb^{2+}$ for the metal ion concentrations ranging from 0.01 to 0.04 ppb, while it changed to $Hg^{2+} > Pb^{2+} > Cd^{2+}$ for the last two concentrations evaluated. The highest percentage inhibition obtained at the highest metal ion concentration of 0.06 ppb evaluated were 42.6% (Cd^{2+}), 44.9% (Pb^{2+}) and 73.1% (Hg^{2+}).

Analysis of the above inhibition results obtained for the three sets of starting metal ion concentrations investigated showed interesting results and similarities. It was observed that the enzyme HRP was inhibited by all three of the metal ions, Cd^{2+}, Pb^{2+} and Hg^{2+}. The inhibition results obtained for Hg^{2+} ions was the highest of the metal ions investigated, clearly indicating the known toxicity of these metal ions. At low to intermediate metal ion concentrations, the results obtained showed a decreasing inhibition trend of $Hg^{2+} > Cd^{2+} > Pb^{2+}$. The highest inhibition results obtained for the Pt/PANI/HRP biosensor was 74.4% for Hg^{2+}, followed by 47.2% for Pb^{2+} and 44.1% for Cd^{2+}.

However, no results for the simultaneous analysis of two or more of either Cd^{2+}, Pb^{2+} or Hg^{2+} were collected. These experiments will be conducted in future with the biosensor system described in this work and published in future papers.

3.4.4. Analytical characteristics of the Pt/PANI/HRP biosensor applied in metal inhibition studies

The amperometric responses of the Pt/PANI/HRP biosensor to various H_2O_2 substrate concentrations were evaluated and compared to that obtained in the presence of selected heavy metal ions of Cd^{2+}, Pb^{2+} and Hg^{2+}. For all three metal ions, decreased biosensor responses were obtained, clearly showing that inhibition was taking place. The analytical characteristics of the Pt/PANI/HRP biosensor were evaluated for each metal ion using various calibration curves to obtain the linear ranges, slopes of the calibration plots, correlation coefficients, limits of detection $(LOD = \frac{3 \times SD}{m})$ and limits of quantification $(LOQ = \frac{10 \times SD}{m})$. For these calculations SD is the standard deviation of the blank signal ($n = 10$) obtained in 0.1 M PB solution, while m is the slope of the calibration curve. The results obtained are listed in Table 2.

Metal ion	Linear range (ppb)	Sensitivity (μA/ppb)	R^2	LOD (ppb)	LOQ (ppb)
Cd^{2+}	1.5 - 4	3.19×10^{-2}	0.992	0.0579	0.193
Pb^{2+}	1.1 - 5	1.90×10^{-2}	0.991	0.0931	0.310
Hg^{2+}	1.5 - 4	1.20×10^{-2}	0.991	0.0268	0.089

Table 2. Results for the analytical characteristics of the Pt/PANI/HRP biosensor from the various calibration curves for the determination of Cd^{2+}, Pb^{2+} and Hg^{2+} heavy metal ions.

Analysis of the results in Table 2 showed that the Pt/PANI/HRP biosensor was very sensitive for the determination of Hg^{2+} ions, followed by Pb^{2+} and Cd^{2+} ions. This sensitivity was confirmed with the lowest LOD obtained for Hg^{2+} ions. The increasing LOD values obtained for this biosensor was $Pb^{2+} < Cd^{2+} < Hg^{2+}$. These results further compared favourably with that obtained in the study by Nomngongo et al. (2011), which used the same biosensor construction for the determination of Cd^{2+}, Pb^{2+} and Cu^{2+} metal ions.

4. Conclusions

The results obtained in this study have showed that an amperometric Pt/PANI/HRP biosensor can be successfully applied for the inhibition determination of selected heavy metals of Cd^{2+}, Pb^{2+} and Hg^{2+}. The inhibition results obtained for the three sets of starting metal ion concentrations investigated showed interesting results and similarities. It was observed that the enzyme HRP was inhibited by all three of the metal ions, Cd^{2+}, Pb^{2+} and Hg^{2+}. The inhibition results obtained for Hg^{2+} ions was the highest of the metal ions investigated, clearly indicating the known toxicity of these metal ions. At low to intermediate metal ion concentrations, the results obtained showed a decreasing inhibition trend of $Hg^{2+} > Cd^{2+} > Pb^{2+}$. The highest inhibition results obtained for the Pt/PANI/HRP biosensor was 74.4% for Hg^{2+}, followed by 47.2% for Pb^{2+} and 44.1% for Cd^{2+}. The analytical features obtained for the HRP biosensor showed high sensitivity for the determination of Hg^{2+} ions, followed by Pb^{2+} and Cd^{2+} ions. The respective LOD values obtained were 0.027 ppb (Hg^{2+}), 0.058 ppb (Cd^{2+}) and 0.093 ppb (Pb^{2+}).

Acknowledgements

This study was financially supported by the Water Research Commission (WRC), Council for Scientific and Industrial Research (CSIR) and the National Research Foundation (NRF) of South Africa. The assistance from researchers at the SensorLab, Chemistry Department, University of the Western Cape, Bellville, South Africa is also acknowledged.

Author details

B. Silwana[1,2], C. van der Horst[1,2], E. Iwuoha[2] and V. Somerset[1*]

*Address all correspondence to: vsomerset@csir.co.za

1 Natural Resources and the Environment (NRE), Council for Scientific and Industrial Research (CSIR), Stellenbosch, South Africa

2 SensorLab, Department of Chemistry, University of the Western Cape, Bellville, South Africa

References

[1] Chen-C, C., & Gu, Y. (2008). Enhancing the sensitivity and stability of HRP/PANI/Pt electrode by implanted bovine serum albumin. Biosensors and Bioelectronics, , 23, 765-770.

[2] Ensafi, A. A., & Zarei, K. (2000). Simultaneous determination of trace amounts of cadmium, nickel and cobalt in water samples by adsorptive voltammetry using ammonium 2-amino-cyclopentene dithiocarboxylate as a chelating agent. Talanta, 52(3), 435-440.

[3] Estevez-Hernandez, O., Naranjo-Rodriguez, I., Hidalgo-Hidalgo de, Cisneros. J. L., & Reguera, E. (2007). Evaluation of carbon paste electrodes modified with 1-furoylthioureas for the analysis of cadmium by differential pulse anodic stripping voltammetry. Sensors and Actuators B, , 123(1), 488-494.

[4] Guascito, M. R., Malitesta, C., Mazzotta, E., & Turco, A. (2008). Inhibitive determination of metal ions by an amperometric glucose oxidase biosensor study of the effect of hydrogen peroxide decomposition. Sensors and ActuatorsB, , 131, 394-402.

[5] Honeychurch, K. C., Hawkins, D. M., Hart, J. P., & Cowell, D. C. (2002). Voltammetric behaviour and trace determination of copper at a mercury-free screen-printed carbon electrode. Talanta, 57(3), 565-574.

[6] Honeychurch, K. C., & Hart, J. P. (2003). Screen-printed electrochemical sensors for monitoring metal pollutants. TrAC, 22(7-8): 456-469.

[7] Hwang, G. H., Han, W. K., Park, J. S., & Kang, S. G. (2008). Determination of trace metals by anodic stripping voltammetry using a bismuth-modified carbon nanotube electrode. Talanta, 76, 301-308.

[8] Iwuoha, E. I., Saenz de, Villaverde. D., Garcia, N. P., Smyth, M. R., & Pingarron, J. M. (1997). Reactivities of organic phase biosensors. 2. The amperometric behavior of horseradish peroxidase immobilised on a platinum electrode modified with an electrosynthetic polyaniline. film. Biosensors and Bioelectronics, , 12(8), 749-761.

[9] Li, H., Lei, L., Zeng, Q., Shib, J., Luo, C. X., Ji, H., Ouyanga, Q., & Chen, Y. (2010). Laser emission from dye doped microspheres produced on a chip. Sensors and Actuators B, , 145(1), 570-574.

[10] Mathebe, N. G. R., Morrin, A., & Iwuoha, E. I. (2004). Electrochemistry and scanning electron microscopy of polyaniline/peroxidase-based biosensor. Talanta, 64, 115-120.

[11] Morrin, A., Guzman, A., Killard, A. J., Pingarron, J. M., & Smyth, M. R. (2003). Characterisation of horseradish peroxidase immobilisation on an electrochemical biosensor by colorimetric and amperometric techniques. Biosensors and Bioelectronics, , 18, 715-720.

[12] Morrin, A., Ngamna, O., Killard, A. J., Moulton, S. E., Smyth, M. R., & Wallace, G. G. (2005). An amperometric biosensor fabricated from polyaniline nanoparticles. Elect. roanalysis , 17, 423-430.

[13] Nomngongo, P. N., Ngila, J. C., Nyamori, V. O., Songa, E. A., & Iwuoha, E. I. (2011). Determination of Selected Heavy Metals Using Amperometric Horseradish Peroxidase (HRP) Inhibition Biosensor. *Analytical Letters*, 44, 2031-2046.

[14] Shyuan, L. K., Heng, L. Y., Ahmad, M., Aziz, S. A., & Ishak, Z. (2008). Evaluation of pesticide and heavy metal toxicity using immobilised enzyme alkaline phosphatase with an electrochemical biosensor. Asian Journal of Biochemistry, , 3(6), 359-365.

[15] Somerset, V., Leaner, J., Mason, R., Iwuoha, E., & Morrin, A. (2010a). Development and application of a poly(2,2′-dithiodianiline) (PDTDA)-coated screen-printed carbon electrode in inorganic mercury determination. Electrochimica Acta, , 55, 4240-4246.

[16] Somerset, V., Leaner, J., Mason, R., Iwuoha, E., & Morrin, A. (2010b). Determination of inorganic mercury using a polyaniline and polyanilinemethylene blue coated screen-printed carbon electrode. International Journal of Environmental Analytical Chemistry, , 90(9), 671-685.

[17] Somerset, V., Baker, P., & Iwuoha, E. (2009). Mercaptobenzothiazole-on-gold organic phase biosensor systems: 1. Enhanced organophosphate pesticide determination. Journal of Environmental Science and HealthB, , 44, 164-178.

[18] Somerset, V. S., Klink, M. J., Baker, P. G. L., & Iwuoha, E. I. (2007). Acetylcholinesterase-polyaniline biosensor investigation of organophosphate pesticides in selected organic solvents. Journal of Environmental Science and HealthB, 42, 297-304.

[19] Sonthalia, P., Mc Gaw, E., Show, Y., & Swain, G. M. (2004). Metal ion analysis in contaminated water samples using anodic stripping voltammetry and a nanocrystalline diamond thin-film electrode. *Analytica Chimica Acta*, 522(1), 35-44.

[20] Ugo, P., Moretto, L. M., & Mazzocchin, G. A. (1995). Voltammetric determination of trace mercury in chloride media at glassy carbon electrodes modified with polycationic ionomers. *Analytica Chimica Acta*, 74 EOF.

[21] Zen-M, J., Yang-C, C., & Kumar, A. S. (2002). Voltammetric behavior and trace determination of Pb^{2+} at a mercury-free screen-printed silver electrode. *Analytica Chimica Acta*, 464, 229-235.

[22] Zhao, J., Henkens, R. W., & Crumbliss, A. L. (1996). Mediator-Free Amperometric Determination of Toxic Substances Based on Their Inhibition of Immobilized Horseradish Peroxidase. *Biotechnology Progress*, 12(5), 703-708.

Biosensing Food

Surface-Enhanced Raman Scattering Liquid Sensor for Quantitative Detection of Trace Melamine in Dairy Products

Mingqiang Zou, Xiaofang Zhang, Xiaohua Qi and
Feng Liu

Additional information is available at the end of the chapter

1. Introduction

Raman spectroscopy has emerged as a fast, non-invasive, analytical method for the detection and quantification of adulterants in many fields (Wong et al., 2007; Weng et al., 2003; Muik et al., 2003; Micklander et al., 2002; Peica et al., 2005; Rubayiza et al., 2005; Ellis et al., 2005; Paradkar et al., 2001; Abalde-Cela et al., 2009; Mulvihill et al., 2008; Zhou et al., 2006). Although signals from conventional Raman spectroscopy are very weak, great progress has been made with the development of surface-enhanced Raman spectroscopy (SERS) as a sensing method. SERS is a powerful spectroscopy technique that can provide ultra-sensitive characterization of adsorbate molecules on roughened metal (e.g., Ag, Au, and Cu) surfaces that produce a large enhancement to the Raman scattering signal (Lin et al., 2008; Lee et al., 1982; Wei et al., 2009; Küstner et al., 2009; Koglin et al., 1996; House et al., 2008; Leopold et al., 2003; Yaffe et al., 2008; Yu et al., 2007; Tiwari et al., 2007; Guingab et al., 2007; Tian et al., 2002; Wang et al., 2005; Chen et al., 2012; Betz et al., 2012). Generally, solid/liquid substrates are necessary to enhance the SERS spectrum to obtain adequate sensitivity.Solid substrates, generally prepared as gold or silver nanoparticles with a silica or alumina shell, have a wide application range. However, only a few examples of liquid substrates have been reported, though they are easily prepared and enhance the analysis some analytes.For example, using a silver colloid, at least a 10^5-fold enhancement of the Raman signal is achieved for the measurement of melamine (Zou et al., 2010).

Presently, there are two commonly accepted sensing mechanisms(Chu et al., Phys. Rev; Campion et al., 1998; Knoll, 1998; Kneipp et al., 1999; Moskovits et al., 1998; Otto et al. 2005):

the electro-magnetic enhancement mechanism, which involves enhancement in the field intensity by plasmon resonance excitation; and the chemical enhancement mechanism, which involves enhancement of the polarizability by chemical effects such as a charge-transfer excited states.The efficiency of the generation of the SERS signal is high enough to observe the Raman spectrum of even a single molecule. With the rapid development of nanofabrication technology, SERS has grown to become a very active field of research in several areas of materials and analytical sciences, such as medicine, the environment, food, gems, cultural relics, and archaeology (Fan et al., 2011; Jun et al., 2010; Deiss et al., 2011).

In the following section, liquid milk melamine detection using a SERS liquid sensor is described as an example of this technique. In the example, liquid milk samples preparation process is very easy, i.e. only diluted with double-distilled water and centrifugation is required. With the aid of silver colloid, at least a 10^5-fold enhancement of the Raman signal was achieved for the measurement of melamine. The limit of detection by this method was 0.01 g mL^{-1} for melamine standard samples. Based on the intensity of the Raman spectroscopy with vibration bands normalized by the band at 928 cm^{-1} (CH2), external standard method was employed for the quantitative analysis. The linear regression square (R^2) of curve was 0.9998, the limit of quantitation using this approach was 0.5 g mL^{-1} of melamine in liquid milk, the relative standard deviation was $\leq 10\%$ and recoveries were from 93 to 109%. The test results for SERS were very precise and as good as those obtained by LC/MS/MS.

2. Background of surface-enhanced Raman scattering liquid sensor for melamine detection

Since 2008, there has been mounting concern about the intentional adulteration of protein ingredients in milk powder with melamine, because milk powder blended with melamine can lead to kidney disease and even death in babies. Thisfear of milk powder tainted with melamine has an important influence on the dairy production of milk powder and cow breeding, as well as an important impact on the food market and industry. Currently, new methods such as high-performance liquid chromatography (HPLC) (Ehling et al., 2007; Muniz-Valencia et al., 2008), liquid chromatography coupled with mass spectroscopy (LC-MS) (Varelis et al., 2008), LC-MS/MS (http://www.cfsan.fda.gov/~frf/lib4421.htm), thin-layer chromatography (TLC) (Broszat et al., 2008), commercial enzyme-linked immunosorbent assay technology (Eric et al., 2008), matrix-assisted laser desorption/ionization mass spectrometry (Tang et al., 2009), and surface desorption atmospheric pressure chemical ionization mass spectrometry (Yang et al.,2009) are the principal analysis techniques used for the detection and quantification of melamine in food. However, these methods are time consuming and cannot satisfy the need for melamine detection in practice because raw milk spoils and must be assayed within 4 h. Moreover, these methods require access to complicated and expensive laboratory facilities, especially in terms of sample preparation and clean-up steps. Therefore, it is of particular importance to develop a simple, quick, cost-effective, and sensitive method for detection of melamine in food.

We demonstrate an approach to detect melamine in liquid milk using surface-enhanced Raman spectroscopy in a silver colloid, which can be used for the rapid and online detection of melamine in dairy products.

2.1. Optimization of the surface-enhanced Raman scattering liquid sensorfor melamine detection

In recent years, gold nanoparticles (Au NPs) and silver nanoparticles (Ag NPs) have been widely used as colorimetric probes for chemical sensing and biosensing of various substances (Zhao, et al., 2008), such as viruses (Niikura et al., 2009), protein (Wang et al., 2008), DNA (Cho, et al., 2008), cancerous cells (Medley et al., 2008), and small molecules (Chen, et al., 2010; Li et al., 2009; Zhang et al., 2008), relying on their unique size-dependent and/or interparticle distance-dependent absorption spectra and solution color. For example, triple hydrogenbonding recognition between melamine and a cyanuric acid derivative grafted on the surfaced of Au NPs can be used for reliable detection of melamine (Ai et al., 2009).

Currently, much attention has been paid to the study of the optical absorption spectra of nanoscale colloidal silver in the quest for SERS enhancement factors. Compared to Au NPs, Ag NPs have some advantages, for example, lower cost of preparation and higher extinction coefficients relative to Au NPs of the same size (Lee, et al, 2007). Therefore, Ag NPs are also good candidates for melamine sensing (Han, et al., 2010; Ping et al., 2012).

Upon considering the influence of temperature, ionic strength, and aggregation behavior of colloids on the SERS spectra band intensity in the presence of adsorbates and the wavelength at which maximum enhancement occurs, the latter shift to higher values with time. In particular, the adsorption of the colloid is strongly influenced by chloride ions (Koglin et al., 1996) and pH (House et al., 2008). Scanning electron microscopy images of a silver colloid before and after addition of reagent A (Sodium chloride aqueous solution or aqueous potassium chloride solutions) and reagent B (Aqueous sodium hydroxide or potassium hydroxide solution) are presented in Figure 1. As shown in Figure 1a, the colloidal silver particles mainly displayed a spherical morphology with a uniform size of ~70-100 nm. After added reagent A (Fig. 1b) or reagent B (Fig. 1c), the silver colloid became aggregated and inhomogeneous. When reagents A and B were added to the colloidal silver at the same time, the morphology of the silver colloid became more dense and uniform (Fig. 1d), which is the best form for SERS enhancement. Thus, this system was chosen as the surfaced-enhancing substrate for further study. Figure 1e shows the SERS spectra of 1 µg mL^{-1} melamine on the corresponding enhancing substrates from (a), (b), (c) and (d) in Figure 1. There are no evident Raman bands of melamine on silver colloid (curve ⊕in Fig. 1e) or on silver colloid with reagent A (curve ⊕in Fig. 1e). However, when reagent B was added to the silver colloid (curve⊕in Fig. 1e), a weak characteristic peak of melamine was observed at 698 cm^{-1}, i.e., the SERS spectra band intensity was affected by pH. After reagents A and B were added to the silver colloid (Curve IV in Fig. 1e), the characteristic peak of melamine was strongly enhanced, with the intensity of the peak at 698 cm^{-1}being the greatest.

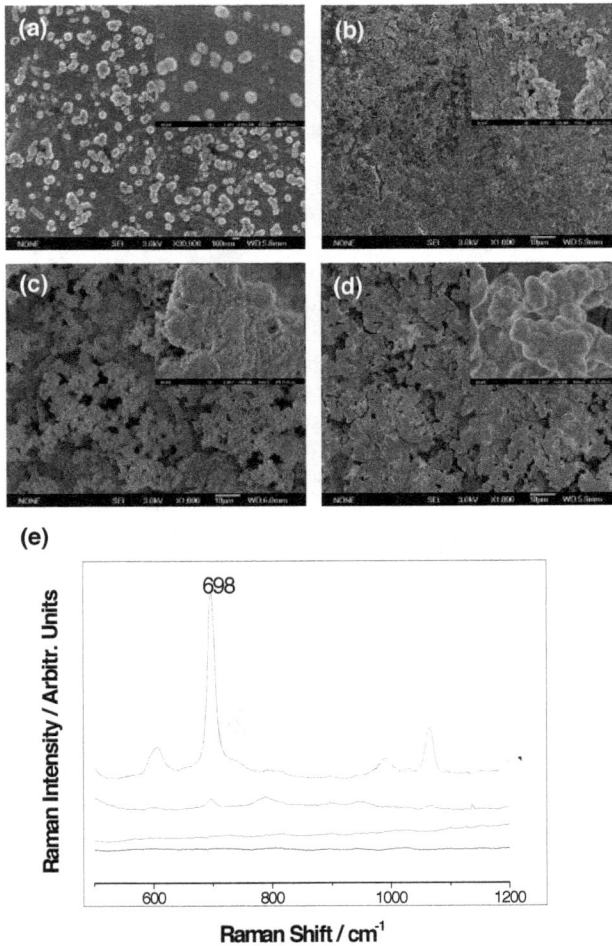

Figure 1. a-d) Scanning electron microscopy images of colloids and(e)SERS spectra of 1 μg mL⁻¹ melamine with the corresponding enhancing substrates.Scanning electron microscopy images of silver colloids (a)before and(b) after addition of reagent A, (c) reagent B, and (d) reagents A and B together.(e) Curves ⊡⊡⊡ and ⊡are SERS spectra of 1 μg mL⁻¹ melamine from the corresponding enhancing substrates from (a), (b), (c), and (d), respectively.

2.2. Description of use of the milk melamine liquid sensor

It is believed that melamine (2,4,6-triamino-1,3,5-triazine) is sometimes intentionally added to food ingredients to make the products appear to contain higher protein levels due to the high nitrogen content of melamine. A safety limit for melamine ingestion is officially set at 2.5 ppm for adult food and 1 ppm infant formula by the US Food and Drug Administration (Zhao et al.,

2009; http://www.fda.gov/NewsEvents/ Newsroom-/ PressAnnouncements/ 2008/ ucm116960.htm.). The maximum residue level of melamine in infant formula is now legally regulated at 1 ppm by the Chinese government after the recent melamine accident (Guo et al., 2010). To achieve this lower limit of detection (LOD), silver colloids are ideal candidates to be used as surfaced-enhancing substrate liquid sensors due to their strong Raman-enhancing effect. Thus, we chose silver colloid as a surfaced-enhancing substrate for the detection of melamine in this study, and the detection process is diagrammed in Figure 2.First, liquid milk was diluted with double-distilled water (Fig. 2a) toobtain a diluted milk sample. Next, the diluted sample was placed into a 1.5-mL conical centrifuge tube and centrifuged for 4 min at 14,000 rpm, and then it was delaminated (Fig 2b). Next, the supernatant was removed from the centrifuge tube and was added to the silver colloid, which was previously prepared with dropwise addition of reagents A andB, and uniformly mixed (Fig. 2c).Finally, the SERS spectra were recorded using a portable Raman spectrometer (Fig. 2d) to collect analytical results.

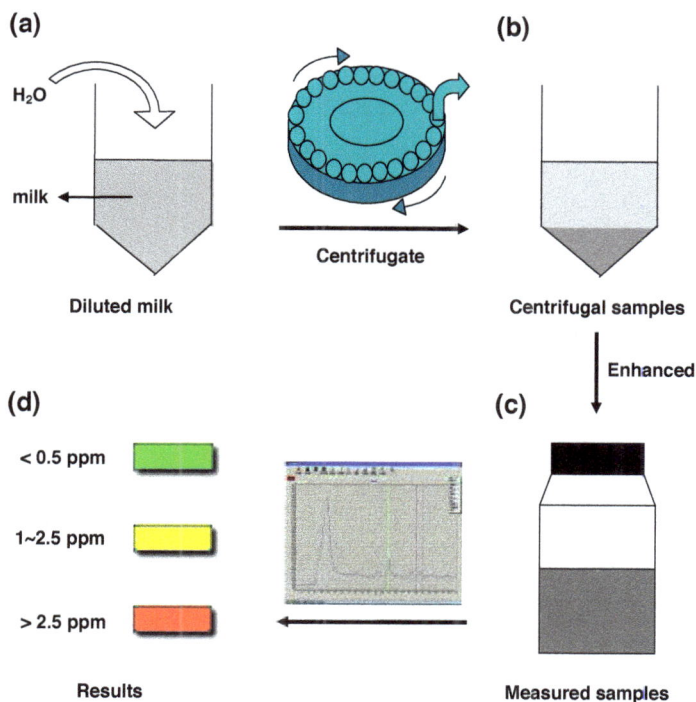

Figure 2. Schematic diagram of the on-line and rapid method for measuring melamine in liquid milk using surface-enhanced Raman spectroscopy.(a) Liquid milk was first diluted with double-distilled water.(b) The diluted sample was then centrifuged and delaminated.(c) The supernatant was addedto the silver colloid.(d) SERS spectra were recorded using a portable Raman spectrometer.

2.3. Optimization of the melamine spectra

Based on these experimental results, the spectra of different concentrations of melamine in solution were investigated from 500–1200 cm^{-1}, as shown in Figure 3.Typical Raman peaks of solid melamine at 382, 584, 678, and 983 cm^{-1} were observed (Fig. 3a).The most intense peak at 678 cm^{-1} is assigned to the ring breathing II mode, which involves in-plane deformation of the triazine ring. And the second most intense peak at 983 cm^{-1} arises from the ring breathing mode I of the triazine ring (Koglin et al., 1996). The peaks at 698 and 1005 cm^{-1}, visible in the SERS spectra of Figure 3b–d, were obtained from melamine samples at concentrations of 5×10^{-1}, 10^{-1}, and 10^{-2} μg mL^{-1}. The Raman spectra of the enhanced substrate, i.e., silver colloid treated with reagents A and B,is shown in Figure 3e.In the absence of melamine, small peaks at 698 and 1005 cm^{-1} were observed, and the other peaks disappeared. Only a small peak at 678 cm^{-1} was observed in the Raman spectra of melamine dissolved in water (Fig. 3f), and no peaks were evident in the spectra obtained from the 10^{3} μg mL^{-1} melamine sample in the absence of the enhancing substrate (Fig. 3g).

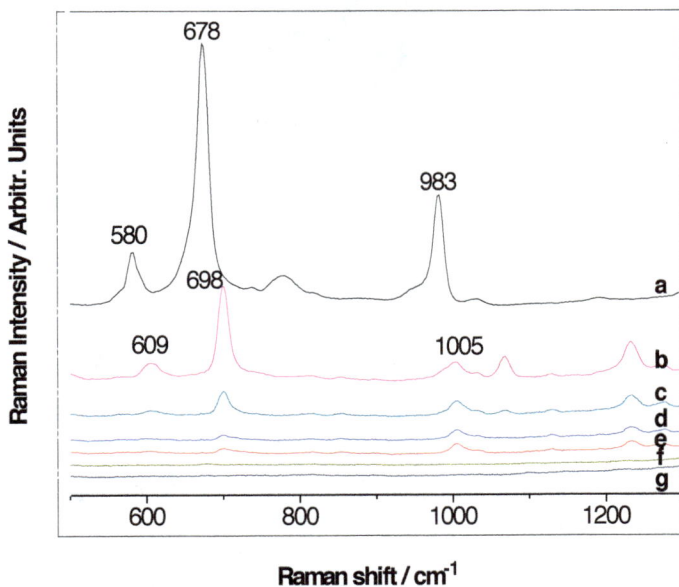

Figure 3. Raman spectra and SERS spectra of melamine at different concentrations.(a) Raman spectra of solid melamine.SERS spectra of melamine solution at (b) 5×10^{-1} μg mL^{-1}, (c) 1×10^{-1} μg mL^{-1}, and (d) 1×10^{-2} μg mL^{-1}.Raman spectra of silver colloid treated with reagents A and B (e) and melamine at different concentrations: (f) ~3.3×10^{3} μg mL^{-1}; and (g) 1×10^{3} μg mL^{-1}.

2.4. Analysis of detection results

To demonstrate the practical application of melamine in liquid milk, we used melamine in raw liquid milk as an example. Various concentrations of melamine in liquid milk were extracted and analyzed by their SERS spectra (Fig. 4). As shown in Figure 4a, seven concentrations (0.5, 1, 2, 2.5, 5, 8, and 10 µg mL^{-1}) of melamine in liquid milk were studied, and the intensity of the melamine peak at 698 cm^{-1} was enhanced with increasing melamine concentration. To eliminate the effects of the matrix and other factors (e.g., temperature, humidity, and focal distance), the intensity of the peak at 928 cm^{-1} was set at 100 for milk, and the Raman peak at 698 cm^{-1} in the absence of melamine had a fixed value. Accordingly, a melamine standard curve was obtained by establishing a plot correlating the melamine concentrations in liquid milk to the intensity of the intense SERS spectral peak of melamine at ~698 cm^{-1}. A linear regression (R^2 = 0.9996) was found between the Raman intensity and melamine concentration (Fig. 4b). The limit of quantification (LOQ) using this approach to detect melamine in liquid milk was also investigated, as shown in Figure 5. We found that this specific approach is reasonable for the detection of melamine in liquid milk because only one prominent peak was present in the melamine SERS spectra, which can be applied to field detection of various liquid milk products.

Moreover, the tests were performed and assessed by the Ministry of Science and Technology of the P. R. China, complying with the general administration quality supervision inspection quarantine of the P. R. China, the Ministry of Agriculture of the P. R. China, the Ministry of Health of the P. R. China, and the National Institute of Metrology P. R. China. The SERS test results were very precise and as good as those obtained by the LC/MS/MS method (Table 1). Forty-nine of 50 test samples results were correct, i.e., melamine was correctly detected in 98% of the test samples (Table 2).The concentration error in the samples was 0.2 ppm, which exceeds the limit of quantification using Raman spectra. The relative standard deviations (RSDs) were ≤ 10%, and the relative measurement deviations (RMD) were ≤ 10%.Therefore, the SERS method is an effective approach for measuring liquid milk melamine, which provides on-line, rapid, and reliable screening.

Sample #	LC/MS/MS (µg mL^{-1})	Quantity	Quality
No.1	2.37	2.6	Positive
No.2	2.37	2.5	Negative
No.3	0.48	1.1	Negative
No.4	7.20	7.5	Positive
No.5	2.37	2.8	Positive
No.6	2.37	2.7	Positive
No.7	≤0.1	0	Negative
No.8	2.37	2.6	Positive
No.9	1.91	2.05	Positive

Table 1. Comparison of results obtained by Raman spectroscopy and LC/MS/MS of liquid milk from the first test.

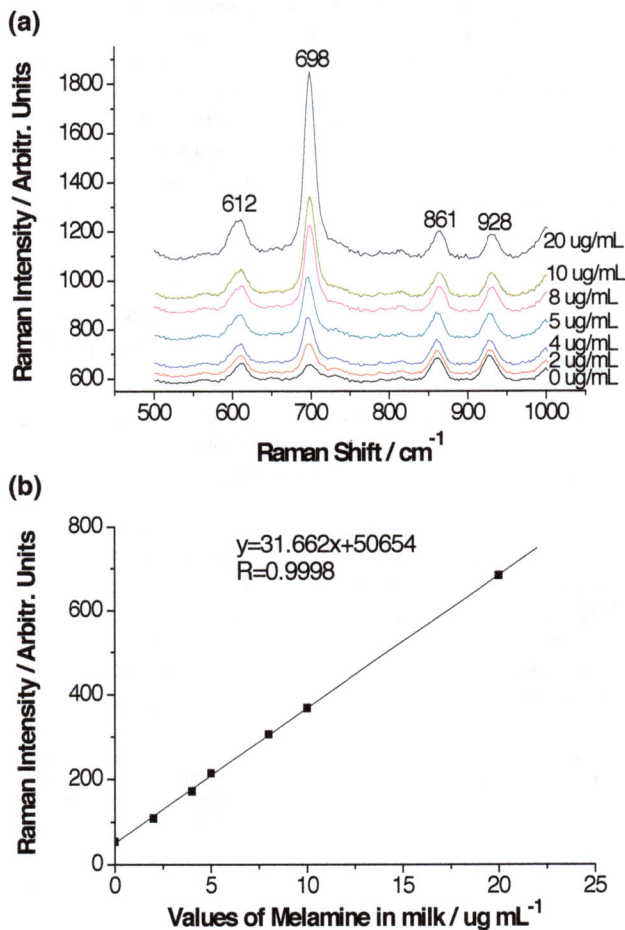

Figure 4. SERS spectra and standard curve of melamine in milk.(a) SERS spectra of different concentrations of melamine in milk. (b) Standard curve of melamine in milk.

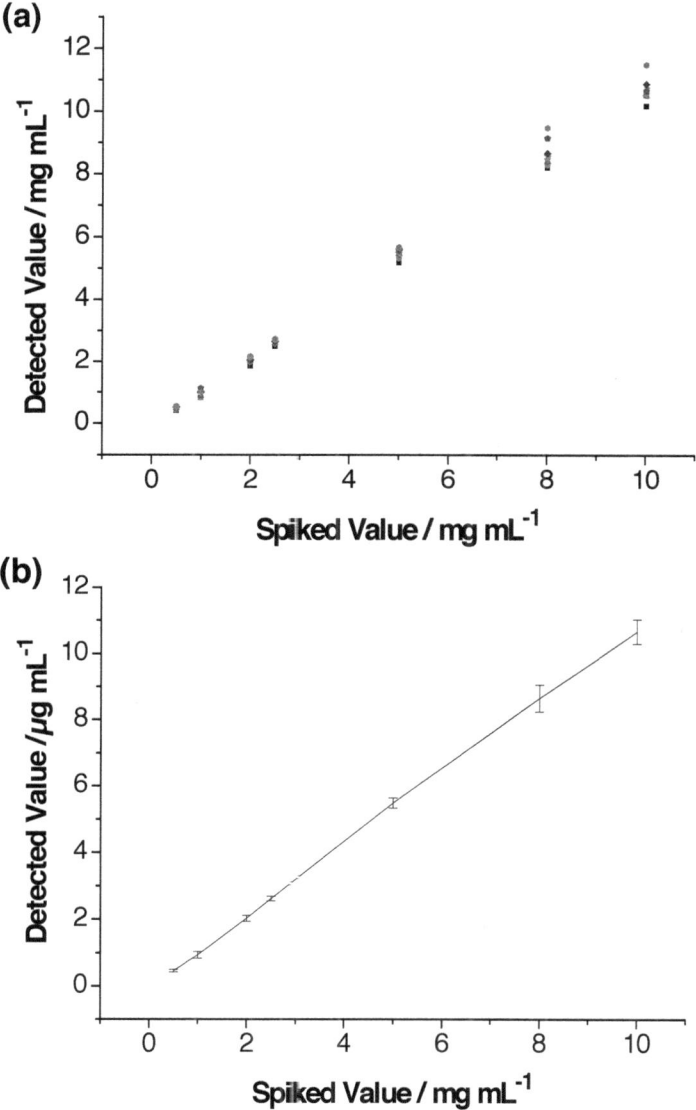

Figure 5. Predicted melamine value (μg mL⁻¹) compared to a spiked melamine value (μg mL⁻¹) using (a) the external standard method and (b) the error line. The spectral region = 1000-1800 cm⁻¹; spectral number n = 63.

Serial number	Random number	Raman (ppm)	LC/MS (μg mL^{-1})	Average value (μg mL^{-1})	RSD (%)	RMD (%)
1	754	0	<0.03			
2	769	0				
3	775	0				
4	781	<0.2				
5	788	0				
6	800	0				
7	695	0.19	0.20	0.25		
8	709	0.29				
9	719	0.24				
10	725	0.16				
11	731	0.28				
12	736	0.51				
13	741	0				
14	751	0.29				
15	658	0.57	0.50	0.55	10	0.10
16	669	0.52				
17	676	0.47				
18	684	0.58				
19	692	0.62				
20	700	0.54				
21	711	0.54				
22	606	1.19	1.02	1.07	10	0.10
23	617	1.08				
24	685	0.98				
25	694	1.01				
26	703	1.06				
27	708	1				
28	721	1.15				
29	451	2.26	2.02	2.23	2	0.10
30	523	2.14				
31	574	2.2				
32	611	2.23				
33	615	2.25				
34	620	2.3				
35	626	2.2				
36	634	2.23				
37	642	2.2				
38	652	2.22				
39	686	2.27				
40	589	28.37	30.25	30.78	6	0.02
41	590	32.69				
42	609	32.31				

Serial number	Random number	Raman (ppm)	LC/MS (µg mL⁻¹)	Average value (µg mL⁻¹)	RSD (%)	RMD (%)
43	621	33.4				
44	625	30.86				
45	644	29.97				
46	646	30.51				
47	691	28.1				
48	457	10.55	10.07	10.67	3	0.07
49	614	10.8				
50	683	11.15				

Table 2. Comparison of results obtained by Raman spectroscopy and LC/MS/MS of liquid milk from the second test.

A method was established to detect melamine in liquid milk using surface-enhanced Raman spectroscopy with the aid of a silver colloid enhancing substrate. An enhancement factor of $\geq 10^5$-fold was achieved in the measurement of melamine on this SERS-active substrate. In addition, the milk sample preparation process used in this technique is easy and time-saving, only requiring four steps: dilution, centrifugation, addition of samples to the enhanced base, and collection of the Raman spectra. The total detection time using SERS to measure a sample was ~3 min, which is starting from the dilution up to the final results. And the Raman spectra were acquired for only 3 s. Based on the calculations of the most intense peak in the melamine SERS spectra at approximately698 cm⁻¹, the LOQ of the SERS spectra achieved a level of 0.01 µg mL⁻¹ for melamine standard samples, which corresponds to 0.5 µg mL⁻¹ melamine in liquid milk. The RSD was ≤ 10 %, and recoveries were from 93-109%.The results from actual sample analyses were very precise and as good as those results obtained by LC/MS/MS.

3. Summary

Melamine, a nitrogen-rich chemical, has recently caused enormous economic losses to the food industry due to instances of milk products being adulterated by melamine, which has led to an urgent need for a rapid and reliable detection method for melamine in food. Here, we used a SERS liquid sensor to detect melamine in dairy products. The preparation processfor the dairy product samples is very easy, i.e.,only dilution with double-distilled water and centrifugation is required.With the aid of a silver colloid, at least a 10⁵-fold enhancement of the Raman signal was achieved for the measurement of melamine. The LOD by this method was 0.01 g mL⁻¹ for melamine standard samples. Based on the intensity of the Raman spectra with vibration bands normalized by the band at 928 cm⁻¹ (CH₂),the external standard method was employed for quantitative analysis. The linear regression (R^2) of the curve was 0.9998, the LOQ using this approach was 0.5 g mL⁻¹ melamine in dairy product samples, the relative standard deviation was ≤ 10%, and the recoveries ranged from 93-109%. The test results for SERS were very precise and as good as those obtained by LC/MS/MS.

Our method is simple, quick (only requiring ~3 min), cost-effective, and sensitive for the detection of melamine in dairy product samples using a SERS liquid sensor. Therefore, Ag NPs are good candidates for melamine sensing and suitable for the detection of melamine in dairy products. We believe that liquid Au NPs and Ag NPs will be widely used as liquid sensing substrates and that SERS will be widely investigated and applied for the analysis of other molecules, including pesticides, herbicides, pharmaceutical chemicals, banned food dyes, explosives, nicotine, and organic pollutants.

Acknowledgements

We are grateful for financial support by the International Science and Technology Cooperation and Exchange Foundation (No. 2008DFA40270), a Strategic Eleventh-five-year Science and Technology Supporting Grant (No. 2009BAK58B01), and Special Funded Projects of the Fundamental Research Funds from the Chinese Academy of Inspection and Quarantine of China (Grant No. 2010JK017).

Author details

Mingqiang Zou, Xiaofang Zhang, Xiaohua Qi and Feng Liu

Chinese Academy of Inspection and Quarantine, China

References

[1] Abalde-Cela S., Ho S., Rodríguez-González B., Correa-Duarte M. A., Álvarez- Puebla R. A., Liz-Marzán L. M., Kotov N. A. (2009). Loading of Exponentially Grown LBL Films with Silver Nanoparticles and Their Application to Generalized SERS Detection.Angew. Chem. Int. Ed., vol. 48, pp. 1-5.

[2] Ai K. L., Liu Y. L., Lu L. H. (2009). Hydrogen-bonding recognition-induced color change of gold nanoparticles for visual detection of melamine in raw milk and ifrant formula.Journal of the American Chemical Society, vol. 131, pp. 9496-9497.

[3] Betz J. F., Cheng Y., Rubloff G. W. (2012). Direct SERS detection of contaminants in a complex mixture: rapid, single step screening for melamine in liquid infant formula. Analyst, vol. 137, pp. 469-470.

[4] Broszat M., Brämer R., Spangenberg B. (2008). A new method for quantification of melamine in milk by absorption diode-array thin-layer chromatography Planar J. Chromatogr., vol. 21, pp. 469-470.

[5]   Campion A, Kambhampati P (1998). Surface-enhanced Raman scattering. Chem. Soc. Rev., vol. 4, pp. 241-250.

[6]   Chen l., Liu Y. (2012). Ag-nanoparticle-modified single Ag nanowire for detection of melamine by surface-enhanced Raman spectroscopy. Journal of Raman Spectroscopy, vol. 43, pp. 986-991.

[7]   Chen X., Parker S. G., Zou G., Eu W., Zhang Q. J. (2010). ß-cyclodextrin-functionalized silver nanoparticles for the naked eye detection of aromatic isomers.ACS Nano, vol. 4, pp. 6387-6394.

[8]   Cho M., Han, M. S., Ban C. (2008).Detection of mismatched DNAs via the dinding affinity of MutS using a gold nanoparticle-based competitive colorimetric method.Chemical Communications, vol. 38, pp. 4573-4575.

[9]   Chu P., Mills D. (2008). Electromagnetic response of nanosphere paris: collective plasmon resonances, enhance field, and laser-induced force. Phys. Rev.vol. B77, pp. 45416-45416.

[10] Deiss F., Sojic N., White D. J., Stoddart P. R. (2010). Nanostructured optical fibre arrays for high-density biochemical sensing and remote imaging. Anal Bioanal Chem, vol. 396, pp. 53–71.

[11] Ehling S., Tefera S., Ho I. P., (2007).High-performance liquid chromatographic method for the simultaneous detection of the adulteration of cereal flours with melamine and related triazine by-products ammeline, ammelide, and cyanuric acid. Food Addit.Contam., vol. 24, pp. 1319-1325.

[12] Ellis D. I., Broadhurst D., Clarkeb S. J., Goodacre R. (2005).Rapid identification of closely related muscle foods by vibrational spectroscopy and machine learning.Analyst, vol. 130, pp. 1648-1654.

[13] Eric A. E., Garber J., (2008). Detection of melamine using commercial enzyme-linked immunosorbent assay technology.Food Prot., vol. 3, pp. 590-594.

[14] Guingab J. D., Lauly B., Smith B. W., Omenetto N., Winefordner J. D., (2007). Stability of silver colloids as substrate for surface-enhanced Raman spectroscopy detection of dipicolinic acid.Talanta, vol. 74, pp. 271-274.

[15] Guo L. Q., Zhong J. H., Wu J. M., Fu F. F., Chen G. N., Zheng X. Y., (2010). Visual detection of melamine in milk products by label-free gold nanoparticles.Talanta, vol. 82, pp. 1654-1658.

[16] Han C. P., Li H. B. (2010).Visual detection of melamine in infant formula at 0.1 ppm level based on silver nanoparticles.Analyst, vol. 135, pp. 583-588.

[17] Haynes C. L., McFarland A. D., VanDuyne R. P., (2005). Surface-Enhanced Raman Spectroscopy.Anal.Chem., vol. 77, pp. 338a-346a.

[18] He L., Liu Y., Lin M., Awika J., Ledoux D. R., Li H., Mustapha A., (2008). A new approach to measure melamine, cyanuric acid, and melamine cyanurate using surface-

enhanced Raman spectroscopy coupled with gold nanosubstrates. Sens. Instrumen. Food Qual., vol. 2, pp. 66-71.

[19] House P. G., Schnitzer C. S., (2008).SERRS and visible extinction spectroscopy of copper chlorophyllin on silver colloids as a function of Ph. J. Colloid Interface Sci., vol. 318, pp. 145-151.

[20] Jun B. H, Kim Noh G., Kang M. S, Kim H., Y. K, Cho M. H, Jeong D. H, Lee Y. S., (2011). Surface-enhanced Raman scattering-active nanostructures and strategies for bioassays.Nanomedicine, vol. 6, pp. 1463–1480.

[21] Kneipp K, Kneipp H., Itzkan I., Dasari R. R., Feld M. S., (1999). Ultrasensitive chemical analysis by Raman spectroscopy. Chem. Rev., vol. 99 (10): 2957-2976.

[22] Kneipp K., Haka A. S., Kneipp H., (2002). Surface-Enhanced Raman Spectroscopy in Single Living Cells Using Gold Nanoparticles.Appl. Spectrosc., vol. 56, pp. 150-154.

[23] Knoll W. (1998). Interfaces and thin films as seen by bound electromagnetic waves. Ann. Rev. Phys. Chem., vol. 49, pp. 569-638.

[24] Koglin E., Kip B. J., Meier R. J., (1996). Adsorption and Displacement of Melamine at the Ag/Electrolyte Interface Probed by Surface-Enhanced Raman Microprobe Spectroscopy.J. Phys. Chem., vol. 100, pp. 5078-5089.

[25] Küstner B., Gellner M., Schütz M., Schöppler F., Marx A., Ströbel P., Adam P., Schmuck C., Schlücker S., (2009). SERS Labels for Red Laser Excitation: Silica-Encapsulated SAMs on Tunable Gold/Silver Nanoshells. Angew. Chem. Int. Ed., vol. 48, pp. 1950-1953.

[26] Lee J. S., Lytton-Jean A. K. R., Hurst S. J., Mirkin C. A., (2007). Silver nanoparticle oligonucleotide conjugates based on DNA with triple cyclic disulfide moieties. Nano Letter, vol. 7, pp. 2112-2115.

[27] Lee P. C., Meisel D.,(1982).Adsorption and Surface-Enhanced Raman of Dyes on Silver and Gold Sols.J Phys.Chem., vol. 86, pp. 3391-3395.

[28] Leopold N., Lendl B., (2003).A New Method for Fast Preparation of Highly Surface-Enhanced Raman Scattering (SERS) Active Silver Colloids at Room Temperature by Reduction of Silver Nitrate with Hydroxylamine Hydrochloride.J. Phys. Chem. B, vol. 107, pp. 5723-5727.

[29] Li L., Li B. X. , (2009). Sensitive and selective detection of cysteine using gold nanoparticles as colorimetric probe.Analyst, vol. 134, pp. 14217-14226.

[30] Lin M., He L., Awika J., Yang L., Ledoux D. R., Li H., Mustapha A., (2008). Detection of Melamine in Gluten, Chicken Feed, and Processed Foods Using Surface-enhanced Raman Spectroscopy and HPLC.J. Food Sci., vol. 73, pp. 129-134.

[31] Medley C. D., Smith J. E., Tang Z., Wu Y., Bamrungsap S., Tan W. H. (2008). Gold nanoparticle-based colorimetric assay for the direct detection of cancerous cells.Analytical Chemistry, vol. 80, pp. 1067-1072.

[32] Meikun Fan, Gustavo F.S. Andrade, Alexandre G. Brolod. (2011). A review on the fabrication of substrates for surface-enhanced Raman spectroscopy and their applications in analytical chemistry.Analytica Chimica Acta, vol. 693, pp. 7–25.

[33] Micklander E., Brimer L., Engelsen S. B., (2002).Noninvasive Assay for Cyanogenic Constituents in Plants by Raman Spectroscopy: Content and Distribution of Amygdalin in Bitter Almond (Prunus amygdalus) .Appl. Spectrosc., vol. 56, pp. 1139-1146.

[34] Moskovit M, Dilella D. Maynard K., (1988). Surface Raman spectroscopy of a number of cyclic aromatic molecules adsorbed on silver: selection rules and molecular reorientation.,Langmuir, vol, 4, pp. 67-76.

[35] Muik B., Lendl B., Molina-Diza A., Ayora-canada M. J., (2003).Direct, reagent-free determination of free fatty acid content in olive oil and olives by Fourier transform Raman spectrometry. Anal.Chim.Acta., vol. 487, pp. 211-220.

[36] Mulvihill M., Tao A., Benjauthrit K., Arnold J., Yang P., (2008).Surface-enhanced Raman spectroscopy for trace arsenic detection in contaminated water.Angew. Chem.Int. Ed., vol. 47, pp. 6456-6460.

[37] Muniz-Valencia R., Ceballos-Magana S. G., Rosales-Martinez D., Gonzalo-Lumbreras R., Santos-Montes A., Cubedo-Fernandez-Trapiella A. R., Izquierdo-Hornillos C., (2008).Method development and validation for melamine and its derivatives in rice concentrates by liquid chromatography. Application to animal feed samples. Anal.Bioanal. Chem., vol. 392, pp. 523-531.

[38] Niikura K., Nagakawa K., Ohtake N., Suzuki T., Matsuo Y., Sawa H., (2009). Gold nanoparticle arrangement on viral particles through carbohydrate recognition: a non-cross-linking approach to optical virus detection. Bioconjugate Chemistry, vol. 20, pp. 1848-1852.

[39] Otto A. (2005).The 'chemical' (electronic) contribution to surface-enhanced Raman scattering. J Raman Spectrosc., vol. 36, pp. 497-509.

[40] Paradkar M. M., Irudayaraj J., (2001).Discrimination and classification of beet and cane inverts in honey by FT-Raman spectroscopy.Food Chem., vol. 76, 231-235.

[41] Peica N., Pavel I., Rastogi V. K., Kiefer W.,(2005).Vibrational characterization of E102 food additive by Raman and surface-enhanced Raman spectroscopy and theoretical studies.J. Raman Spectrosc., vol. 36, pp. 657-666.

[42] Rubayiza A. B., Meurens M., (2005). Chemical Discrimination of Arabica and Robusta Coffees by Fourier Transform Raman Spectroscopy. J. Agric. Food Chem., vol. 53, pp. 4654-4659.

[43] Tang H., Ng K., Chui S. S., Che C., Lam C., Yuen K., Siu T., Lan L. C., Che X., (2009). Analysis of melamine cyanurate in urine using matrix-assisted laser desorption/ionization Mass Spectrometry. Anal. Chem., vol. 81, pp. 3676-3680.

[44] Tian Z., Ren B., Wu D., (2002).Surface-enhanced Raman scattering: from noble to transition metals and from rough surfaces to ordered nanostructures J. Phys. Chem. B, vol. 106, pp. 9463-9483.

[45] Tiwari V.S., Oleg T., Darbha G.K., Hardy W., Singh J.P., Ray P.C., (2007). Non-resonance SERS effects of silver colloids with different shapes. Chem. Phys. Lett., vol. 446, 77-82.

[46] Varelis P., Jeskelis R., (2008). Preparation of [13C3]-melamine and [13C3]-cyanuric acid and their application to the analysis of melamine and cyanuric acid in meat and pet food using liquid chromatography-tandem mass spectrometry.Food Addit.Contam., vol. 25, 1208-1215.

[47] Wang H., Levin C. S., Halas N. J., (2005). Nanosphere Arrays with Controlled Sub-10-nm Gaps as Surface-Enhanced Raman Spectroscopy Substrates.J. Am. Chem. Soc. 127, 14992-14993.

[48] Wang Y. L., Li D., Ren W., Liu Z. J., Dong S. J., Wang E. K. (2008). Ultrasensitive colorimetric detection of protein by aptamer-Au nanoparticles conjugates based on a dot-blot assay. Chemical Communications, vol. 22, pp. 2520-2522.

[49] Wei W., Li S., Millstone J. E., Banholzer M. J., Chen X., Xu X., Schatz G. C., Mirkin C. A., (2009).Surprisingly long-range surface-enhanced Raman scattering (SERS) on Au–Ni multisegmented nanowires.Angew.Chem. Int. Ed., vol. 48, pp. 4210-4212.

[50] Weng Y. M., Weng R. H., Tzeng C. Y., Chen W. L., (2003). Structural analysis of triacylglycerols and edible oils by near-infrared Fourier transform Raman spectroscopy. Appl. Spectrosc., vol. 57, pp. 413-418.

[51] Wong H., Phillips D. L., Ma C., (2007). Raman spectroscopic study of amidated food proteins.Food Chem., vol. 105, 784-792.

[52] Yaffe N. R., Blanch E. W., (2008). Effects and anomalies that can occur in SERS spectra of biological molecules when using a wide range of aggregating agents for hydroxylamine-reduced and citrate-reduced silver colloids.Vib. Spectrosc., vol. 48, 196-201.

[53] Yang S., Ding J., Zheng J., Hu B., Li J., Chen H., Zhou Z., Qiao X., (2009). Detection of melamine in milk products by surface desorption atmospheric pressure chemical ionization Mass Spectrometry. Anal.Chem., vol. 81, pp. 2426-2436.

[54] Yu D., (2007).Formation of colloidal silver nanoparticles stabilized by Na^+–poly(γ-glutamic acid)–silver nitrate complex via chemical reduction process. Colloids Surf. B: Biointerfaces, vol. 59, pp. 171-178.

[55] Zhang J., Wang L. H., Pan D., Song S. P., Boey F. Y. C., Zhang H., (2008). Visual cocaine detection with gold nanoparticle and rationally engineered aptamer structures.Small, vol. 4, pp. 1196-1200.

[56] Zhang X., Zou M., Qi X., Liu F., Zhu X., Zhao B. (2010). Detection of melamine in liquid milk using surface-enhanced Raman scattering spectroscopy, J. Raman Spectrosc., vol. 41, pp. 1365–1370.

[57] Zhao B., Liu Z. L., Liu G. X., Li Z., Wang J. X., Dong X. T. (2009). Silver microspheres for application as hydrogen peroxide sensor. Electrochemistry Communications, vol. 11, pp. 1707-1710.

[58] Zhao W., Brook M. A., Li Y. F. (2008). Design of gold nanoparticle-based color-imetric biosensing assays. Chembiochem, vol. 9, pp. 2363-2371.

[59] Zhou Q., Li X., Fan Q., Zharg X., Zheng J.,(2006).Charge transfer between metal nanoparticles interconnected with a functionalized molecule probed by surface-enhanced Raman spectroscopy. Angew.Chem.Int. Ed., vol. 45, pp. 3970-3973.

Chapter 8

Multiplexing-Capable Immunobiosensor Based on Competitive Immunoassay

Yanfei Wang, Mingqiang Zou, Ping Yao and Yuande Xu

Additional information is available at the end of the chapter

1. Introduction

As biological receptors, biosensors offer enormous potential for detecting a wide range of analytes in the food industry and in drug residue monitoring. For immunobiosensors, a probe is required to conjugate with a protein to monitor a specific interaction, and these probes possess reactive functionalities convenient for covalent linkage to antibodies and other protein molecules. As the usage of these probes becomes standardized, an increasing number of instruments have become widely available for measuring their signals. The development of multiplexing-capable immunobiosensors offers the possibility to simultaneously measure many different analytes in a small sample volume. The advantages of a multiplexing-capable immunobiosensor increase with each additional analyte analyzed and can be translated into substantial savings in the cost of reagents and time.

In this chapter, we review the main applications of multiplexing-capable immunobiosensors and illustrate their multiplexing capabilities by example of a suspension array system. First, the principle of a suspension array system is introduced. Microspheres are the common carrier in suspension array systems. Microspheres with different sizes or colors or a single microsphere with different fluorescent probes are used for multiplexed immunoassays (IAs).

Second, we list the main applications of suspension array systems. Suspension array technology was initially widely used in nucleic acid detection and genotyping. Recently, it has been gradually introduced for the detection and quantification of many viruses and antibodies in various samples based on IAs. However, there are also quite a few reports on multiplexed detection of low molecular weight compounds, such as drug residues, because this is more difficult for competitive IAs than for noncompetitive (sandwich) ones to meet the sensitivity and specificity requirements.

In the following section, a rapid, sensitive, multiplex, and competitive IA model based on a suspension array system is described Antigens are covalently bound to different coded functional carriers to compete for antibodies with analytes in sample, and the fluorescent probe is used as a transducer. As the fluorescence signals are measured, the analytes can be screened simultaneously.

2. Multiplexing-capable suspension array system

2.1. History of high-throughput technology

High-throughput screening (HTS) has been developed as an advanced technique to perform large-scale screening of samples to analyze biological or chemical activity (Burbaum & Sigal, 1997; Hill, 1998). Descriptions of this technology first appeared in the scientific literature as early as the 1980s. Over the past 30 years, the developments in the scale, efficiency, and technical level of HTS have rendered it a key technology in the areas of international drug research, genomics, proteomics, and analytical techniques (Burbaum, 1998; Eggeling et al., 2003; Haber et al., 2005).

HTS technology can be used to construct a micro-biochemical analysis system by integrating micromachining technology with microelectronics technology. With a large number of nucleic acid fragments with specific sequence or proteins fixed on a functional carrier, as well as labeled reactants (nucleic acids or protein) reacting in the reaction system, HTS can quantitatively determine the target analytes in samples by detecting the intensity of fluorescent reporter signals and can screen for compounds, nucleic acids, proteins, cells, and other biological components. Multiplexing and miniaturization are current trends in the development of HTS technology. Assays performed on large numbers of candidates in small sample volumes and using target analytes (e.g., antigens, antibodies, nucleotides, and peptides) can be screened. HTS has important application prospects in the fields of drug discovery, drug screening, and diagnostics, and suspension array technology is one of the representatives in a variety of new high-throughput analysis methods.

2.2. Suspension array system and its detection principles

The suspension array system is based on encoded microspheres, which are both the encoded carrier and analysis platform for the probes. Compared with traditional flat biochips, the advantages of suspension biochips are as follows (Meza, 2000; Battersby, 2000; Grondahl, 2000):

- Liquid phase conditions are conducive for maintaining the natural conformation of protein molecules, and the use of microspheres as the solid carrier in the assay leads to reaction kinetics approaching phase conditions;

- Differently coded microspheres conjugated to the capture antibody or target molecules are required for each reaction and can be flexibly pooled together for a multiplexed assay and separated later during data acquisition, reducing the analysis cost;

- These "no-wash" assays eliminate the need for washing steps; and

- Suspension array systems offer an approach for large-scale screening, which is an important breakthrough in the number of analytes compared with traditional flat chips.

Of course, suspension array system has some shortcomings: optimization of multiple liquid phase reaction conditions is more difficult, and the cross reaction and its elimination is more complicated which would require thorough analysis. In addition, on-line monitoring of analytes could not be done using suspension array system.

In spite of this, with widespread prospects for application,suspension array technology is becoming a research focus.

Microsphere-based suspension array technology consists of a detection platform and microspheres. Such technology has a long history of simplifying the separation of biological systems, the positioning of biological interactions, and signal amplification in the biomedical field (Brown, 1987; Deleo, 1991). After decades of development, microsphere-based analysis techniques have now been developed into a wide range of analysis systems to meet specific needs. However, despite their various forms, the basic principle of these assays is microspheres pre-encoded by chemical, spectral, electronic, or physical methods. Each microsphere encodes specific identifying information, and these microspheres offer binding sites for bioconjugation. Before testing, differentially coded microspheres are coupled with the capture antibody or target molecules (e.g., antigens, oligonucleotides, receptors, or peptides) that are required for each reaction. Then, a reporter molecule labeled with a fluorescent marker (e.g., phycoerythrin or Alexa 532) binds to the analytes captured on the microspheres in the sample solution. The bound microspheres are passed through the detection channel and separated during data acquisition.

It can be concluded that microspheres-based suspension array technology includes two key factors: 1) coding and decoding a large number of microspheres, and 2) conjugating capture antibodies or target molecules with functional microspheres for the identification of different analytes.

As a typical multiplexed microsphere-based suspension array platform, the Luminex®xMAP™ system is a flow-based dual-laser system capable of analyzing and reporting up to 100 different analytes in a single reaction vessel in just a few seconds per sample (Dubar, 2006).The target molecules are fixed on the carrier via non-covalent or covalent bonds. Usually, reactive functionalities on microspheres provide the means to specifically label certain target groups on capture antibodies or target molecules in the form of a covalent bond. The common covalent bonds include amide, ester, and ether bonds (Deleo, et al. 2000; Mehnaaz, et al. 2003).

3. Competitive model based on IA

3.1. Immunoassay (IA)

Immunoassay technology is a type of analysis method based on antigen-antibody specificity, identification, and binding. Depending on the available antigen or antibody reagents, one can achieve qualitative and quantitative detection of a specific analyte.

Immunoassay is a very popular method, has been rapidly developed by many research laboratories,and many versions are commercially available due to their rapid, simple, specific detection capabilities. Different types ofIA with specific characteristics listed in Table 1 would give a comparison. In the early 1990s, IA, together with liquid chromatography (LC) and gas chromatography (GC), were called three of the most advanced analytical techniques by the Association of small molecule residue determination mainly uses competitive IA.

type	characteristics
Radioimmunoassay,RIA	Reagent with low cost, high sensitivity; complex operation, radioactive pollution, short duration
enzyme immunoassay,EIA	Reagent with low cost, simple operation; low sensitivity, suitable for qualitative and semi-quantitative determination
chemiluminescent enzyme immunoassay,CIZIA	Simple operation, low cost, high sensitivity, stable reagent; working curve drift with time
fluorescence immunoassay,FIA	High sensitivity, stable reagent; complex operation, high cost of reagents, the background interference

Table 1. Different types ofIA with specific characteristics

3.2. Operational process of competitive IA

The basic operational process of the competitive model includes fixing the antigen on a solid support, pooling the sample (antigen) and antibody in the reaction vessel together for competitive interaction, and adding a reporter molecule labeled with a marker to the sample solution to bind to the analyte captured by the antigen fixed on the solid carrier. Competitive IA can quantitatively determine the analyte in a sample indirectly by detecting the intensity of the reporter signals because the detection signal is inversely proportional to the concentration of the analyte in a competitive model. Small molecular weight compounds, such as drug residues, and theconsist of preparation of the immunogen (artificial antigen), production of the antibody, and optimization of reaction conditions.

Small molecular weight compounds enzyme-linked immunosorbent assay became an inevitable trend, and multiplexed assays have been further simplified by introducing immunosensors, which are available for HTS and automation. Remarkably, optical immunosensors are widely employed to their full potential.

4. Applications of a suspension array system based on competitive IA

4.1. Main applications of a suspension array system

Microsphere-based suspension array technologies, such as the Luminex® xMAP™ system, offer a flexible technical platform based on Flexible Multi-Analyte Profiling (xMAP) technology, which was recently developed (Morgan, et al. 2004). Such technology combines encoded microspheres with flow cytometry. Two lasers analyze the microspheres in a flow stream. The first laser indentifies each microsphere-associated analyte according to the fluorescent signature of the microsphere, and the second measures the reporter molecules attached to the analytes quantify the amount of analyte (Nolan &Sklar, 2002).

Many studies have been performed in the field of clinical diagnosis and genomic research by combining flow-based technology with IA (Yan et al. 2004; Fuja et al. 2004). Beads of different sizes/colors or a single bead with different fluorescent probes are used for multiplexed IA. Previous studies using multiplexed assay technology include detection of thyroxine and thyrotropin from blood spot samples for congenital hypothyroidism (Bellisario, et al. 2000), antibodies to West Nile virus in human serum and cerebrospinalfluid (Wong et al. 2004), and Ig classes in serum and stool samples (Dasso, et al. 2002), as well as identifying response patterns in hyperinflammatory diseases (Hsu, et al. 2008) and cytokines in human serum (Ray et al. 2005). As an open platform, xMAP was gradually introduced for the detection and quantification of many virus and drug residues in food and agriculture samples. For example, a multiplex microsphere IA was developed to monitor seed potatoes for potato viruses X, Y, and PLRV (Bergervoet et al. 2008), and a flow cytometric IA was developed to detect sulfonamide residues in milk (de Keizer et al. 2008).

Notably, the multiplexed IAs using suspension array technology are currently more prevalent for analyzing viruses and antibodies than detecting small molecular weight compounds such as drug residues.

A recent study reports the use of a suspension array system for the detection of drug residues, highlighting its simplicity, low cost, specificity, and multiplexing capabilities (Zou et al. 2008). Fig. 1 shows the competitive IA strategy and the detection principles of the Luminex® xMAP™ system.

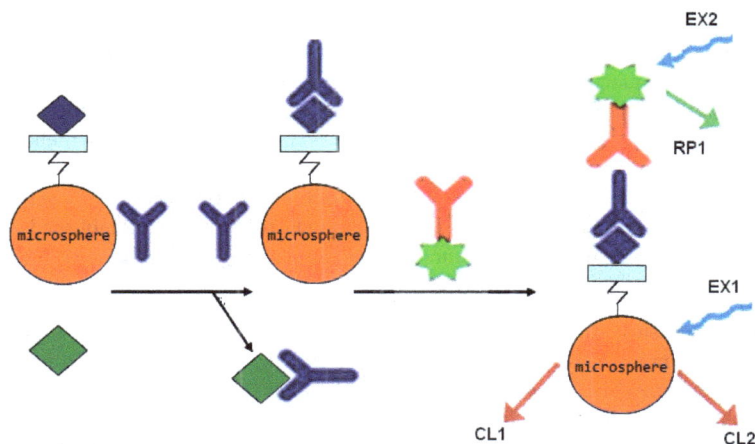

Figure 1. Microsphere-based competitive IA strategy. EX: excitation laser; RP: reporter laser; CL: clarification laser

4.2. Detection of low molecular weight compounds using an immunobiosensor based on competitive IA

Low molecular weight compounds are widely utilized as animal drugs, food additives, and pesticides to pursue maximum productivity and profits directly or indirectly in food products. However, residues of such low molecular weight compounds in food products are detrimental to human health. For instance, clenbuterol (313.65u), a β-adrenergic agonist growth promoter, is associated with a series of severe food poisoning outbreaks around the world. Therefore, quantitative analysis of such low molecular weight compounds is essential to safeguardpublic health, prevent their illicit use, and facilitate governmental regulation and surveillance.

In comparison to conventional ELISAs, a microsphere-based competitive fluorescent IA (MCFI) has been established to analyze low molecular weight compounds in food as described in the literature (Zou et al. 2008). This method utilizes an artificial antigen bound to 4-μm polystyrene beads together with an FITC-conjugated monoclonal antibody. In this indirect IA, the fluorescent conjugated antibody is bound to an artificial antigen fixed on the polystyrene beads and subsequently analyzed by multiparameter flow cytometry. Quantification is based on the intensity of the fluorescent reporter signal: the higher the hazardous low molecular weight compound concentration, the lower the fluorescence intensity.

It is worth pointing out that the entire process is specially designed for aqueous systems that are bio-compatible with proteins' native structures. To illustrate the feasibility of this novel strategy, clenbuterol was selected as a practical analyte. Compared to traditional IAs, such as ELISA, microsphere-based competitive fluorescent IA by flow cytometry offers comparable or higher sensitivity, better reproducibility, and a greater dynamic range.

4.3. Simultaneous detection of antibiotics in raw milk

Antibiotics, as conventional veterinary medicine, are widely used for the treatment and pre-vention of microbial infections. Antibiotic residues can enter the body and are metabolized by the liver. Further, antibiotics and their metabolites can impair the nervous and psychiat-ric systems, circulatory system, and even brain function before elimination by the kidneys. Thus, every country has established a maximum residue limit (MRL) for the total amount of antibiotics in various food products (e.g., raw milk) to bolster public health and the security of international trade in food products. For example, the European Union established a MRL of 100 µg/L for the total amount of gentamycin in milk and, 150µg/L for the total amount of kanamycin in milk.

As previously published (Wang, et al. 2011), we applied suspension array technology to de-velop an indirect multiplexed competitive IA for simultaneous detection of antibiotic resi-dues, such as kanamycin and gentamycin, in raw milk. First, we successfully produced a monoclonal antibody (McAb) against gentamycin. Then, with the coating antigens of kana-mycin and gentamycin attached to differentially encoded beads and R-phycoerythrin-conju-gated goat anti-mouse IgG as the fluorescent probe, an indirect bead-based competitive fluorescent IA using a Bio-Plex™ 200 suspension array system was developed. To develop the actual application, raw milk samples spiked with kanamycin or gentamycin were ana-lyzed using ELISA and our Bio-Plex™ assay for comparison.

A McAb against gentamycin was successfully obtained, and an indirect bead-based compet-itive fluorescent IA was developed for rapid detection of small molecule drug residues in animal-derived food products, such as raw milk. Notably, even with the matrix's back-ground interaction with milk, the fluorescent intensities detected from raw milk samples are still strong enough and sufficient to detect the lowest allowable quantity of the antibiotics in milk at the MRL level. In fact, cephalosporins and sulfonamides, which are used worldwide in cows and are also tested in milk, could be added to the assay because the suspension ar-ray technology permits up to 100 different assay combinations. Compared to ELISA, the multiplexed assay had a similar limit of detection,but its ability to simultaneously measure several analytes in a single sample is superior to ELISA. Details of limit of detection, detec-tion time, etc. are described in the literature (Wang, et al. 2011).

With a particular focus on applications in real samples containing complex matrices, kana-mycin was intraperitoneally administered to three cows at a dose of 10 mg/kg (twice per day for 3 days), and gentamycin was administered in another three cows in the same way.

Sixty raw milk samples were collected from these six cows. Fig. 2 and Fig. 3 show the com-parison of the detection results obtained by the bead-based competitive fluorescent IA and ELISA. The analysis shows that there is good correlation between the results of the multi-plexed assay and those of ELISA. Basically, the kanamycin concentrations in raw milk de-creased gradually. Seven hours after the last injection, the residues were present at 190 ng/ml, and 20 h later, the residues were reduced to 40 ng/ml, which is lower than the MRL provided by the European Union. The gentamycin concentrations in raw milk decreased in a similar manner.

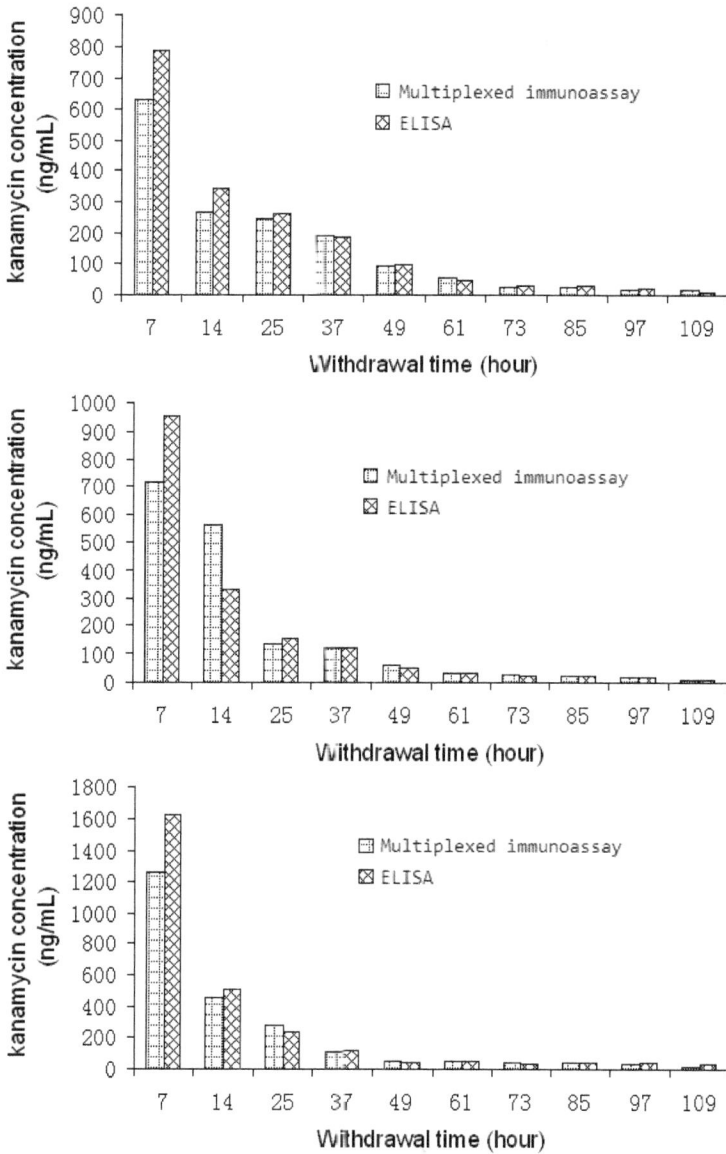

Figure 2. Raw milk depletion profile of kanamycin after intramuscular administration to cows.

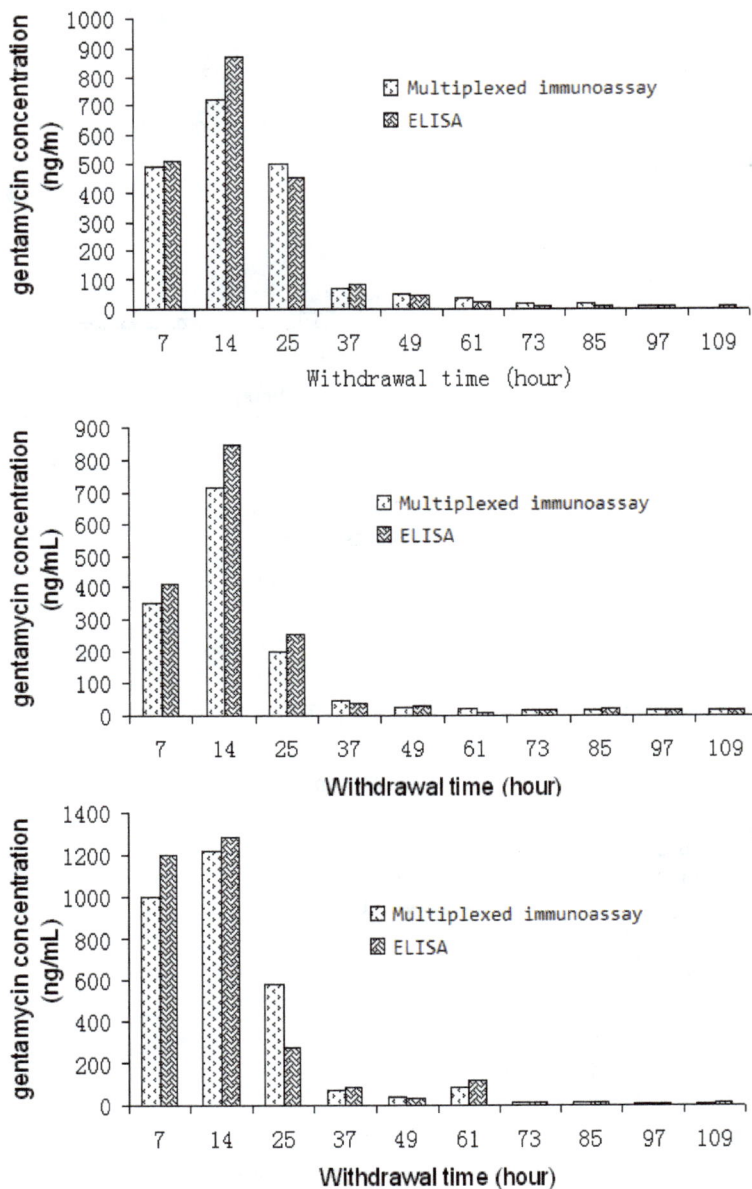

Figure 3. Raw milk depletion profile of gentamycin after intramuscular administrationto cows.

4.4. Other applications using immunobiosensors based on competitive IA

To illustrate the feasibility of this competitive strategy, we present an example of the simultaneous detection of antibiotic residues like kanamycin and gentamycin, though other related reports exist.

Sulfonamides are a class of antibiotics used in veterinary and human medicine for the treatment and prevention of microbial infections. The European Union (EU) established a MRL of 100 µg/L for the total amount of sulfonamides in milk. To detect sulfonamides in raw milk, a multi-sulfonamide flow cytometric IA (FCI) was developed using the xMAP technology as described in the literature (de Keizer, et al. 2008). In this automated FCI, a previously developed biotinylated multi-sulfonamide mutant antibody (M.3.4) was applied in combination with fluorescent beads, directly coated with a sulfathiazole derivative, and streptavidin–phycoerythrin (SAPE) for detection. Because of differences in sensitivity toward different sulfonamides, the FCI was considered and validated as a qualitative screening assay.

Simultaneous detection for three kinds of veterinary drugs: chloramphenicol (CAP), clenbuterol, and 17-beta-estradiol has also been reported (Liu, et al. 2009). Conjugates of chloramphenicol and clenbuterol coupled with bovine serum albumin were synthesized and purified. Probes for the suspension array were created by coupling the three conjugates on fluorescent microspheres, and the microstructures on the microspheres' surfaces were observed by scanning electron microscopy, which was direct confirmation for successful conjugate coupling. The addition of conjugates and the amounts of antibodies were optimized and selected, respectively. The suspension array is specific and has no significant cross-reactivity with other chemicals. Meanwhile unknown samples were analyzed by suspension array and ELISA in comparison with each other. The errors found for the detection of the unknown samples were relatively small in both methods, but the detection ranges of the suspension array are broader and more sensitive than that of traditional ELISA.

Further research investigated the simultaneous quantitative determination of five antibiotics (tylosin, tetracycline, gentamycin, streptomycin, and chloramphenicol) in milk in the same research laboratory (Su, et al. 2011). A novel treatment of milk samples for the suspension array with diethyl ether was performed, which greatly reduced the interference of the disturbing components in milk on the reaction results with no significant effect on detection sensitivity. Compared to using a biotin-labeled monoclonal antibody, using a secondary biotinylated antibody further increased the detection sensitivity. Thus, suspension assay technology is powerful for the rapid, quantitative analysis of multiple drug residues.

5. Conclusions

In this chapter, the competitive IA method with immunobiosensor as a detection platform was first used with a single analyte in simple buffer for its sensitivity, specificity, and repeatability, and then complex substrates such as milk were introduced. Thousands of actual samples were screened for confirmation of this multiplexing-capable immunobiosensor based on competitive IA. As an easy, effective, and time-saving method, the multiplexing-capable immunobiosensor based on competitive IA has the potential to detect small molecular weight compounds in various actual applications. Previous research has laid a firm foundation for simultaneous screening of multiple analytes in food products. Future research should focus on adding more analytes to the assay and establishing a reliable multiplexed assay to detect analytes in additional samples, such as tissue and fat. It is a flexible technical platform with multiplexing capabilities.

Acknowledgments

This work was partially supported by a Strategic Eleventh Five-Year Science and Technology Supporting Grant No: 2009BAK61B04 and Special Funded Projects of the Fundamental Research Funds from the Chinese Academy of Inspection and Quarantine of China No: 2010JK016.

Author details

Yanfei Wang[1], Mingqiang Zou[1], Ping Yao[2] and Yuande Xu[3]

1 Chinese Academy of Inspection and Quarantine, China

2 The People's Hospital of Juxian, China

3 COFCO Corporation, China

References

[1] Burbaum, J. J., & Sigal, N. H. (1997). New technologies for high-throughput screening.Curr.Opin. Chem. Biol. , 1, 72-78.

[2] Hill, D. C. (1998). Novel screen methodologies for identification of new microbial metabolites with pharmacological activity. Adv. Biochem. Eng. Biotechnol. , 59, 73-121.

[3] Burbaum, J. J. (1998). Miniaturization technologies in HTS: how fast, how small, how soon? Drug Discov. Today , 3, 313-322.

[4] Eggeling, C., Brand, L., Ullmann, D., et al., & (2003, . (2003). Highly sensitive fluorescence detection technology currently available for HTS.Drug Discov.Today. , 8, 632-641.

[5] Haber, C., Boillat, M. ., & Schoct, B. V. D. (2005). Precisenanoliter fluid handling system with integrated high-speed flow sensor. Assay Drug Dev. Technol. , 3, 203-212.

[6] Meza, M. B. (2000). Bead-based HTS applications in drug discovery.Drug Discov.Today (Suppl.). , 5, 38-41.

[7] Battersby, B. J., Bryant, D., Meutermans, W., et al. (2000). Toward larger chemical libraries: encoding with fluorescent colloids in combinatorial chemistry. J. Am. Chem. Soc. , 122, 2138-2139.

[8] Grondahl, L. B., Bronwyn, J., Bryant, D., et al. (2000). Encoding combinatorial libraries: a 104novel application of fluorescent silica colloids. Langmuir. , 16, 9709-9715.

[9] Brown, W., & (1987, . (1987). Microparticle-capture membranes: Application to Test Pack hCG-urine. Clin. Chem. , 33, 1567-1568.

[10] Deleo, D. T., & (1991, . (1991). Particle-enhanced turbidimetric IA of sex-hormone-binding globulin in serum.Clin. Chem. , 37, 527-531.

[11] Jane, A., Ferguson, F. J., Walt, D. R., & (2000, . (2000). High-density fiber-optic DNA random microsphere array.Anal. Chem. , 72, 5618-5624.

[12] Mehnaaz, F. A., Romy, K., Adrian, P. G., et al. (2003). DNA hybridization and discrimination of single-nucleotide mismatches using chip-based microbead arrays. Anal. Chem. , 75, 4732-4739.

[13] Bellisario, R., Colinas, R. J., Pass, K. A., & (2000, . (2000). Simultaneous Measurement of Thyroxine and Thyrotropin from Newborn Dried Blood-Spot Specimens Using a Multiplexed Fluorescent Microsphere IA.Clini. Chem. , 46, 1422-1424.

[14] Bergervoet, J. H. W., Peters, J., van Beckhoven, J. R. C. M., van den, Bovenkamp. G. W., Jacobson, J. W., & van der Wolf, J. M. (2008). Multiplex microsphere immuno-detection of potato virus Y, X and PLRV. J. Virol. Methods. , 149, 63-68.

[15] Dasso, J., Lee, J., Bach, H., Mage, R. G., & (2002, . (2002). A comparison of ELISA and flow microsphere-based assays for quantification of immunoglobulins. J. Immunol. Methods. , 263, 23-33.

[16] de Keizer, W., Bienenmann-Ploum, M. E., Bergwerff, A. A., Haasnoot, ., & , W. (2008). Flow cytometric IA for sulfonamides in raw milk. Anal.Chim.Acta. , 620, 142-149.

[17] Fuja, T., Hou, S., Bryant, P., & (2004, . (2004). A multiplex microsphere bead assay for comparative RNA expression analysis using flow cytometry.J Biotechnol. , 108, 193-205.

[18] Hsu, H. Y., Wittemann, S., Schneider, E. M., Weiss, M., Joos, ., & , T. O. (2008). Suspension microarrays for the identification of the response patterns in hyperinflammatory diseases. Med. Eng. Phys. . 30, 976-983.

[19] Liu, N., Su, P., Gao, Z. X., Zhu, M. X., Yang, Z. H., Pan, X. J., Fang, Y. J., & Chao, F. H. (2009). Simultaneous detection for three kinds of veterinary drugs: Chloramphenicol,

clenbuterol and 17-beta-estradiol by high-throughput suspension array technology. Anal.Chim.Acta. , 632, 128-134.

[20] Morgan, E., Varro, R., Sepulveda, H., Ember, J. A., Apgar, J., Wilson, J., Lowe, L., Chen, R., Shivraj, L., Agadir, A., Campos, R., Ernst, D., & Gaur, A. (2004). Cytometric bead array: a multiplexed assay platform with applications in various areas of biology. Clin.Immunol. , 110, 252-266.

[21] Nolan, J. P. ., Sklar, L. A., & (2002, . (2002). Suspension array technology_evolution of the flat-array paradigm.TRENDS Biotechnol. , 20, 9-12.

[22] Ray, C. A., Bowsher, R. R., Smith, W. C., Devanarayan, V., Willey, M. B., Brandt, J. T., & Dean, R. A. (2005). Development, validati. on, and implementation of a multiplex IA for the simultaneous determination of five cytokines in human serum. J. Pharmaceut. Biomed. , 36, 1037-1044.

[23] Wong, S. J., Demarest, V. L., Boyle, R. H., Wang, T., Ledizet, M., Kar, K., Kramer, L. D., Fikrig, E., Koski, ., & , R. A. (2004). Detection of Human Anti-Flavivirus Antibodies with a West Nile Virus Recombinant Antigen Microsphere IA. J. Clin. Microbiol. , 42, 65-72.

[24] Yan, X. M., Tang, A. J., Schielke, E. G., Hang, W., & Nolan, J. P. (2004). On the development of a microsphere-based multiplexed IA for influenza virus typing and subtyping by flow cytometry. International Congress Series. , 1263, 342-345.

[25] Zou, M. Q., Gao, H. X., Li, J. F., Xu, F., Wang, L., & Jiang, J. Z. (2008). Rapid determination of hazardous compounds in food based on a competitive fluorescence microsphere IA.Anal.Biochem. , 374, 318-324.

[26] Wang, Y. F., Wang, D. N., Zou, M. Q., Jin, Y., Yun, C. L., Gao, ., & , X. W. (2011). Application of suspension array for simultaneous detection of antibiotic residues in raw milk.Analytical letters. , 44, 2711-2720.

[27] Dubar, S. A. (2006). Applications of Luminex® xMAP™ technology for rapid, high-throughput multiplexed nucleic acid detection. Clin.Chim.Acta. , 363, 71-82.

[28] Su, P., Liu, N., Zhu, M. X., Ning, B. A., Liu, M., Yang, Z. H., Pan, X. J., Gao, ., & , Z. X. (2011). Simultaneous detection of five antibiotics in milk by high-throughput suspension array technology. Talanta. , 85, 1160-1165.

Chapter 9

Biosensors for Contaminants Monitoring in Food and Environment for Human and Environmental Health

Lívia Maria da Costa Silva,
Vânia Paula Salviano dos Santos,
Andrea Medeiros Salgado and Karen Signori Pereira

Additional information is available at the end of the chapter

1. Introduction

Environmental security is one of the fundamental requirements of our well-being. However, it still remains a major global challenge. on account of the increasing number of potentially harmful pollutants (chemical compounds, toxins and pathogens) discharges into the environment [1]. In this context, the detection and monitoring of environmental pollutants in soil, water and air is very important in the overall safety and security of humans, other animals and plants.

The requirements for application of most traditional analytical methods to environmental pollutants analysis often constitute an important impediment for their application on a regular basis. These analysis calls for fast and cost-effective analytical techniques to be used in extensive monitoring programs. So, the need for disposable systems or tools for environmental applications has encouraged the development of new technologies and more suitable methodologies. In this context, biosensors appear as a suitable alternative or as a complementary analytical tool. Biosensors can be considered as a subgroup of chemical sensors in which a biological mechanism is used for analyte detection [2,3,4].

A biosensor (Figure 1) is defined by the International Union of Pure and Applied Chemistry (IUPAC) as a self-contained integrated device that is capable of providing specific quantitative or semi-quantitative analytical information using a biological recognition element (biochemical receptor), which is retained in contact direct spatial with a transduction element [5].

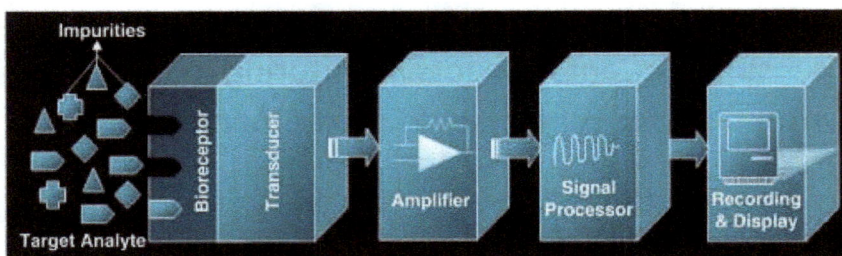

Figure 1. Biosensor scheme [6].

Based on the principle of specific biological-recognition measurements coupled with a signal transducer, biosensor technologies have developed over the past several decades, resulting in commercial production of versatile and portable instruments for many applications that greatly impacts the bioprocess control, food quality control, agriculture, environment, military and mainly medicine and clinical analysis. Overall, there are three so-called generations of biosensors: first generation biosensors operates on electrical response, second generation biosensors function involving specific mediators between the reaction and the transducer for generation of a improved response, and in third generation biosensors the reaction itself causes the response and no product or mediator diffusion is directly involved [7].

In the 21st century, with the progression in the sciences, nanobiosensors with superbly dedicated miniature sensors with highly miniaturization were developed. In this context, nanomaterials transducer modification and genetic engineering of the biocomponents are the main strategies to overcome the reported drawbacks of low sensitivity and reusability/regeneration of working electrode [7,8].

The main classes of bioreceptor elements that are applied in environmental analysis are microbial whole-cell, enzymes, antibodies and DNA. Additionally, electrochemical transducers are used in most of the biosensors described in the literature for environmental applications [5]. In recent years, the number of investigations on biosensors has been very high, which reflects the considerable interest in the theme. Nevertheless, there is a lag between the high level of technological and scientific development and the limited use of these devices in the real environmental and agro-food analysis [9, 10].

So, given the applicability of biosensors, this chapter reviews the development and use of some biosensors in the contaminants monitoring in food and environment for human and environmental health, describing the three main areas of application: fertilizers, pesticides and quality and safety of food.

2. Biosensors for nitrogen compounds and fertilizers determination

Because nitrogen compounds are pollutant found in several industrial effluents its determination is of extreme importance for the environment. Several methods are used to urea determination, including spectrophotometry, fluorimetry, potentiometry and amperometry. But some of these require a pretreatment or are unsuitable for monitoring in situ. For this reason there has been growing interest in the development of biosensors for these determinations.

Urea is a nitrogen compound widely used, such as fertilizer in agriculture and as a nitrogen source for animal feeding stuff additive. The determination of urea is of great importance in areas such as clinical analysis, food industry, cosmetics and environmental assessment. In the environmental area, the large amount of urea in wastewater and water bodies can encourage the eutrophication process.

In this context, the first urea potentiometric biosensor ever made was built by Guilbault and Montalvo [11], and measured urea through its enzyme-catalyzed hydrolysis. This first device was consisted of urease immobilization on polyacrylamide gel on the surface of an ammonium ion-selective electrode. Furthermore, in 1984, Arnold and Glazier [12], developed a urea potentiometric biosensor based on jack bean extract immobilization on the surface of an ammonia gas sensitive electrode. Since then, several instruments were developed and Table 1 shows some biosensors for urea quantification with different transduction system.

Urease is abundant enzyme in plants and, moreover, it can be found at numerous of eukaryotic microorganisms and bacteria. The bacterial and plant ureases have high sequence similarity, suggesting that they have similar three-dimensional structures and a conserved catalytic mechanism. Ureases (urea amidohydrolase, EC 3.5.1.5) catalyzes the hydrolysis of urea to yield ammonia and carbamat, the latter compound decomposes spontaneously to generate a second molecule of ammonia and carbon dioxide [13].

Biological component	Transductor	Reference
Urease immobilized on gelatin beads via cross-linking with glutaraldehyde	Potentiometry	[14]
Urease immobilized on electrosynthesized polymer	Potentiometry	[15]
Urease immobilized on air stable lipid films	Optical	[16]
Urease immobilized, through entrapping, onto the ion sensitive membrane using a polymer matrix of poly(carbamoylsulphonate) and polyethyleneimine	Potentiometry	[17]
Urease immobilized on platinum electrode both by chemical binding and electropolymerization	Conductometry	[18]
Urease immobilized on electropolymerized toluidine blue film	Amperometry	[19]
Urease immobilized with sol-gel approach onto a nylon membrane	Colorimetry	[20]

Table 1. Biosensors for urea quantification.

Other nitrogen compounds such as nitrate and nitrite are used in fertilizers. Nitrate is the most ubiquitous chemical contaminant in groundwater and soil. The increasing levels of nitrate found in groundwater and surface water concern because they can harm the aquatic environment. In line with this, the regulations for treatment of urban wastewater in order to reduce pollution, including pollution by nitrates from sewage treatment works of industrial and domestic have been implemented [21]. Spectrophotometric methods for nitrate determination have been developed over the past several decades. Popular methods for nitrate analysis utilize ion-exchange chromatography combined with spectrometric, conductometric or electrochemical detection and are suitable for a wide range of environmental samples [22]. On the other hand, nitrites are widely used for food preservation and for fertilization of soils. However, continuous consumption of these ions can cause serious implications on human health, particularly because it can react irreversibly with hemoglobin [23].

Moorcroft et al. [23] had written a review article on the current strategies employed to facilitate the detection, determination and monitoring of nitrate and/or nitrite presence. This review shows that the increasing demand for rapid on-site analysis will ensure the continued development of both spectroscopic and electrochemical methods, which are more applicable to miniaturization and remote operation. Thereby the biosensors development goes against to the conclusion of Moorcroft et al [23]. Moreover, for the analysis of these compounds, biosensors based on nitrate reductase enzyme have been developed. Truly amperometric enzyme sensors for this nitrogen compounds were first described in 1994. Nitrate reductase (NR) (EC 1.7.1.3) catalyzes NAD(P)H reduction of nitrate to nitrite.

Silveira et al. [24] had developed a non-mediated electrochemical biosensor for nitrite determination. The instrument was based on the stable and selective cytochrome c nitrite reductase (ccNiR) from *Desulfovibrio desulfuricans* (ATCC 27774), which has both high turnover and heterogeneous electron transfer rates. This biological element performs the fast six electron reduction of nitrite to ammonia. This biosensor had 120nM as a limit detectable concentration and keeping a stable response up to two weeks. Other nitrite biosensor was developed by Rosa et al. [25]. It was an optical biosensor based on cytochrome cd1 nitrite reductase, from the aerobic denitrifier *Paracoccus pantotrophus*, immobilized in controlled pore glass beads. The developed biosensor operates by measuring the optical reflectance of nitrite reductase, which shows spectroscopic changes when nitrite reversibly binds to the reduced form and oxidizes the enzyme. A biosensing device was developed to detect nitrite in water and obtained detection limit of 0.93μM (nitrite concentration lower than the permissible in water imposed by European Community regulations that is 2.2μM).

A highly sensitive, fast and stable conductimetric enzymatic biosensor for the determination of nitrate in waters was described in Wang et al. [26-27]. Conductimetric electrodes were modified by methyl viologen mediator mixed with nitrate reductase from *Aspergillus niger* by cross-linking with glutaraldehyde in the presence of bovine serum albumin and Nafion® cation-exchange polymer, allowing retention of viologen mediator. A linear calibration curve in the range of 0.02 and 0.25 mM with detection limits of 0.005 mM nitrate was obtained. When stored in pH 7.5 phosphate buffer, the sensors showed good stability over two weeks.

3. Biosensors for pesticides determination

A pesticide, as defined by the EPA, is any substance or mixture of substances intended for preventing, destroying, repelling, or lessening the damage of any pest [28]. Of all the environmental pollutants, pesticides are the most abundant, present in water, atmosphere, soil, plants, and food [3].

Pesticides (insecticides, fungicides and herbicides) are used worldwide due to their wide range of activity. They are released intentionally into the environment and, through various processes, end up contaminating it. The presence of pesticide residues and metabolites in food, water and soil currently represents one of the major issues in environmental chemistry research [29]. Due to their increasing use in agriculture, pesticides are among the most important environmental pollutants. So, the continuous monitoring for low pesticide levels in food, water, and air has become a key activity in respect to human health [30].

Three of the main classes of pesticides that pose a serious problem are organophosphates (OPs), organochlorines and carbamates. OPs are usually esters, amides or thiol derivatives of phosphoric, phosphonic or phosphinic acids. These compounds commonly used includes parathion, malathion, methyl parathion. chlorpyrifos, diazinon, dichlorvos, phosmet, fenitrothion, tetrachlorvinphos and azinphos methyl [8].

While pesticides are associated with many health effects, there is a lack of monitoring data on these contaminants. Traditional chromatographic methods, as High Performance Liquid Chromatography (HPLC), are effective for the analysis of pesticides in the environment, but have limitations and prevent adequate monitoring [31]. Due to the restrictions in conventional methodologies, the development of biosensors for direct and indirect pesticide detection is of particular interest.

Different types of pesticides used in food production can accumulate in fatty tissue in animals – including humans, while the excessive use of fertilizers contaminates ground water with nitrates, nitrites and phosphates. The majority of the insecticides used are acetylcholinesterase (AChE) inhibitors, 55% of them belong to the group of organophosphates and 11% to carbamates, whereas the others are pyrethroids, chlorinated hydrocarbons or other insecticides [32]. The acetylcholinesterase (EC 3.1.1.7) catalysed hydrolysis of acetylthiocholine generates the electroactive product thiocholine. The current of its oxidation is recorded amperometrically at a potential of +0.80 V/SCE [33].

As mentioned, applied to the monitoring of pesticides in the environment and in many foods, the biosensors typically employ the enzyme acetylcholinesterase, monitoring the occurrence of enzymatic inhibition by organophosphates and carbamates, as these chemical compounds bind to the active center of the enzyme, preventing the hydrolysis reaction of acetylcholine into choline and acetate [1].

For the detection of herbicides such as phenyl urea and triazines, which inhibit photosynthesis, biosensors have been designed with membrane receptors of thylakoid and chloroplasts or complete cells such unicellular alga, for which mainly amperometric and optical transductors have been employed [10]. Furthermore, enzymes like cholinesterase (AChE, BChE), organo-phosphorus-hydrolase (OPH), and urease are used in the design of electrochemical biosensors for pesticides detection. These cholinesterase enzymes have different substrates: AChE preferentially hydrolyzes acetyl esters, such as acetylcholine, whereas BChE hydrolyzes butyrylcholine.

A great number of research studies report the development of biosensors for detecting pesticides based on AChE enzymatic inhibition as shown in Table 2.

Biological component	Transductor	Analyte	Limit detection	Reference
AChE immobilized on multiwalled carbon nanotubes	Amperometry	Carbaryl	4.0µM	[34]
AChE-choline oxidase on a gold-platinum bimetallic nanoparticles	Electrochemical impedance spectroscopy	Paraoxon ethyl, aldicarb and sarin	150-200nM, 40-60 µM and 40-50nM	[35]
AC1.W2.R1/ACCHE sensors with the help of Biosensor Toxicity Analyzer	Amperometry	Residual pesticides (organophosphorous and carbamate) on cotton	-	[36]
AChE immobilized on polyaniline and multiwalled carbon nanotubes	Chronoamperometry	Carbaryl and methomyl	around 10µM	[30]

Table 2. Biosensors used in pesticides detection based on the inhibition of enzymatic activities.

Sassolas et al. [37] had developed a review of biosensors for pesticide detection. The authors discussed that the molecular imprinted polymers (MIPs) are innovative affinity-based recognition elements that are exploited for the development of environmental sensors. MIPs have been used as artificial recognition elements of biosensors for pesticide detection. These synthetic materials can mimic the function of biological receptors but with less stability constraints and can provide high sensitivity and selectivity while maintaining excellent thermal and mechanical stability. In this context, Jenkins et al. [38] had constructed a pesticide sensor with detection limits less than 10ppt and a response time of less than 15 min.

4. Biosensor applications in the food industry

The food industry needs suitable analytical methods for process and quality control. The determination of chemical and biological contaminants in foods is of paramount importance to the health of food because, unlike the contamination of a physical nature, they cannot be displayed. Apart from a few important analytes, such as sugars, alcohols, amino acids, flavours and sweeteners, food applications mainly focus on the determination of contaminants. Therefore, it is necessary to invest in the development of biosensors to the analysis of the quality of food, since they have proven to be an extremely viable alternative to traditional analytical techniques such as chromatography. However, very few biosensors play a prominent role in food processing or quality control. Considerable effort must be made to develop biosensors that are inexpensive, reliable, and robust enough to operate under realistic conditions [39].

The potential uses of biosensors in agriculture and food transformation are numerous and each application has its own requirements in terms of the concentration of analyte to be measured, required output precision, the necessary volume of the sample, time required for the analysis, time required to prepare the biosensor or to reuse it and cleanliness requirements of the system [9].

In the area of food the interest in the development of biosensors mainly focuses on analysis of food security (detection of compound contaminants, allergens, toxins, pathogens, and additives etc.) Food composition and online process control (Table 3) [40].

Food safety		
Xenobiotic compounds	Bacterial toxins:	Pathogenic microorganisms:
• Additives	• Mycotoxins	• Virus
• Drugs	• Marine Toxins	• Bacteria
• Pesticides and fertilizers		• Protozoa
• Other contaminants: dioxins, PCB's, PAH's, heavy metals and biotoxins		
Food quality		
Composition of food:	Lifetime:	
• Sugars	• Polyphenols and fatty acids (rancid)	
• Amino acids	• Sugars and organic acids (maturation)	
• Alcohols	• Biogenic Amines (index freshness)	
• Organic acids	• Aliina (garlic and onions)	
• Cholesterol		
Process control		
• Sugars (fermentation and pasteurization) • Amino acids (fermentation)		
• Lactic acid (cheese production) • Alcohols (fermentation)		
Other applications		
• GMO's • Animal Reproductive Cycle		

Table 3. Main areas applying biosensors technologies in food industry [40].

Quality control is of paramount importance in food and beverage industries. In food and fermentation processes, quick and reliable analytical methods are required to analyze sugars like glucose, fructose and sucrose for better process efficiency and economy Also, the recent demands for high quality food products to meet the customer needs have opened up newer and improved sensor technologies that are coupled with production processes for quality control and consumer assurance.

The concept of food safety involves ensuring the production and marketing of harmless food, and by that way ensure the health of the consumer. The quantity and types of food additives incorporated into food products are regulated by the legislation of each country, their detection and quantification are important to prevent fraud and malpractice by manufacturers, allergies and other adverse effects to determined groups of the population [10]. Because of this, special attention has been given the way to detect the presence of contaminants, such as residues of heavy metals and components antinutritional.

On the other hand, foods can naturally present anti-nutritional compounds that can generate disorders in the consumer, given that they hinder absorption and metabolize distinct nutrients causing them to have a deficiency. Antinutritional components (oxalate and glycoalkaloids) or allergen (gluten) can be contained naturally in foods. First mentioned are mostly detected by enzymatic amperometric biosensors, while for allergens are described imunosensors [40]. Table 4 presents some examples of biosensors used in the detection of anti-nutrients.

Biological component	Transductor	Analyte	Reference
oxalate oxidase immobilized on chitosan	Potenciometric	Oxalate	[41]
oxalate oxidase and peroxidase	Amperometry	Oxalate in urine	[42]
oxalate oxidase immobilized on gold nanoparticles	Amperometry	Oxalate	[43]
β-glucosidase	Potenciometric	amygdalin	[44]
Peroxidase	Potenciometric	amygdalin	[45]

Table 4. Biosensors used in anti-nutrients detection.

Oxalic acid is of great importance in food industries and clinical analysis. An increase in oxalate excretion through urine indicates hyperoxaluria, renal failure, kidney lesions and pancreatic insufficiency. The ingestion of a large quantity of food rich in oxalic acid can cause loss of calcium in the blood as well as injury to the kidneys [43,46]. Many methods have been recommended for oxalate determination in clinical laboratory analyses but some of them are time-consuming (as chromatographic and spectrophotometric) while some others need a chemically pre-treated sample [42].

Cyanogenic glycosides, such as amygdalin, are found in a wide variety of plants. Although cyanogenic glycosides are not toxic as it is, cyanide liberated from them as a result of hydrolysis has acute toxicity as is well known [45]. The design of a simple D-amygdalin biosensor is important for its applications, namely in the analytical monitoring of this cyanogenic glyco-side. Several dried fruits like bitter almonds or kernels etc. contain amygdalin. Cyanoglyco-sides yield glucose, benzaldehyde and hydrocyanic acid when hydrolysed in vitro by mineral acids or in vivo by enzymes. Unexpectedly, efluents of food and feed production can also contain cyanide [44].

4.1. Commercially available biosensors for food industry

Despite the large number of publications on biosensors used in food analysis, only a few systems are commercially available (Table 5). Among some limitations that must be overcome are the limited lifespan of biological components, mass production, as well as convenience in handling. However, these problems can be managed in a near future, since the biosensors provide unique solutions for food analysis in terms of specificity and time saving [47].

Company	Biosensor	Country
Oriental electric	Fish deterioration tracking	China
Massachusetts Institute of Technology	Detection of *Escherichia coli* 0157:H7 in lettuce (Canary)	USA
Michigan State University's Electrochemical Biosensor	Detection of *Escherichia coli* O157:H7 and *Salmonella* in meat products in USA	USA
Georgia Research Tech Institute	Detection of *Salmonella* and *Campylobacter* in pork industry	USA
Naval Research Laboratory	Detection of *Staphylococcal* enterotoxin B and *Botuminum* toxin A in tomatoes, sweet corn, beans and mushrooms	USA
Universitat Autònoma de Barcelona in collaboration with CSIC	Detection of atrazine traces	Spain
Molecular Circuitry Inc.	*Escherichia coli* 0157, *Salmonella, Listeria* and *Campylobacter*	USA
Research International	Proteins, toxins, virus, bacteria, spores and fungi (simultaneous analysis)	USA
Universal Sensors	Ethanol, methanol, glucose, sucrose, lactose, L-aas, glutamine, ascorbic acid and oxalate	USA
Texas Instruments Inc.	Penaut Al ergens, antibiotics	USA
Yellow Springs Instruments Co	Glucose, sucrose, lactose, L-lactate, galactose, L-glutamate, ethanol, H_2O_2, starch, glutamine and choline	USA
Affinity Sensors	*Staphylococcus aureus* and cholera toxin	UK
Ambri Limited	Pathogens such as *Salmonella e Enterococcus*	USA
Biacore AB	Water soluble vitamins, chemical veterinary residues and mycotoxins	Sweden
BioFutura Srl	Glucose, fructose, malic acid and lactic acid (fermentation)	Italy
Biomerieux	Microorganisms	France
Biosensor Systems Desing	Microorganisms and toxic substances	USA
Biosensores S.L.	Toxic substances	Spain
Chemel AB	Glucose, sacarose, ethanol, methanol and lactose	Sweden
IVA Co Ltd	Heavy metals	Rusia
Motorola	Microorganisms and GMO's	Japan
Inventus Bio Tec	Ascorbic acid	Germany

Table 5. Commercial biosensors for food industry. Modified from [40,47].

5. Biosensors for benzoic acid detection

The concept of food security implies the production and marketing of food products that offer no risk to consumer health. The use of additives has become an increasingly common practice in the food industry seeking greater lifetime favoring storage and long distance transport [48].

Due to the growing demand for processed food, the use of preservatives has been gaining importance in modern food technology. Benzoic acid as well as salts, benzoates Na and K, are among the most widely used preservatives to inhibit microbial growth, depending on the cost-benefit [49]. Benzoates constitute very important group of food additives to protect the consumer from microbiological risks of some bacteria, fungi and yeasts which may be responsible for poisoning [50].

Given the wide use of benzoic acid and its salts (benzoates) as preservatives in the food industry, detection and quantification of these are of great importance in controlling product quality in order to prevent fraud and improper manufacturing practices, considering the possible adverse effects those including preservatives, exacerbation of symptoms of chronic rhinitis, asthmatic reactions, hyperactivity in children, genotocixidade, clastogenicity and mutagenicity (in human lymphocytes) [51-54]. Furthermore benzoates may undergo decarboxylation in the presence of ascorbic acid under certain conditions to form benzene beverages [55-56].

Although considered safe for health, studies suggest that the consumption of these preservatives is related to a number of adverse health effects, especially in children and susceptible people. The acute toxicity is low benzoates. However, cases of urticaria, rhinitis, asthma and anaphylaxis have been reported after oral, dermal or inhalation. In sensitive people, even at doses lower than 5 mg / kg may cause immunologic reactions do not contact (pseudo-allergy) [50].

In the case of food additives, it is important to note that the toxicity or the benefits depend on the extent to which this food components has adsorption, metabolism and excretion affected as a whole, since there are synergistic interactions among them, the limits for human consumption can be changed [10]. Some studies have reported cases of allergic cross-reactions benzoates and other additives such as sunset yellow, for example [53].

Aiming to make the analysis of benzoates in food, faster, cheaper and simpler, there has been development of various methodologies described in the literature in terms of biosensors as promising alternatives to conventional methods (Table 6). Thus the immense development opportunities and market potential, has driven research analysis methods to the area of food in terms of biosensors [65].

Amperometric biosensors based on enzymes have emerged in the last decade, with the possibility of very promising application in the food and beverage industry. These devices are usually highly selective, sensitive, relatively inexpensive and easy to handle system integration and continuous analysis. Successful application of these sensors for industrial purposes, however, requires a sensor design which meets the specific needs monitoring of the target

Biosensor/principle	Detection range/sample	References
Mushroom tissue homogenate/ enzyme inhibition	25–100 µM (non-alcoholic)	[57]
PPO-PANI/enzyme inhibition	0.0366 mg/L (non-alcoholic)	[58]
PPO-PANI- Pan/enzyme inhibition	2×10^{-7}M (milk, yogurth andnon-alcoholic beverages)	[59]
PPO-Teflon/grafite/enzyme inhibition	9.0×10^{-7} M (mayonnaiseand non-alcoholic beverages)	[60]
Carbon Electrode modified with PPO- nano-CaCO3/enzyme inhibition	5.6×10^{-7} - 9.2×10^{-5} M (yogurth, and non-alcoholic beverages)	[61]
Glassy Carbon Electrodemodified with PPO/CaHPO$_4$enzyme inhibition	˙ 19.6-132 mg/L (non-alcoholic beverages, mayonnaise)	[62]
PPO-Gel made of Titanium modifiedwith carbon nanotubes /enzyme inhibition	0.33 mM – 1.06 mM (non-alcoholic beverages)	[63]

Table 6. Biosensors for detection benzoate/benzo c acid in food samples.

analyte in the particular application, since each individual application requires different operating conditions and characteristics of sensor [64].

Addition of enzyme activity, enzyme inhibition can also be used as analytical signal to be monitored in biosensors thus the activity is measured before and after inhibition promoted by a specific inhibitor in a given time range [66]. The percentage inhibition of enzyme is sustained by quantitatively related to the concentration of the inhibitor [67]. Generally, biosensors developed for detection of benzoic acid in food, are based on the principle of enzyme inhibition [67].

The characteristics of biosensors are dependent biological element of the recognition and signal transducer used as well as the communication between these two elements, which means that although promising, have limitations and drawbacks as well as any other methods of analysis [68].

6. Biosensors for heavy metals

Heavy metals are the most dangerous environmental contaminants, which present a serious threat to human health, even in trace quantities. Contamination of soils due to discharge of industrial effluents is one of the most significant problems faced by man. Heavy metals are widely existent in these contaminated environments. For example, many places are considerably polluted with chromium from tannery waste waters. In these areas, chromium exists in both the hexavalent and the trivalent forms. The plants grown in such areas can accumulate chromium ions. These ions have certain threshold levels for essential functions of living organism and man, but cause toxic actions if the tolerance levels exceed [69].

Moreover, fertilizer has become one of the polluting sources of heavy metals. So repetitive applications of commercial fertilizers and pesticides continually for agriculture have contributed to a continuous accumulation of heavy metals in soils. The trace metal content of commercial fertilizers is also highly variable, depending mainly on the phosphate rock source and the fertilizer production process. The heavy metals in fertilizer can endanger the human body by the crop containing heavy metals [70-71].

The majority of existing techniques used for trace analysis of heavy metals includes spectroscopic, voltammetric and chromatographic methods, which can detect species at low concentrations or even in single elements. However, all of these traditional methods are generally expensive and can hardly be used for in situ analysis. Recently, the ability to detect heavy metal contaminants using biosensors for in situ analysis has gained much interest [72].

Soldatkin et al. [73] presented by a biosensor composter a differential pair of planar thin-film interdigitated electrodes, deposited on a ceramic pad, (used as a conductometric transducer) together with the three-enzyme system (invertase, mutarotase, glucose oxidase), immobilized on the transducer surface, (used as a bioselective element). The developed biosensor demonstrated the best sensitivity toward ions Hg^{2+} and Ag^+.

Ravikumar et al. [74], designed and applied molecular biosensor for heavy metals, zinc and copper, for use in bioremediation strategies. Bacteria utilize two component systems to sensor change in the environment by multi signal components incluing heavy metals and control gene expression in response to changes in signal molecules.

7. Conclusions

The modern environmental and food analysis requires sensitive, accurate, and express methods. The growing field of the biosensors represents an answer to this demand. Unfortunately, most biosensor systems have been tested only on distilled water or buffered solutions, but more biosensors that can be applied to real samples have appeared in recent years. In this context, biosensors for potential environmental and food applications continue to show advances in areas such as genetic modification of enzymes and microorganisms, improvement of recognition element immobilization and sensor interfaces.

Acknowledgements

The authors thank the financial support of the National Council for Scientific and Technological Development (CNPq), CAPES (Coordination for the Improvement of Higher Level Personnel) and the Foundation for Research of the State of Rio de Janeiro (FAPERJ).

Author details

Lívia Maria da Costa Silva[1], Vânia Paula Salviano dos Santos[1], Andrea Medeiros Salgado[1*] and Karen Signori Pereira[2]

*Address all correspondence to: andrea@eq ufrj.br

1 Laboratory of Biological Sensors, Biochemical Engineering Department, Chemistry School, Technology Center, Federal University of Rio de Janeiro, Ilha do Fundão, Rio de Janeiro, Brazil

2 Laboratory of Food Microbiology, Biochemical Engineering Department, Chemistry School, Technology Center, Federal University of Rio de Janeiro, Ilha do Fundão, Rio de Janeiro, Brazil

References

[1] Silva LMC, Melo AF, Salgado A.Biosensors for environmental applications, (2011). Environmental Biosensors, Vernon Somerset (Ed.), 978-9-53307-486-3InTech.

[2] Rogers, K. R, & Gerlach, C. L. Environmental biosensors: A status report, Environ. Sci. Technol. 30 ((1996). , 486-491.

[3] Rodriguez-mozaz, S, & Marco, M-P. Alda MJL, Barceló D. Biosensors for environmental applications: future development trends, Pure Appl. Chem. (2004). , 2004(76), 723-752.

[4] Rogers, K. R. Recent advances in biosensor techniques for environmental monitoring. Anal. Chim. Acta (2006). , 568, 222-231.

[5] Thévenot, D. R, Toth, K, Durst, R. A, & Wilson, G. S. Electrochemical biosensors: Recommended definitions and classification, Pure Appl. Chem. 71 ((1999). , 2333-2348.

[6] Korostynska, . , Monitoring of nitrates and phosphates in wastewater: current technologies and further challenges, International Journal on Smart Sensing and Intelligent Systems, v.5, n.1, pp.149-176, 2012

[7] Rai, . , Implications of nanobiosensors in agriculture, Journal of Biomaterials and Nanobiotechnology, v.3, pp. 315-324, 2012

[8] Gahlaut, . , Electrochemical biosensors for determination of organophosphorus compounds: review, Open Journal of Applied Biosensor, v.1, pp. 1-8, 2012

[9] Velasco-garcía, M. N, & Mottram, T. (2003). Biosensor technology addressing agricultural problems, Biosyst. Eng, 84, , 1-12.

[10] Cock, L. S. Arenas, AMZ, Aponte, AA. Use of enzymatic biosensor as quality indices: a synopsis of present and future trends in the food industry, Chilean Joural of Agricultural Research (2009 6). , 2009(69), 2-270.

[11] Guilbault GG Montalvo, J. Urea specific enzyme electrode. Journal of the American Chemical Society, (1969). , 91, 2164-2569.

[12] Arnold, MA Glazier, SA Jack. Bean meal as biocatalyst for urea biosensors. Biotechnology Letters, n. 5, (1984). , 6, 313-318.

[13] Takishima, K, Suga, T, & Mamiya, G. (1988). The structure of jack bean urease. European Journal of Biochemistry, , 175, 151-165.

[14] Panpae, K, Krintrakul, S, & Chaiyasit, A. Development of a urea potentiometric biosensor based on gelatin-immobilized urease. Kasetsart Journal (Natural Science) (2006). , 2006(40), 74-81.

[15] Chirizzi, D, & Malitesta, C. Potentiometric urea biosensor based on urease immobilized by an electrosynthesized poly (o-phenylenediamine) film with buffering capability. Sensors and Actuators B: Chemical (2011 1). , 2011(157), 1-211.

[16] Nikoleli, G-P, Nikolelis, D. P, & Methenitis, C. Construction of a simple optical sensor based on air stable lipid film with incorporated urease for the rapid detection of urea in milk. AnalyticaChimicaActa (2010). , 2010(675), 58-63.

[17] Trivedi, U. B, Lakshminarayana, D, Kothari, I. L, Patel, N. G, Kapse, H. N, Makhija, K. K, Patel, P. B, & Panchai, C. J. Potentiometric biosensor for urea determination in milk. Sensors and Actuators B, (2009). , 140, 260-266.

[18] Hedayatollah, G, Ahmad, M. R, & Hossein, E. A conductometric urea biosensor by direct immobilization of urease on Pt electrode. Iranian Journal of Chemistry & Chemical Engineering, n. 2, (2004). , 23, 55-63.

[19] Vostiar, I, Tkac, J, Sturdik, E, & Gemeiner, P. Amperometric urea biosensor based on urease and electropolymerized toluidine blue dye as a pH-sensitive redox probe. Bioelectrochemistry, n. 1-2, (2002). , 56, 113-115.

[20] Verma, N, Kumar, R, & Kumar, M. S. Simple, qualitative cum quantitative, user friendly biosensor for analysis of urea, Advances in Applied Science Research, (2012). , 135-141.

[21] Rodriguez-mozaz, S, Alda, M. J, Marco, M. P, & Barceló, D. A global perspective: Biosensors for environmental monitoring, Talanta. 65 ((2005). , 291-297.

[22] Cho, S-J, Sasaki, S, Ikebukuro, K, & Karube, I. A simple nitrate sensor system using titaniumtrichloride and an ammonium electrode, Sensors and Actuators B, (2002). , 85, 120-125.

[23] Moorcroft, M. J. Davis J Compton RG, Detection and determination of nitrate and nitrite: A review. Talanta (2001). , 2001(54), 785-803.

[24] Silveira, C. M, Gomes, S. P, & Araújo, A. N. Montenegro MCBSM, Todorovic S, Viana AS, Silva RJC, Moura JJG, Almeida MG, An efficient non-mediated amperometric biosensor for nitrite determination, Biosensors and Bioelectronics, (2010). , 25, 2026-2032.

[25] Rosa, C. C, Cruz, H. J, Vidal, M., & Oliva, A. G. Optical biosensor based on nitrite reductaseimmobilised in controlled pore glass, Biosensors & Bioelectronics, (2002). , 17, 45-52.

[26] Wang, X, Dzyadevych, S. V, Chovelon, J. M, Jaffrezic-renault, N, Ling, C, & Siqing, X. Conductometric nitrate biosensor based on Methyl viologen/Nafion®/Nitrate reductaseinterdigitated electrodes. Talanta 69 ((2006). , 450-455.

[27] Wang, X, Dzyadevych, S. V, Chovelon, J. M, Jaffrezic-renault, N, Ling, C, & Siqing, X. Development of conductometric nitrate biosensor based on Methyl viologen/ Nafion® composite film, Electrochem. Commun. 8 ((2006). , 201-205.

[28] Tothill, L. E. Biosensors developments and potential applications in the agricultural diagnosis sector. Comput. Electron. Agric. 30 ((2001). , 205-218.

[29] Mostafa, G. A. Electrochemical biosensors for the detection of pesticides, The Open Electrochemistry Journal, (2010). , 2, 22-42.

[30] Cesarino, I, & Moraes, F. C. Lanza MRV, Machado SAS, Electrochemical detection of carbamate pesticides in fruit and vegetables with a biosensor based on acetylcholinesteraseimmobilised on a composite of polyaniline-carbon nanotubes, Food Chemistry, (2012). , 135, 873-879.

[31] Van Dyk, J. S. Pletschkeb. Review on the use of enzymesfor the detection of organochlorine, organophosphate and carbamatepesticides in the environment. Chemosphere, Elmsford, (2011). , 82, 291-307.

[32] Schulze, H, Scherbaum, E, Anastassiades, M, Vorlova, S, Schmid, R. D, & Bachmann, T. T. Development, validation, and application of an acetylcholine esterase biosensor test for the direct detection of insecticide residues in infant food. Biosensors and Bioelectronics (2002). , 2002(17), 1095-1105.

[33] Stoytcheva, M, Zlatev, R, Ovalle, M, Velkova, Z, Gochev, V, & Vladez, B. Electrochemical biosensors for food quality control, (1998).

[34] Cai, J, & Du, D. A disposable sensor based on immobilization of acetylcholinesterase to multiwall carbon nanotube modified screen-printed electrode for determination of carbaryl, Journal of Applied Electrochemistry, n. 9, (2008). , 38, 1217-1222.

[35] Upadhyay, S, Rao, G. R, Sharma, M. K, Bhattacharya, B. K, Rao, V. K, & Vijayaraghavan, R. Immobilization of acetylcholineesterase-choline oxidase on a gold-platinum bimetallic nanoparticles modified glassy carbon electrode for the sensitive detection of organophosphate pesticides, carbamates and nerve agents, Biosens. Bioelectron., n.4, (2009). , 25, 832-838.

[36] Hassan SZU, Militky J,Acetylcholinesterase based detection of residual pesticides on cotton, American Journal of Analytical Chemistry, (2012). , 3, 93-98.

[37] Sassolas, A, Prieto-simón, B, & Marty, J-L. Biosensors for pesticide detection: new trends, American Journal of Analytical Chemistry, (2012). , 3, 210-232.

[38] Jenkins, A. L, Yin, R, & Jensen, J. L. Molecularly imprinted polymer sensors for pesticide and insecticide detection in water, Analyst, n. 6, (2001). , 126, 798-802.

[39] Luong JHT, Groom CA, Male KB.The potential role of biosensors in the food and drink industries. Biosens. and Bioelectron. (1991). , 1991(6), 547-554.

[40] Rumayor, V. G, Iglesias, E. G, Galán, O. R, & Cabezas, L. G. Aplicaciones de biosensores en la industria agroalimentaria. Madrid: Comunidad de Madrid y la UniversidadComplutense de Madrid; (2005).

[41] Benavidez, T. E, Alvarez, C, & Baruzzi, A. M. Physicochemical properties of a mucin/chitosan matrix used for the development of an oxalate biosensor. Sensors and Actuators B (2010). , 2010(143), 660-665.

[42] Milardovic, S, & Kerekovic I, . . A novel biamperometric biosensor for urinary oxalate determination using flow-injection analysis. Talanta 2008 (77) 222-228.

[43] Pundir, C. S, & Chauhan, N. Rajneesh, Verma M, Ravi. A novel amperometric biosensor for oxalate determination using multi-walled carbon nanotube-gold nanoparticle composite. Sensors and Actuators B (2011). , 2011(155), 796-803.

[44] Merkoçi, A, Braga, S, Fábregas, E, & Alegret, S. A potentiometric biosensor for D-amygdalin based on a consolidated biocomposite membrane. AnalyticaChimicaActa (1999). , 1999(391), 65-72.

[45] Tatsuma, T, Tani, K, Ogawa, T, & Oyama, N. Interference-based amygdalin sensor with emulsin and peroxidase. SensorsandActuators B (1998). , 1998(49), 268-272.

[46] Fiorito, P. A, & Cordoba, S. I. Otimized Multilayer oxalate Biosensor, Talanta, (2004). , 62, 649-654.

[47] Furtado, R. F. Dutra RAF, Alves CR, Pimenta MGR, Guedes MIF. Aplicações de Biossensores na Análise da Qualidade de Alimentos. Fortaleza: EMBRAPA; (2008). novembro. Report (1677-1915), 1677-1915.

[48] Kochana, J, Kozak, J, Skrobisz, A, & Wozniakiewicz, M. Tyrosinase biosensor for benzoic acid inhibition-based determination with the use of a flow-batch monosegmented sequential injection system. Talanta. (2012). December; 96(2).

[49] Mello, L. D, & Kubota, L. T. Review of the use of biosensors as analytical tools in the food and drink industries. Food Chemistry. (2002). dezembro 2; 77: , 237-256.

[50] Gustavsson, E, Bjurling, P, & Sternesjö, A. Biosensor analysis of penicillin G in milk based on the inhibition of carboxypeptidase activity. Analytica Chimica Acta. (2002). julho 02;(468): , 153-159.

[51] Leonard, P, Hearty, S, Brennan, J, Dunnea, L, Quinn, J, Chakraborty. T, et al. Advances in biosensors for detection of pathogens in food and water. Enzyme and Microbial Technology. (2003). agosto;(32): , 3-13.

[52] Mutlu, M. Biosensors in Food Processing, Safety, and Quality Control. 1st ed. Mutlu M, editor.: CRC Press; (2010).

[53] Metcalfe, D. D, Sampson, H. A. & Simon, R. A. Food Allergy: Adverse Reactions to Foods and Food Additives. 4th ed. Malden: John Wiley & Sons; (2009).

[54] Shan, D, Li, Q, Xue, H, & Cosnier, S. A highly reversible and sensitive tyrosinase inhibition-based amperometric biosensor for benzoic acid monitoring. Sensors and Actuators B: Chemical. (2008). julho; I(134): , 1016-1021.

[55] Csöoregi, E, Gáspñr, S, Niculescu, M, Mattiasson, B, & Schuhmann, W. Physics and chemistry basis of biotechnology Amsterdan: Springer Netherlands; (2002).

[56] Viswanathan, S, Radecka, H, & Radecki, J. Electrochemical biosensors for food analysis. MonatsheftefürChemie. (2009). Abril; I(140): , 891-899.

[57] Sezgintürk, M. K, Göktug, T, & Dinçkaya, E. Detection of Benzoic Acid by an Amperometric Inhibitor Biosensor Based on Mushroom Tissue Homogenate. Food Technology and Biotechnology. (2005). setembro; IV(43): , 329-334.

[58] Li, S, Tan, Y, Wang, P, & Kan, J. Inhibition of benzoic acid on the polyaniline-polyphenol oxidase biosensor. Sensors and Actuators B: Chemical. (2010). agosto; I(144): , 18-22.

[59] Shan, D, Shi, Q, Zhu, D, & Xue. H. Inhibitive detection of benzoic acid using a novel phenols biosensor based on polyaniline-polyacrylonitrile composite matrix. Talanta. (2007). fevereiro; I(72): , 1767-1772.

[60] Morales, M. D, Morante, S, Scarpa, A, González, M. C, Reviejo, A. J, & Pingarrón, J. M. Design of a composite amperometric enzyme electrode for the control of the benzoic acid content in food. Talanta. (2002). Abril; II(57): , 1189-1198.

[61] Shan, D, Li, Q, Xue, H, & Cosnier, S. A highly reversible and sensitive tyrosinase inhibition-based amperometric biosensor for benzoic acid monitoring. Sensors and Actuators B: Chemical. (2008). julho; II(132): , 1016-1102.

[62] López MSPLópez-Ruiz B. Inhibition Biosensor Based on Calcium Phosphate Materials for Detection of Benzoic Acid in Aqueous and Organic Media. Electroanalysis. (2011). outubro; I(23): , 264-271.

[63] Kochana, J, Kozak, J, Skrobisz, A, & Wozniakiewicz, M. Tyrosinase biosensor for benzoic acid inhibition-based determination with the use of a flow-batch monosegmented sequential injection system. Talanta. (2012). December; 96(2).

[64] Gardner, L. K, & Lawrence, G. D. Benzene Production from Decarboxylation of Benzoic Acid in the Presence of Ascorbic Acid and a Transition-Metal Catalyst. Journal of Agricultural and Chemistry Food. (1993). maio; 41(5): , 693-695.

[65] Moutinho IDLBertges LG, Assis RVC. Prolonged use of the food dye tartrazine (FD&C yellow n° 5) and its effects on the gastric mucosa of Wistar rats. Brazilian Journal of Biology. (2007). fevereiro; 67(1): , 141-145.

[66] Stadle, R. H, & Lineback, D. R. Process-Induced Food Toxicants: Occurrence, Formation, Mitigation, and Health Risks. 1st ed. New Jersey: John Wiley & Sons; (2008).

[67] Marques PRBdOYamanaka H. Biossensores baseados no processo de inibição enzimática. Química Nova. (2008). setembro 9; VII(31): , 1791-1799.

[68] Ene, C. P, & Diacu, E. High-performance liquid chromatography method for the determination of benzoic acid in beverages. U.P.B. Sci. Bull. (2009). , 81-88.

[69] Ntihuga, J. N. Biosensor to detect heavy metals in waste water. Proceedings from the International Conference on Advances in Engineering and Technology, (2006). , 159-166.

[70] Atafar, Z, Mesdaghinia, A, Nouri, J, Homaee, M, Yunesian, M, Ahmadimoghaddam, M, & Mahvi, A. H. Effect of fertilizer application on soil heavymetal concentration, Environ. Monit. Assess., (2010). , 160, 83-89.

[71] Frost, H. L. Jr LHK, Trace metal concentration in durum wheat from application of sewage sludge and commercial fertilizer, Advances in Environmental Research, (2000). , 4, 347-355.

[72] Castillo, J, Gáspár, S, Leth, S, Niculescu, M, Mortari, A, Bontidean, I, Soukharev, V, Dorneanu, S. A, Ryabov, A. D, & Csoregi, E. Biosensors for life quality: Design, development and applications, Sensors and Actuators B: Chemical, n.2, (2004). , 102, 179-194.

[73] Soldatkin, O. O, Kucherenko, I. S, Pyeshkova, V. M, Kukla, A. L, Affrezic-renault, N, Skaya, A. V, Dzyadevych, S. V, & Soldatkin, A. P. Novel conductometric biosensor based on three-enzyme system for selective determination of heavy metal íons. Bioelectrochemistry (2012).

[74] Ravikumar, S, Ganesh, I, Yoo, I, & Hong, S. H. Construction of a bacterial biosensor for zinc and copper and its application to the development of multifunctional heavy metal adsorption bactéria. Process Biochemistry (2012). , 2012(47), 758-765.

Chapter 10

Novel DNA-Biosensors for Studies of GMO, Pesticides and Herbicides

Magdalena Stobiecka

Additional information is available at the end of the chapter

1. Introduction

The modern intensive and highly-efficient agricultural technologies impose strong demands on the crop protection against pests and weeds. Among others, there are two main ways, often controversial, to achieve an efficient crop protection: to use the pesticides and herbicides and/or to modify the crop genome so that it can become immune or less susceptible to the invaders' attacks. This approach is widely applied in industrial countries and is gaining popularity in the rest of the world. Due to the unavoidable toxicity of pesticides and herbicides, including possible genotoxicity and carcinogenicity, there is an urgent need to extend a pollution control and to monitor the levels of pesticides and herbicides in ground water and soil. On the other hand, the genetically modified organisms (GMOs) are the products of a new technology and the effects of genome modification on human health are largely unknown. Hence the GMOs are not permitted or are restricted in some countries. In European Union, the content of GMO's in foodstuff is restricted to 0.9%. Since the Roundup Ready corn GMO, for instance, is cultivated in USA already on more than 80% of the fields, the analysis of food products for GMO becomes necessary. In this Chapter, the progress in the development of new inexpensive analytical sensors for pesticides, herbicides, and GMO is presented and discussed in view of the necessity of environmental pollution control, as concerns to pesticides and herbicides, and in view of the GMO content in foodstuff, as required by the mandated restrictions. The common feature of the sensors discussed is the strong dependence of the analytical signal of the sensors on the interaction of DNA molecules immobilized on the sensor surface with the analyte. In the case of GMOs, this provides a straightforward means for biorecognition and very sensitive identification of the GMO genes. In the case of pesticides and herbicides, it enables testing formulations that include adjuvants in addition to the main pesticides or herbicides, and their effect on DNA, thereby probing the affinity of analytes to DNA and possible DNA damage.

2. Effect of pesticides and herbicides on DNA

The pesticides and herbicides are designed to either kill or disable pests and weeds. They often act directly onto the DNA of pests or weeds and while they do not appear to be immediately harmful to larger animals, the evidence suggest teratogenicity and in some cases they clearly exhibit genotoxic and carcinogenic properties.

The pre-mutagenic DNA modifications leading to DNA damage (strand scission, mutations) often begins with nitrogen base oxidation. It has recently been found [1] that catechol-containing compounds in the presence of copper(II) or iron(II) ions, may induce a Fenton cascade leading to reactive oxygen species (ROS) generation which are potent enough to damage DNA. Currently, more than one half of the world production of catechol is consumed by the pesticide and herbicide industry. The oxidative damage to DNA structure is of critical importance for many biological processes, including aging, mutation, and carcinogenesis [1-4]. A number of pesticides and herbicides interact with DNA and can cause DNA damage [5-12].

In this Chapter, we describe the extent of DNA damage done by herbicide paraquat (PQ) and the effect of another herbicide, atrazine (Atz), on DNA helix using electrochemical DNA-biosensor method [4, 11, 13, 14]. Also we present DNA-based piezoelectric sensors developed for the detection of genetically modified foodstuff [15].

Paraquat is a broadly used herbicide which is highly toxic and acts nonselectively. This herbicide can cause fatal intoxication in humans and animals [16] since it targets the dopaminergic neurons. It has been reported that paraquat may also induce neurodegenerative diseases such as Parkinson's disease [17-22]. There are known extreme cases of PQ causing widespread damage to many organs [23, 24]. We have demonstrated [4] that paraquat can initiate the formation of ROS, such as HO$^•$, O$^{2•-}$, in the presence of H$_2$O$_2$ and induce DNA damage. PQ-mediated DNA damage was found also by Yamamoto and Mohanan [25] and Ali et al. [24]. Tokunaga et al. [26] have reported that paraquat caused oxidation of guanine and increased the 8-hydroxy-deoxyguanosine amount in heart, brain, and lung. Schmuck et al. [27] have shown that PQ can induce an oxidative stress in rat cortical neurons and astrocytes in vitro, leading to the dopaminergic cell death in the nigrostriatum.

Atrazine is another commonly applied herbicide. It is known as an inhibitor of photosystem II (PSII) in plants [28]. Several studies indicate atrazine genotoxicity [7, 8, 29-32]. Because of this, its use has been regulated in many countries [9]. Other studies have shown no toxicity of pure atrazine. There is a growing consensus that atrazine alone may act as a sensitizer [11] increasing the DNA susceptibility to a damage inflicted by various adjuvants of herbicide preparations which then become cytotoxic [32].

3. Biosensors for pesticide/herbicide pollution monitoring and screening of their interaction with DNA

The extensive use of highly toxic herbicides and pesticides and serious risk for the environment and human health compel the development of pollution control. The Food and Agriculture Organization (FAO) and the World Health Organization (WHO) have established maximum residue limits for pesticides in food [33-36].

The standard laboratory procedures for routine analysis of pesticides and herbicides are based on gas chromatography, mass spectrometry and high-performance liquid chromatography (HPLC). The application of these methods requires sophisticated and costly instrumentation, laborious sample preparation, and highly trained personnel. Therefore, there is a pressing necessity for the advancement of new analytical platforms able to assist in the rapid and inexpensive field-deployable testing. It becomes apparent that biorecognition-inspired biosensors based on DNA, antibody, and whole cells can fill the gap:

i.   *Electrochemical DNA-biosensors.* In Hepel's laboratory, the interactions of herbicides and pesticides with DNA have recently been widely studied [11, 12], including among herbicides: atrazine (Atz), paraquat (PQ), glufosinate ammonium (GA), and 2,4-dichydrophenoxyacetc acid [2,4-D), and among pesticides: diflubenzuron (DFB), carbofuran (CF), paraoxon-ethyl (PE). The DNA electrochemical biosensor was also used by Mascini to the determination of intercalating and groove-binding drugs and pollutants including daunomycin, polychlorinated biphenyls (PCBs), aflatoxin B1, cisplatin, atrazine, and hydrazine [37]. Moreover, an electrochemical DNA-biosensor has been used to investigate the interactions between DNA and derivatives of 1,3,5-triazine herbicides [38].

ii.   *Piezoimmunosensors.* The next kind of biosensors used for highly-sensitive, quantitative detection of herbicides has been the antibody based quartz crystal nanobalance biosensor. Halamek et al. [39] have developed a piezoelectric immunosensor for the detection of 2,4-dichlorophenoxyacetic acid [2,4-D) and Pribyl et al. [40, 41] for the detection of atrazine and polychlorinated biphenyls (PCB).

iii. *Cell-based biosensors.* Immensely significant in the determination of environmental pollutants, toxic chemicals, pesticides or water quality assessment has been the development of cell-based biosensors. These excellent analytical tools based mainly on bacteria, algae or yeast have been widely investigated in recent years [42-46]. Li and coworkers have described fluorescence and bioluminescence bacterial biosensors for the determination of petroleum products such as benzene, toluene, ethylbenzene, and xylenes (BTEX) in groundwater and soil samples as an alternative to conventional HPLC and GC-MS methods of BTEX measurement [47]. Naessens et al. [28] have developed a new algal-based fluorescence biosensor for the detection of inhibitor of photosystem II (PSII) herbicides: atrazine, diuron and simazine with the detection limit of 0.1 µg/L.

4. GMO alternative to pesticides and herbicides

Area of the genetically modified cultivations including: soybean, maize, cotton, and rapeseed increases in recent years and more and more of genetically modified plants or their derivatives are involved into the food industry. Mainly, a soybean and a maize are used in the production of food and feed. Many countries, such as Korea, Japan, and Australia, have developed laws controlling the marketing of the GMOs. The European Union (EU) and polish legislation (EC Directive 18/2001, 1829/2003, 1830/2003, 1946/2003] imposes a duty to control GMO by qualitative and quantitative assays. The food products and ingredients, containing transgenic material in percentage higher than 0.9 %, 3%, and 5%, have to be labeled in EU, Korea, and Japan, respectively. However, the labeling of GM foods is not compulsory in the United States and Canada. The genetic modification confers plants novel characteristics which improve their agronomic properties (eg. response to herbicides), quality (taste, maturation, shelf live, color), and pest resistance, including viral, fungal, insect, and parasite resistance. The most common of the genetically modified plants inoculated against herbicide glyphosate are Roundup Ready soybean, maize, and cotton. The Roundup Ready genetic insert contains a portion of the cauliflower mosaic virus (CaMV) 35 promoter, the Petunia hybrid 5-enolpyruvylshikimate-3-phosphate synthase (EPSPS), chloroplast transit peptide (CTP), the CP4 EPSPS coding sequence, and a portion of the 3' non-translated region of the nopaline synthase gene terminator (NOS) [48]. The microbial CP4 gene introduces a glyphosate resistance to the plants. Glyphosate is a broad-spectrum herbicide which controls plants by inhibiting enzyme EPSPS, an essential enzyme in the shikimate pathway which plays a role in the biosynthesis of aromatic amino acids, used in the protein synthesis, cell wall formation, and pathogen defense. Glyphosate is toxic to plants because it prevents the production of tryptophan, tyrosine, and phenylalanine.

5. GMO detection technologies

Presently, the main assays used for GMO detection are DNA- and protein-based methods. In these methods, the genetic modification such as the inserted/altered gene is detected or the product resulting from the genetic modification is identified. DNA-based detection method relies on the inherent ability of the complementary strands to form a double-helix of a double-stranded DNA and may utilize either a Southern Blot or a Polymerase Chain Reaction (PCR) technique [49-55]. For GMO quantification, the Real-Time PCR is used [56, 57]. The methods based on liquid chromatography - mass spectrometry (LC-MS) enable precise analytical measurements. The protein-based detection method relies on finding proteins coded by the transgene. The following techniques are used for the determination of proteins from GMOs: one- and two-dimensional gel electrophoresis, Western Blot, Enzyme Linked Immunosorbent Assay (ELISA) and the lateral flow strip [54, 55]. Rogan et al. have employed immunological methods to measure the 5-enolpyruvylshikimate-3-phosphate synthase (CP4 EPSPS) protein derived from the *Agrobacterium* sp. strain CP4 in the major processed fractions derived from Roundup Ready soybean [58]. All of these methods are very sensitive but costly, requiring very expensive equipment and reagents, and highly experienced personnel. Therefore, the mandated monitoring of GMOs on the market calls for the development of new, fast, and inexpensive analytical biosensing platforms enabling field-testing of crop, foods and feeds.

Recently investigated biosensors based on DNA hybridization [2] may become a viable alternative. The DNA-biosensors are relatively cheap and easy to use. These devices are used in many fields of research including clinical, environmental, and food industryand many reviews evaluating the progress in DNA-biosensors have been published [59-62]. Some of the electrochemical sensors have utilized inherent electroactivity of nucleic acid bases which undergo electrooxidation processes at carbon and mercury electrodes [63-66]. In others, changes in peak current or potential for redox-active probe-molecules, which selectively bind to DNA grooves, electrostatically interact with negative chain, or intercalate into dsDNA helix, are monitored [67-69].

The electrochemical methods of GMO detection based on DNA biosensors have been utilized in the investigations of the Filipiak's group [70-72]. The Authors have been able to detect a specific *bar* gene coding for the resistance to herbicide, phosphinotricin, by using the electrochemical hybridization indicator – [2,2'-bipirydyl]cobalt(III) [68]. This indicator intercalates into the double-stranded DNA after the completion of hybridization of the immobilized 19-mer or 21–mer single-stranded probe DNA (pDNA) with target-gene single-stranded DNA (tDNA). The cathodic signal of $Co(bpy)_3^{3+}$ was significantly higher after the formation of a DNA double-helix from the *bar* gene target and pDNA. Also, the Authors have detected genetically modified plants with a transgenic coding for resistance to kanamycin (nptII). In their investigations, they used an organic dye, methylene blue (MB), which shows considerable affinity toward guanine bases in DNA. After the interaction of the probe with complementary target sequence of nptII gene, the electrochemical signal of this indicator has decreased. Meric et al. have been able to detect the most common insert in GMOs, nopaline synthase terminator (NOS) using DNA biosensor [67]. The Authors have based their investigations on an MB intercalator probe as the hybridization indicator. They tested their sensor with short synthetic oligonucleotides and DNA fragments obtained by PCR amplification. We have investigated the dye Nile Blue, used in DNA staining in carcinoma-cell tumors [73], to evaluate the effect of herbicides and pesticides on DNA [13], [14]. The next kinds of biosensors which are promising for the determination of GMOs are based on the surface plasmon resonance and piezoelectric sensors. The Mascini's group has investigated the promoter [35S] and the terminator elements (T-NOS), which are widely used for the production of many transgenic commercially available vegetables. They have performed a hybridization study using short-oligonucleotide of DNA samples isolated from certified reference materials (CRM), soybean powder, real samples of different dietetic products, which were amplified by PCR, as well as the genomic and plasmidic DNA samples non-amplified by PCR [74-76]. Stobiecka at al. [15], have designed a DNA hybridization biosensor for the determination of genetically modified soybean Roundup Ready using a chemically-modified gold piezoelectrode with single-stranded probe DNA immobilized in sensory films using avidin-biotin binding system.

6. Materials and methods

6.1. Chemicals

All chemicals used for investigations were of analytical grade purity. Avidin, 6-mercapto-1-hexanol (MCH), N-[2-hydroxyethyl]piperazine – N'-[2-ethanesulfonic acid] (HEPES),

$K_3[Fe(CN)_6]$, paraquat (PQ), redox active dye Nile Blue A (NB),mercaptopropionic acid (MPA), N-[3-dimethylaminopropyl)-N'-ethylcarbodiimide (EDC), and ethanolamine were purchased from Sigma-Aldrich Chemical Company (St. Louis, MO, U.S.A. or Poznań, Poland). Atrazine was purchased from Supelco (Bellefonte, PA, U.S.A.). The short synthetic oligonucleotides used in investigations with atrazine and paraquat were obtained from Eurofins MWG/Operon (Huntsville, AL, U.S.A.). Synthetic biotinylated oligodeoxynucleotides used as a probe to the detection of the genetically modified soybean and unbiotinylated oligodeoxynucleotides complementary or noncomplementary to probes were synthesized in the Laboratory of DNA Sequencing and Oligonucleotides Synthesis, IBB PAS, Poland. Samples of fragments of DNA amplified by PCR and genomic DNA were prepared in the Genetic Modifications Analysis Laboratory, IBB PAS, Poland. 3,3'-dithiodipropionic acid di(N-succinimidyl ester) was from Fluka-Sigma-Aldrich (Poznań, Poland). Aqueous solutions were prepared using Millipore Milli-Q deionized water (conductivity $\sigma = 55$ nS/cm) (Billerica, MA, U.S.A.) or Simplicity® 185 Water System (Molsheim, France).

6.2. Quartz crystal nanobalance measurements

For nanogravimetric measurements, a Model EQCN-700 Electrochemical Quartz Crystal Nanobalance from Elchema (Potsdam, NY, U.S.A.) and a Model CHI-410 Time-Resolved Electrochemical Quartz Crystal Microbalance (CH Instruments, U.S.A.) were used. Quartz crystals coated with gold on both sides with resonant frequency of 9.975 MHz or 7.995MHz were used as the substrates for working electrodes, and were obtained from Elchema or CH Instruments, respectively. The geometric surface area of the working electrode was 0.1963 cm² and the apparent-mass changes Δm were related to the fundamental frequency shift Δf using the equation: $\Delta f = 0.8673\Delta m$ or $\Delta f = 1.34\Delta m$ ng, respectively, based on Sauerbrey equation [77-80]:

$$\Delta f = - \frac{2\Delta mn f_0^2}{A\sqrt{\mu_q d_q}} \qquad (1)$$

where, Δf is the change in the resonant oscillation frequency, Δm is the change in the interfacial mass, A is the piezoelectrically active area, n is the overtone number and f_0 is the fundamental frequency which depends on the quartz properties (density, $d_q = 2.648$ g cm⁻³, and shear modulus, $\mu_q = 2.947 \times 10^{11}$ g cm⁻¹ s⁻²) and resonator thickness (here: 0.166 mm)). All experimental variables influencing the resonant frequency [77] of the EQCN electrodes such as the temperature, pressure, viscosity and density of the solution, were kept constant in the apparent mass change measurements.

6.3. Electrochemical measurements

Cyclic voltammetric (CV) measurements were performed with a standard electrochemical setup - a Potentiostat/Galvanostat Model PS-205B with a Data Logger and Control System, Model DAQ-716v, operating under Voltscan 5.0 data acquisition and processing software from Elchema (Potsdam, NY, U.S.A.) or with the Time-Resolved Electrochemical Quartz Crystal Microbalance (CH Instruments, U.S.A.). Potentials were measured versus the double-junction saturated Ag/AgCl reference electrode. As the working electrodes, gold disk electrodes with an area of 1 mm^2, gold coated quartz crystal piezoresonators with a real surface area of 0.264 cm^2 (f_0 = 9.975 MHz) obtained from Elchema, and quartz crystals coated with gold on both sides (f_0 = 7.995 MHz) obtained from CH Instruments, were used. A platinum wire was used as the counter electrode. First, the surfaces of gold electrodes were polished on a flat pad with two kinds of alumina, 0.3 and 0.05 μm dia., in wet alumina slurry (Coating Service Department, Indianapolis, U.S.A.). Next, the electrodes were cleaned electrochemically in deoxygenated solutions of 1M KOH and 0.1 M H$_2$SO$_4$ until the cyclic voltammograms showed no further change. The solutions were deoxygenated by purging with argon. Quartz crystal piezoelectrodes were also cleaned electrochemically.

6.4. Molecular dynamic simulations and quantum mechanical calculations

The molecular dynamics (MD) simulations and quantum mechanical calculation (QC) of electronic structure for a model DNA molecule and herbicides for the analysis of interactions of atrazine and paraquat with DNA, were performed using procedures embedded in Wavefunction Spartan 6 (Irvine, CA, U S.A.).

7. Design of DNA-based biosensors

In Figure 1, a schematic of the biosensor films with DNA immobilized on a gold electrode is presented for:

a.   Au/MPA/dsDNA$_{20\text{-bp}}$ film and

b.   Au/DASE/avidin/RR-gene oligonucleotide film,

where the basal self-assembled monolayer (SAM) film is composed of either the mercaptopropionic acid (MPA) or the 3,3'-dithiodipropionic acid di-(N-succinimidyl ester) (DASE).

In the first kind of biosensor (a), a clean gold electrode was modified with 10 mM mercaptopropionic acid for 1 h. After the activation of carboxyl group in MPA witha 0.1 M N-[3-Dimethylaminopropyl)-N'-ethylcarbodiimide (EDC) solution, the sensor was incubated for 1h with a NH$_2$-modified oligonucleotide probe 5'NH$_2$C$_6$H$_{12}$-ATTCGACAGGGATAGTTC-GA3' with final concentration of 1 μM,to attach it to the thiol film. The hybridization process was performed by injecting 1 μM (final concentration) solution of complementary oligonucleotide 5'TCGAACTATCCCTGTCGAAT3'to PBS solution (pH = 7.4) for 1h. The DNA biosensor prepared in this way was rinsed with 0.02 M PBS buffer (pH = 7.4) and used for testing of DNA damage caused by two herbicides: paraquat (PQ) and atrazine (Atz) using a redox dye Nile Blue (NB) as the probe intercalator and marker of DNA damage. After the interactions of ds DNA helix with herbicides, the DNA based biosensors were discarded and a new modification of gold piezoelectrodes was prepared for next experiments due to the expected DNA damage by atrazine and paraquat. The reproducibility of subsequent modifications was very good (ca. 5 %).

In the second kind of a DNA biosensor (b), a clean gold piezoelectrode was modified with 5 mM solution of 3,3'-dithiodipropionic acid di-(N-succinimidyl ester) (DASE) in chloroform for 2 h. Then, 1 mM solution of 6-mercapto-1-hexanol (MCH) in ethanol was used to block the remaining free surface of the gold electrode for 1 h. Next, the electrodes were modified with 0.2 mg/mL aqueous solution of avidin for 1 h and additionally for 1 h with 1 mM aqueous solution of 2-aminoethanol (pH 8.00). The biotinylated probes 5'-biotin-ATC CCA CTA TCC TTC GCA AGA-3' were injected with final concentration 300 nM. The DNA biosensor prepared in this way was then tested with short complementary and non-complementary synthetic oligonucleotides and PCR products. Following the tests, the sensor was used for the detection of genetically modified soybean Roundup Ready. Sensors were regenerated by immersing the quartz crystal electrode into the HEPES denaturation buffer pH 8.00, for 10 min at 95 °C, and then cooled down on ice or washing thrice with 10 mM NaOH for 2 min. The reproducibility of the hybridization process in the samples non-amplified by PCR, expressed by the average coefficient of variation, was 20%.

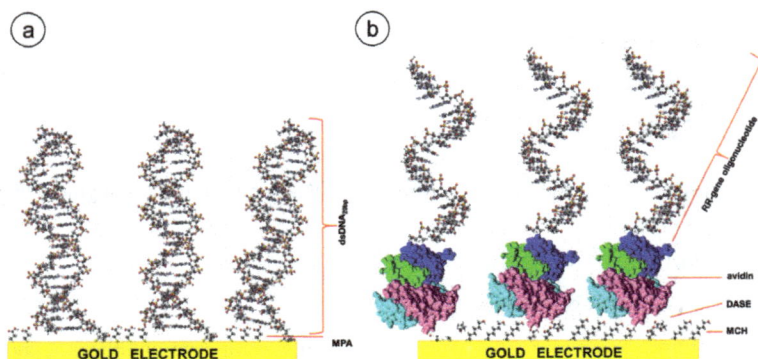

Figure 1. Design of functionalized DNA biosensors for (a) the determination of DNA damage by herbicides (Au/MPA/dsDNA$_{20\text{-bp}}$ film) and (b) the detection of the genetically modified soybean Roundup Ready (Au/DASE/avidin/RR-gene oligonucleotide film).

7.1. Studies of DNA damage by atrazine and paraquat using intercalation redox probe

In Figures 2 and 3, cyclic voltammetric response of the Nile Blue redox dye probe incorporated in sensory films of the DNA-biosensors is presented. This electroactive probe was used for the determination of DNA damage by herbicides atrazine and paraquat (in Figure 2 and 3, respectively). Voltammograms were obtained in pure PBS buffer without NB dye. The sensor was incubated first in a 100 µM solution of NB dye for 10 min, carefully washed in PBS buffer, and tested in the same buffer (curve 2). The NB molecules intercalate into the double-helix structure of DNA. A couple of well resolved redox peaks with cathodic and anodic peak potentials at E_{pc} = -0.41 V and E_{pa} = -0.38 V vs. Ag/AgCl is observed. The linear dependence of the peak currents, I_{pc} and I_{pa}, on the potential scan rate v indicates that the redox peaks correspond to the surface bound NB species [14]. Then, the sensor was incubated in 100 µM solution of herbicide (atrazine or paraquat, respectively) and characterized in a

PBS solution without intercalative dye (curve 3). Next, the sensor was soaked once again in the solution of NB, washed in PBS, and characterized in pure PBS buffer (curve 4). The changes in cathodic peak current (I_{pc})of a Nile Blue probe before and after the interactions of herbicides with DNA were quantified. It is seen that after the interaction of Atz or PQ with DNA, the I_{pc} of a NB bound to DNA has changed considerably. The DNA damage φ was measured as the relative current increase/decrease in the NB uptake after interaction of DNA with herbicide:

$$\varphi = 100 \left(I_{pc,herbicide} - I_{b1}\right) / \left(I_{pc,0} - I_{b0}\right) \tag{2}$$

where, $I_{pc,0}$ and $I_{pc,herbicide}$ are the cathodic peak currents for the reduction of NB intercalated in dsDNA helix on electrode before and after interactions of DNA with herbicides, respectively; and I_{b0} and I_{b1} are the background currents.

7.2. Interaction of atrazine with DNA

After the incubation of a dsDNA sensor in atrazine solution, the cyclic voltammetric characteristic of NB, recorded in pure PBS buffer solution without redox dye, shows a considerable increase of the cathodic current of NB (Figure 2, curve 4) in comparison to the cathodic current of NB before interaction of dsDNA with the herbicide (Figure 2, curve 2) [13],[11].

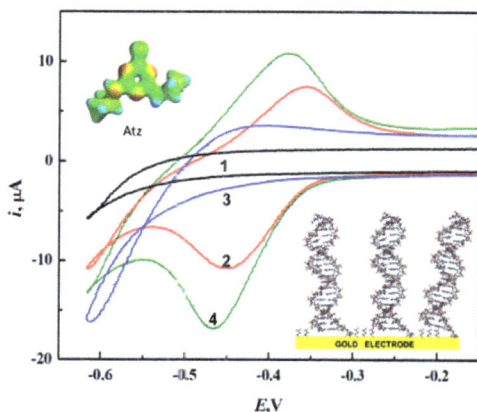

Figure 2. Cyclic voltammograms for an Au/MPA/dsDNA$_{20\text{-bp}}$ film after subsequent treatments: [1] PBS only, [2] after 10 min soaking in 100 µM NB solution; [3] after 40 min soaking in 100 µM atrazine solution; [4] after 10 min soaking in 100 µM NB solution; scan rate v = 400 mV/s, solution: 0.02 M PBS, pH = 7.4. Modified from reference [13] with permission. Copyright 2010 The Electrochemical Society.

It indicates that more molecules of the electroactive dye were intercalated into the DNA helix. It is reasonable to conclude that the higher capacity of dsDNA towards the Nile Blue probe is associated with B-DNA structure altering caused by the herbicide. It is interesting that the second incubation of DNA sensor with NB dyes, after the interaction of dsDNA sensor with atrazine, resulted in so large an increase in the NB uptake (φ = 65%). Separate experiments performed with not fully matched complementary oligonucleotides (C-A mismatched oligonucleotides) have also led to the increased NB uptake but only by φ = 17% [14]. This indicates a high sensitivity of the DNA biosensor proposed. Molecular dynamic simulations have confirmed that atrazine molecules cause underwinding of double-stranded helix of DNA and increase the uptake of Nile Blue redox probe due to the increase in the inter-base spacing in the base stacks.

7.3. Effect of paraquat on DNA biosensor responses

The interactions of paraquat with dsDNA, immobilized on a sensor surface, have been investigated [4]. Changes in cyclic voltammograms of a Nile Blue intercalative redox probe have been observed after the incubation of a DNA-biosensor in a paraquat solutions. For the incubation times of a DNA-biosensor in a paraquat solution for up to 55 min and soaking the biosensors for 10 min in solution of a Nile Blue (100 µM), the peak current of NB reduction was observed to increase in comparison to the peak current of NB before the interactions of DNA with paraquat. The uptake of NB intercalated into the dsDNA helix after the interactions with PQ increased on average by φ = 13.9% (for interaction times of paraquat with dsDNA of 10, 35, and 55 min). This indicates the unwinding of the dsDNA helix, similar to the effect of atrazine on dsDNA, resulting in the increased uptake of NB molecules into the DNA duplex. However, after longer interaction time of paraquat with dsDNA (t = 80 min), the peak current of NB reduction has been found to decrease (Figure 3, curve 4) in comparison to the peak current of the probe in an undamaged dsDNA (Figure 3, curve 2). It clearly indicates on a break and fragmentation of dsDNA caused by long-lasting action of paraquat which leads to the observed diminished uptake of NB probe into the DNA helix, with φ = -32.1%.

7.4. Characterization of the biofilm and testing of EQCN-based DNA biosensor

Each modification step during the construction of a DNA biosensor used for probing the affinity of atrazine to DNA and possible DNA damage was monitored by quartz crystal nanobalance technique (Figure 4). For instance, the immobilization of the mercaptopropionic acid (MPA) on a gold surface of a piezoelectrode has led to a total resonant frequency shift of Δf = 24.3 Hz, corresponding to the apparent mass change of Δm = 21.1 ng and surface coverage γ_{MPA} = 0.76 nmol/cm^2 (A_{geo} = 0.22 cm^2, roughness factor R = 1.2, and real surface area: A = 0.264 cm^2). Next, after activation of the carboxylic groups of the thiol film with EDC, the NH$_2$-modified oligonucleotide probes (pDNA) and complementary to them target oligonucleotides (tDNA) were attached. After the injection of oligonucleotides, the frequency shifts Δf = 73.8 Hz and Δf = 72.3 Hz were observed, for pDNA and tDNA, respectively. The main mass increase was observed within 45 min which was considered as the time necessary to attain stable and full-monolayer coverage of dsDNA biofilm. The dsDNA surface coverage γ_{DNA}= 34 pmol/cm^2 determined from the experimental mass of DNA film was close to the theoretical value for hexagonal close packing and corresponds to the loading of 0.68

nmolbp/cm^2 [13],[11]. The DNA-biosensor prepared in this way was subsequently used for the investigations of the behavior of atrazine and possible DNA damage detection.

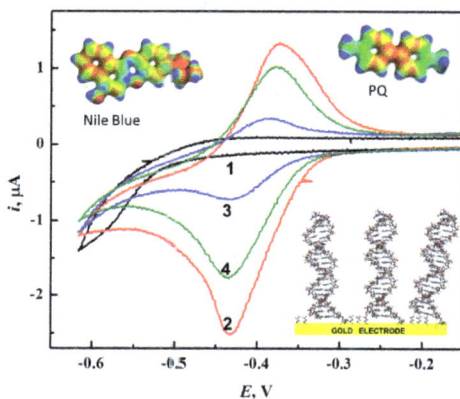

Figure 3. Cyclic voltammograms for an Au/MPA/ dsDNA$_{20\text{-}bp}$ film after subsequent treatments: [1] PBS only, [2] 10 min soaking in NB solution (100 μM); [3] after 80 min soaking in PQ solution (100 μM); [4] after soaking in PQ and 10 min soaking in NB solution (100 μM); scan rate v = 100 mV/s, solution: 0.02 M PBS, pH = 7.4. Modified from reference [4]. Adapted with permission from Antioxidant Effectiveness in Preventing Paraquat-Mediated Oxidative DNA Damage in the Presence of H$_2$O$_2$, M. Stobiecka, A. Prance, K. Coopersmith, M. Hepel, 2011, 211-233, in: S. Andreescu, M. Hepel (Ed.) Oxidative Stress: Diagnostics, Prevention, and Therapy. Copyright 2011 American Chemical Society.

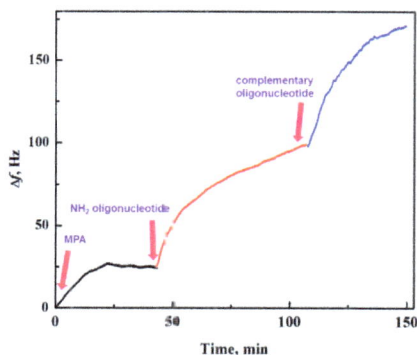

Figure 4. Resonant frequency shift recorded for a gold piezosensor after injection of mercaptopropionic acid (MPA), NH$_2$-modified oligonucleotide (5′NH$_2$C$_6$H$_{12}$-ATTCGACAGGGATAGTTCGA3′) (pDNA) and a complementary oligonucleotide (5′TCGAACTATCCCTGTC GAAT3′) (tDNA); solution: 0.02 M PBS, pH = 7.4; C$_{MPA}$ = 10 mM, C$_{pDNA}$ = 1 μM, C$_{tDNA}$ = 1 μM, all concentrations are the final concentrations Modified from reference [13] with permission. Copyright 2010 The Electrochemical Society.

8. Detection of genetically modified soybean Roundoup Ready by quartz crystal nanogravimetric technique

A DNA biosensor for the determination of genetically modified soybean Roundup Ready was examined using quartz crystal nanobalance technique [15], [69]. In Figure 5, cyclic voltammograms for ferricyanide redox probe obtained after each step of the modification of a gold piezoelectrode are presented. The voltammogram obtained for a bare gold electrode (Figure 5, curve 1) shows a couple of well-developed redox peaks for the marker ion with peak separation, ΔE_p = 111 mV. After forming the 3,3'-dithiodipropionic acid di-(N-succinimidyl ester) (DASE) self-assembling monolayer on a gold surface, a repulsion of the $[Fe(CN)_6]^{3-/4-}$ ions from the film was observed and the peak separation of the redox probe couple in voltammetric characteristics has increased to ΔE_p = 135 mV for the ester-modified gold piezoelectrode (Figure 5, curve 2).

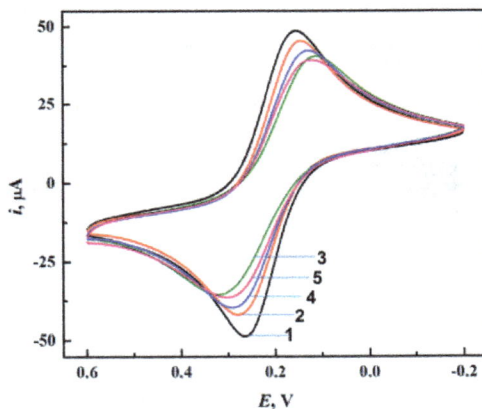

Figure 5. Cyclic voltammograms for a $K_3Fe(CN)_6$ test solution recorded after subsequent steps of the DNA biosensor construction: [1] bare gold piezoelectrode, [2-5] gold piezoelectrode after immobilization of successive layers of [2] 3,3'-dithiodipropionic acid di-(N-succinimidyl ester) (DASE) (5 mM), [3] avidin (0.2 mg/mL), [4] aminoethanol (AET) (1 mM), [5] biotin-oligonucleotide (300 nM), v = 100 mV/s, C_{compl} = 80 nM, $C_{non-compl}$ = 72 nM; the solution composition: 0.1 M phosphate buffer pH = 7.2, 0.1 M KCl and 0.001 M $K_3Fe(CN)_6$, scan rate 0.1V/s. Modified from reference [69], unpublished data

The addition of an avidin solution, has led to further decrease in the marker signal and an increase in the peak separation for ferricyanide ions to ΔE = 208 mV (Figure 5, curve 3). After the immobilization of the protein, the electron transfer of the $Fe(CN)_6^{3-/4-}$ couple has decreased due to the formation of a blocking avidin layer, reduced surface accessibility, and steric hindrance. The immobilization of aminoethanol (AET) molecules promoted an electron transfer between the redox molecules and the electrode surface. Hence, the immobilization of AET on a gold electrode surface resulted in the increase of the redox marker reaction reversibility and a dramatic decrease of the peak separation of the redox probe to ΔE_p = 163 mV (Figure 5, curve 4). The attachment of the biotinylated oligonucleotides resulted in a decrease of the redox response of the electroactive marker and an increase in the peak separation for ferricyanide ions to $\Delta E = 182$ mV. The decrease of the ferricyanide probe signal after the immobilization of oligonucleotides was expected due to repulsions between negatively charged DNA chains and negative ferricyanide ions.

A freshly prepared DNA-biosensor was first functionalized with short synthetic oligonucleotides using quartz crystal nanobalance technique [15] for film formation monitoring. In Figure 6, the hybridization process of short synthetic oligonucleotides, complementary and non-complementary, to the oligonucleotide probe immobilized on the sensor surface is presented. The probe was related to the 5-enolpyruvylshikimate – phosphate synthase (*EPSPS*) gene, which is an active component of an insert integrated into a Roundup Ready soybean. Next, the DNA biosensor was used for testing with a 169 base-pair fragment of *EPSPS* gene extracted from Roundup Ready soybean genome and amplified by PCR, and with a non-complementary 138 base-pair fragment amplified by PCR on maize alcohol dehydrogenase gene template. Finally, the DNA biosensor was employed for the detection of *EPSPS* sequence in PCR non-amplified DNA samples extracted from animal feed containing 30% of the genetically modified soybean RR. The sensor was able to distinguish between a transgene sequence of the modified and unmodified soybean DNA at the genomic DNA quantities used in the analysis: 3.6, 4.6 and 5.4 µg. The detection limit was in the range of 4.7×10^5 genome copies in 200 µL of a QCN cell [15].

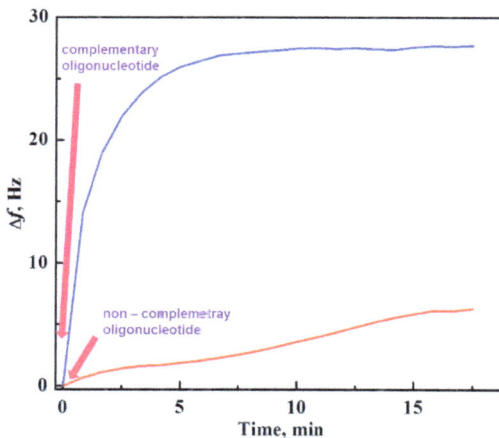

Figure 6. Resonant frequency shift recorded for an Au/DASE/avidin film after injection of a complementary and non-complementary oligonucleotides, $C_{compl} = 80$ nM, $C_{non\text{-}compl} = 72$ nM; the solution composition: 27 mM HEPES, 55 mM NaCl, pH 7.5; total volume in the QCM cell: 200 µl. Modified from reference [15] Copyright 2007 M. Stobiecka , J.M. Cieśla , B. Janowska , B. Tudek, H. Radecka; licensee MDPI, Basel, Switzerland.

9. Molecular dynamics

To better understand the interactions of herbicides with dsDNA, further investigations were carried out using molecular dynamics (MD) simulation and quantum mechanical calculation (QC) of electronic structures. In MD simulations, atrazine and Nile Blue molecules were: [1] docked on the sugar-phosphate chain and at minor and major grooves of DNA to evaluate hydrogen bonding and electrostatic interactions and [2] were inserted into a gap between two stacked bases in model structures of ssDNA and dsDNA [11],[14],[13]. In investigations carried out for the herbicide paraquat, three molecules of the herbicide were placed between the bases in a model dsDNA consisting of 11 nucleic base-pairs [4]. In these investigations, we have found that the intercalation of Atz and PQ in dsDNA helix leads to a conformational alterations of the DNA structure. The herbicides caused an increase in the interbase distance, longitudinal helix expansion and unwinding of the dsDNA confirmed by the higher binding capacity of DNA toward the Nile Blue intercalative redox probe and a higher uptake of the probe after incubation of DNA with herbicides. Long time interactions of paraquat with model dsDNA resulted in the denaturation of the double-stranded helix to single strands decreasing the NB uptake. The molecular dynamics simulations strongly corroborate cyclic voltammetry and fluorimetric measurements carried out by our groups. In Figure 7, an example of a simulation of the unwinding process of a dsDNA during its interactions with atrazine is presented. A clear elongation of the DNA duplex is readily discernible.

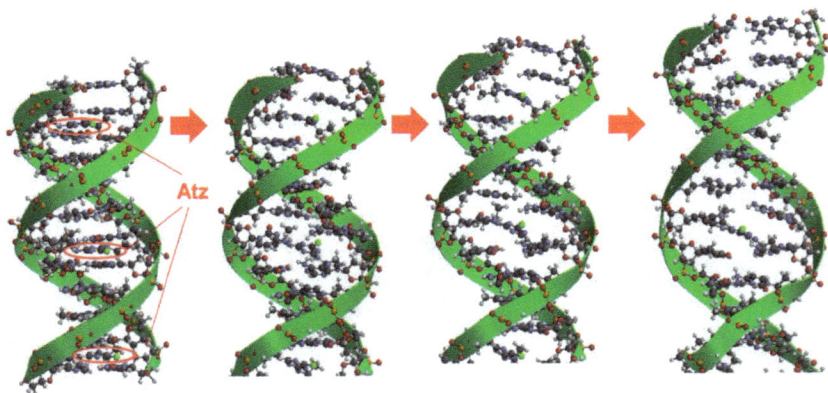

Figure 7. Molecular dynamics simulation of the interactions of atrazine (Atz) with ds DNA helix.

10. Conclusions

In this Chapter, the principles of sensory film designs for DNA-biosensors using nanogravimetric and voltammetric signal transduction techniques were presented. The biosensing platforms we have studied include the electrochemical biosensing with intercalating redox probe (NB) and travelling redox probe (ferricyanide ions), as well as the EQCN monitoring of conformational film canges correlated with herbicide-induced DNA alterations. Applications of DNA-biosensors for probing the affinity of atrazine and paraquat to DNA, the interactions of herbicides with DNA, assessment of DNA damage, and detection of the genetically modified soybean RR, have been reviewed. We have shown that the interactions of herbicides with DNA caused an evident alteration of the B-DNA conformation including unwinding of the helix and an increase in the interbase distance. The unwinding results in an increase in the capacity of dsDNA to bind an intercalative electrochemical probe and higher uptake of Nile Blue molecules into the DNA duplex. At longer interaction times of DNA with paraquat, strand breaks and DNA fragmentation has been observed as evidenced by the lower Nile Blue uptake into the DNA helix. The piezoelectric DNA-biosensor based assay developed was utilized as an alternative, highly sensitive, inexpensive, and simple method for the detection of genetically modified soybean in samples that do not require PCR amplification.

List of abbreviations

AET - aminoethanol

Atz- atrazine

CaMV - cauliflower mosaic virus 35 promoter

CRM - certified reference materials

CV - cyclic voltammetric measurements

DASE - 3,3'-dithiodipropionic acid di-(N-succinimidyl ester)

EDC - N-(3-dimethylaminopropyl)-N'-ethylcarbodiimide

ELISA - Enzyme Linked Immunosorbent Assay

EPSPS-5-enolpyruvylshikimate-3-phosphate synthase gene

EQCN - Electrochemical Quartz Crystal Nanobalance

GMOs -genetically modified organisms

HEPES -N-[2-hydroxyethyl]piperazine – N'-[2-ethanesulfonic acid]

HPLC - high-performance liquid chromatography

I_{pc} -cathodic peak current

LC-MS -liquid chromatography - mass spectrometry

MCH - 6-mercapto-1-hexanol

MD - molecular dynamics

MPA -mercaptopropionic acid

NB - redox active dye Nile Blue A

NOS - nopaline synthase terminator

PCR- Polymerase Chain Reaction technique

pDNA - single-stranded probe DNA

PQ - paraquat

QC -quantum mechanical calculation

ROS -reactive oxygen species

RR soybean - genetically modified soybean Roundup Ready

tDNA - target-gene single-stranded DNA

Author details

Magdalena Stobiecka

Department of Biophysics, Warsaw University of Life Sciences SGGW,Warsaw, Poland

References

[1] Hepel M, Stobiecka M, Peachey J, Miller J. Intervention of glutathione in pre-muta-genic catechol-mediated DNA damage in the presence of copper(II). Mutation Res. 2012;735:1-11.

[2] Halliwell B. Oxidative stress and cancer: have we moved forward? Biochem J 2007;401:1–11.

[3] Tandon VR, Sharma S, Mahajan A, Bardi GH. Oxidative Stress : A Novel Strategy in Cancer Treatment. JK Science. 2005;7 1-3.

[4] Stobiecka M, Prance A, Coopersmith K, Hepel M. Antioxidant effectiveness in pre-venting paraquat-mediated oxidative DNA damage in hte presence of H_2O_2. In: An-dreescu S, Hepel M, editors. Oxidative Stress: Diagnostics, Prevention and Therapy. Washington, DC: American Chemical Society; 2011. p. 211-33.

[5] Clements C, Raph S, Petras M. Genotoxicity of selected herbicides in Rana catesbei-ana tadpoles using alkaline single-cell gel DNA electrophoresis (comet) assay. Envi-ron Mol Mutagen. 1997;29:277-88.

[6]   Croce CD, Morichetti E, Intorre L, Soldani G, Bertini S, Bronzetti G. Biochemical and genetic interactions of two commercial pesticides with the monooxygenase system and chlorophyllin. J Environ Pathol Toxicol Oncol. 1996;15:21-8.

[7]   Ribas G, Ferenzilli G, Barale R, Marcos R. Herbicide induced DNA damage in human lymphocytes evaluated by single-cell gel electrophoresis (SCGE) assay. Mutat Res. 1995;344:41-54.

[8]   Meisner LF, Roloff BD, Belluck DA. In vitro effects of N-nitrosoatrazine on chromosome breakage. Arch Environ Contam Toxicol. 1993;24:108-12.

[9]   Surralles J, Catalan J, Creus A, Norppa H, Xamena N, Marcos R. Micronuclei induced by alachlor, mitomycin-C and vinblastine in human lymphocytes: Presence of centromeres and kinetochores and influence of staining technique. Mutagenesis. 1995;10:417-23.

[10] Fan W, Yanase T, Morinaga H. Atrazine-induced aromatase expression is SF-1 dependent: implications for endocrine disruption in wildlife and reproductive cancers in humans. Environ Health Perspect. 2007;115:720-7.

[11] Hepel M, Stobiecka M. Interactions of Herbicide Atrazine with DNA. New York: Nova Science Publishers; 2010.

[12] Nowicka AM, Kowalczyk A, Stojek Z, Hepel M. Nanogravimetric and voltammetric DNA-hybridization biosensors for studies of DNA damage by common toxicants and pollutants. Biophys Chem. 2010;146:42-53.

[13] Stobiecka M, Coopersmith K, Cutler S, Hepel M. Novel DNA-Hybridization Biosensors for Studies of DNA Underwinding Caused by Herbicides and Pesticides, in: "Electrochemical Nano/Bio Systems 2". ECS Trans. 2010;28 (34):1-12.

[14] Prance A, Coopersmith K, Stobiecka M, Hepel M. Biosensors for the Detection of DNA Damage by Toxicants. ECS Transactions. 2010;33:3-15.

[15] Stobiecka M, Cieśla JM, Janowska B, Tudek B, Radecka H. Piezoelectric sensor for determination of genetically modified soybean Roundup Ready in samples not amplified by PCR. Sensors. 2007;7:1462-79.

[16] Melchiorii D, Reiter RJ, Sewerynek E, Hara M, Chen L, Nistico G. Paraquat Toxicity and Oxidative Damage Reduction by Melatonin. Biochem Pharm. 1996;51:1095-9.

[17] Chen Q, Niu Y, Zhang R, Guo H, Gao Y, Li Y, et al. The toxic influence of paraquat on hippocampus of mice: Involvement of oxidative stress. NeuroToxicology. 2010;31:310-6.

[18] Peng J, Stevenson FF, Hsu M, Andersen JK. The herbicide paraquat induces dopaminergic nigral apoptosis through sustained activation of the JNK pathway. J Biol Chem. 2004;279:32626–32.

[19] Peng J, Stevenson FF, Oo ML, Andersen JK. Iron-enhanced paraquat-mediated dopaminergic cell death due to increased oxidative stress as a consequence of microglial activation. Free Radical Biol Med. 2009;45:312–20.

[20] Shimizu K, Matsubara K, Ohtald K, Saito O, Shiono H. Paraquate induces long-lasting dopamine overflow through the excitotoxic pathway in the striatum of freely moving rats. Brain Res 2003;976:243–52.

[21] Dinis-Oliveira RJ, Remiao F, Carmo H, Duarte JA, Sanchez Navarro A, Bastos ML. Paraquat exposure as an etiological factor of Parkinson's disease. Neurotoxicology. 2006;27:1110–22.

[22] Di Monte DA, Lavasani M, Manning-Bog AB. Environmental factors in Parkinson's disease. Neurotoxicology. 2002;23:487–502.

[23] Hafez AM. Antigenotixic Activity of Melatonin and Selenium Against Genetic Damage Induced by Paraquat. Australian Journal of Basic and Applied Sciences. 2009;3:2130-43.

[24] Ali S, Jain SK, Abdulla M, Athar M. Paraquat induced DNA damage by reactive oxygen species. Biochemistry and Molecular Biology International. 1996;67:63-7.

[25] Yamamoto H, Mohanan PV. Effects of melatonin on paraquat or ultraviolet light exposure-induced DNA damage. JPinealRes. 2001;31:308-13.

[26] Togunaga I, Kubo S-i, Mikasa H, Suzuki Y, Morita K. Determination of 8-hydroxydeoxyguanosine formation in rat organs: assessment of paraquat-evoked oxidative DNA damage. Biochemistry and Molecular Biology International. 1997;43:73-7.

[27] Schmuck G, Rohrdanz E, Tran-Thi QH, Kahl R, Schluter G. Oxidative stress in rat cortical neurons and astrocytes induced by paraquat in vitro. Neurotox Res. 2002;4:1–13.

[28] Naessens M, Leclerc JC, Tran-Minh C. Fiber Optic Biosensor Using Chlorella vulgaris for Determination of Toxic Compounds. Ecotoxicology and Environmental Safety. 2000;46:181-5.

[29] Freeman JL, Rayburn AL. In vivo genotoxicity of atrazine to anuran larvae. Mutation Res. 2004;560:69-78.

[30] Greenlae AR, Ellis TM, Berg RL. Low-dose agrochemicals and lawn-care pesticides induce developmental toxicity in murine preimplantation embryos. Environ Health Persp. 2004;112:703-9.

[31] Hopenhayn-Rich C, Stump ML, Browning SR. Regional assessment of atrazine exposure and incidence of breast and ovarian cancers in Kentucky. Arch Environ Contam Toxicol. 2002;42:127-36.

[32] Zeljezic D, Garaj-Vrhovac V, Perkovic P. Evaluation of DNA damage induced by atrazine and atrazine-based herbicide in human lymphocytes in vitro using a comet and DNA diffusion assay. Toxicology in Vitro. 2006;20:923-35.

[33] Environmental Health Criteria. Geneva: 1986.

[34] Velasco-Garcia MN, Mottram T. Biosensor Technology addressing Agricultural Problems. Biosystems Engineering. 2003;84:1–12.

[35] Palchetti I, Cagnini A, Del Carlo M, Coppi C, Mascini M, Turner APF. Determination of anticholinesterase pesticides in real samples using a disposable biosensor. Anal Chim Acta. 1997;337:315-21.

[36] Del Carlo M, Mascini M, Pepe A, Diletti G, Compagnone D. Screening of food samples for carbamate and organophosphate pesticides using an electrochemical bioassay. Food Chem. 2004;84:651-6.

[37] Mascini M. Affinity electrochemical biosensors for pollution control. Pure Appl Chem. 2001;73:23-30.

[38] Oliveira-Brett AM, Silva LAd. A DNA-electrochemical biosensor for screening environmentl damage caused by s-triazine derivatives. Anal Bioanal Chem. 2002;373:717-23.

[39] Halamek J, M.Hepel, Skladal P Investigation of highly sensitive piezoelectric immunosensors for 2,4-dichlorophenoxyacetic acid. Biosens Bioelectron. 2001;16:253-60.

[40] Pribyl J, Hepel M, Halamek J, Skladal P. Development of piezoelectric immunosensors for competitive and direct determination of atrazine. Sensors Actuators B. 2003;91(1-3):333-41.

[41] Pribyl J, Hepel M, Skladal P. Piezoelectric immunosensors for polychlorinated biphenyls operating in aqueous and organic phases. Sensors and Actuators B. 2006;113:900–10.

[42] Lagarde F, Jaffrezic-Renault N. Cell-based electrochemical biosensors for water quality assessment. Anal Bioanal Chem. 2011;400:947–64.

[43] Shin HJ. Genetically engineered microbial biosensors for in situ monitoring of environmental pollution. Appl Microbiol Biotechnol 2011;89:867–77.

[44] Lei Y, Chenb W, Mulchandani A. Microbial biosensors. Analytica Chimica Acta 2006;568:200–10.

[45] Tkac J, Stefuca V, Gemeiner P. Biosenors with immobilised microbial cells using amperometric and thermal detection principles. In: Nedovic V, Willaert R, editors. Applications of Cell Immobilisation Biotechnology. Netherlands: Springer; 2005. p. 549-66.

[46] Ron EZ. Biosensing environmental pollution. Current Opinion in Biotechnology 2007;18:252–6.

[47] Li Y-F, Li F-Y, Ho C-L, Liao VH-C. Construction and comparison of fluorescence and bioluminescence bacterial biosensors for the detection of bioavailable toluene and related compounds. Environmental Pollution. 2008;152:123-9.

[48] Windels P, Taverniers I, Depicker A, Bockstaele E, Loose M. Characterisation of the Roundup ready soybean insert. EurFood ResTechnol. 2001;213:107.

[49] Hupfer C, Hotzel H, Sachse K, Engel KH. Detection of genetically modified insect-resistant Bt maize by means of polymerase chain reaction. Zeitschrift für Lebensmitteluntersuchung und -Forschung. 1997;205:442-5.

[50] Minunni M, Mascini M, Cozzani I. Screening methodologies for genetic modified organisms (GMOs). Analytical Letters. 2000;33:3093-125.

[51] Holst-Jensen A, Rønning SB, Løvseth A, Berdal KG. PCR technology for screening and quantification of genetically modified organisms (GMOs). Anal Bioanal Chem. 2003;375:985–93.

[52] Lau L-T, Collins RA, Yiu S-H, Xing J, Yu ACH. Detection and characterization of recombinant DNA in Roundup Ready soybean insert. Food Control. 2004;15:471-8.

[53] Forte VT, Pinto AD, Martina C, Tantillo GM, Grasso G, Schena FP. A general multiplex-PCR assay for the general detection of genetically modified soya and maize. Food Control. 2005;16:535-9.

[54] Asfaw A, Tewodros M. Detection and quantification of genetically engineered crops. Journal of SAT Agricultural Research. 2008;6:1-10.

[55] Miraglia M, Berdal KG, Brera C, Corbisier P, Holst-Jensen A, Kok EJ, et al. Detection and traceability of genetically modified organisms In the ford production chain. Food and Chemical Toxicology. 2004;42:1157-80

[56] Berdal KG, Holst-Jensen A. Roundup Ready® soybean event-specific real-time quantitative PCR assay and estimation of the practical detection and quantification limits in GMO analyses. Eur Food Res Technol. 2001;213:432–8.

[57] Vaïtilingom M, Pijnenburg H, Gendre F, Bringon P. Real-time quantitative PCR detection of genetically modified maximizer maize and Roundup Ready soybean in some representative foods. J Agric Food Chem. 1999;47:5261-6.

[58] Rogan GJ, Dudin YA, Lee TC, Magin KM, Astwood JD, Bhakta NS, et al. Immunodiagnostic methods for detection of 5-enolpyruvylshikimate-3-phosphate synthase in Roundup Ready soybeans. Food Control. 1999;10:407-14

[59] Lucarelli F, Marrazza G, Turnet APF, Mascini M. Carbon and gold electrodes as electrochemical transducers for DNA hybridization sensors. Biosens Bioelectron. 2004 19:515-30.

[60] Rivas GA, Pedano ML, Ferreyra NF. Electrochemical DNA biosensors. Electrochemical biosensors for sequence-specific DNA detection. Anal Lett. 2005;38:3653.

[61] Drummond TG, Hill MG, Barton JK. Electrochemical DNA sensors. Nat Biotechnol. 2003;21:1192-9.

[62] Fan CH, Plaxco KW, Heeger AJ. Electrochemical DNA sensors. Trends Biotechnol. 2005;23:186-92.

[63] Palecek E. From polarography of DNA to microanalysis with nucleic acid-modified electrodes. Electroanalysis. 1996;8:7-14.

[64] Palecek E, Jelen F, Trnkova L. Cyclic voltammetry of DNA at a mercury electrode: an anodic peak specific for guanine. Gen Physiol Biophys. 1986;5:315-29.

[65] Fojta M, Stankova V, Palecek E. A supercoiled DNA-modified mercury electrode based biosensor for the detection of DNA strand cleaving agents. Talanta. 1998;46:155-61.

[66] Fojta M. Electrochemical Sensors for DNA Interactions and Damage. Electroanalysis. 2002;14:1449–63.

[67] Meric B, Kerman K, Marrazza G, Palchetti I, Mascini M, Ozsoz M. Disposable genosensor, a new tool for detection of NOS-terminator, a genetic element present in GMOs. Food Control. 2004;15:621-6.

[68] Ligaj M, M.Tichoniuk, Filipiak M. Detection of bar gene encoding phosphinothricin herbicide resistance in plants by electrochemical biosensor. Bioelectrochemistry. 2008;74:32–7.

[69] Stobiecka M, H.Radecka, J.Radecki, Tudek B, Cieśla J. Immobilization and Hybridization of Short Strand Oligonucleotides on Gold Electrodes - Preliminary Results. VIth Conference of the Polish Supramolecular Chemistry Network, Krutyń, Poland, 2004.

[70] Tichoniuk M, Ligaj M, Filipiak M. Application of DNA Hybridization Biosensor as a Screening Method for the Detection of Genetically Modified Food Components. Sensors. 2008;8:2118-35.

[71] Ligaj M, Janowska J, Musiał WG, Filipiak M. Covalent attachment of single-stranded DNA to carbon paste electrode modified by activated carboxyl groups. Electrochimica Acta 2006;51:5193–8.

[72] Ligaj M, Oczkowski T, Janowska J, Musiał WG, Filipiak M. Electrochemical genosensors for detection of L.Monocytogenes and genetically-modified components in food. Polish Journal of Food and Nutrition Sciences. 2003;12:61.

[73] Lin CW, Shulok JR, Wong YK, Schanbacher CF, Cincotta L, Foley JW. Photosensitization, uptake, and retention of phenoxazine Nile Blue derivatives in human bladder carcinoma cells. Cancer Res. 1991;51:1109-16.

[74] Mannelli I, Minunni M, Tombelli S, Mascini M. Quartz crystal microbalance (QCM) affinity biosensor for genetically modified organisms (GMOs) detection. Biosensors and Bioelectronics. 2003;18:129-40.

[75] Minunni M, Tombelli S, Pratesi S, Mascini M, Piatti P, Bogani P, et al. A piezoelectric affinity biosensor for genetically modified organisms (GMOs) detection. Analytical Letters. 2001;34:825-40.

[76] Minunni M, Tombelli S, Fonti J, Spiriti MM, Mascini M, Bogani P, et al. Detection of fragmented genomic DNA by PCR-free piezoelectric sensing using a denaturation approach. JAmChemSoc. 2005;127:7966-7.

[77] Hepel M. Electrode-Solution Interface Studied with Electrochemical Quartz Crystal Nanobalance In: Wieckowski A, editor. Interfacial Electrochemistry Theory, Experiment and Applications. New York: Marcel Dekker, Inc.; 1999. p. 599-630.

[78] Bruckenstein S, Shay M. Experimental Aspects of Use of the Quartz Crystal Microbalance in Solution. Electrochim Acta. 1985;30:1295-300.

[79] Sauerbrey G. Verwendung von Schwingquarzen zur Wagung dunner Schichten und Microwagung. Z Phys. 1959;155:206–22.

[80] Hepel M, Redmond H, Dela I. Electrochromic WO3-x films with reduced lattice deformation stress and fast response time. Electrochimica Acta. 2007;52(11):3541-9.

Biosensing Disease Biomarkers

Chapter 11

Amperometric Biosensor for Diagnosis of Disease

Antonio Aparecido Pupim Ferreira,
Carolina Venturini Uliana,
Michelle de Souza Castilho,
Naira Canaverolo Pesquero, Marcos Vinicius Foguel,
Glauco Pilon dos Santos, Cecílio Sadao Fugivara,
Assis Vicente Benedetti and Hideko Yamanaka

Additional information is available at the end of the chapter

1. Introduction

Our main interest is discussing amperometric biosensors with application in certain disease diagnosis. These biosensors are based on the affinity reaction between antigen/antibody (immunosensor) or DNA/DNA (genosensor) or enzymatic catalytic reaction. The selective interactions will be also discussed in this chapter. In this first part, the central goal is to present and discuss some aspects of working electrode (WE) surface preparation and characterization, electrochemical cell arrangements and (chrono)amperometry as a simple electrochemical technique to evaluate some types of biosensors.

It is well known that in chemical sensors the chemical information is transformed into useful analytical signal. The chemical information can be associated with the concentration of a specific component present in the sample. In a simple way, a molecular receptor in series with a physico-chemical transducer characterizes what is called chemical sensor [1]. When the molecular receptor involves a biochemical component, a biosensor is obtained [2]. In a biosensor, the biological component is responsible for the selectivity while the characteristics of the electrochemical detector determine the sensitivity. It means that the electrochemical detector (transducer) must be carefully selected and prepared. In its selection, the mechanism nature of the biosensor must be known. This mechanism depends basically on the type of active components involved and on the mode of signal transduction. For instance, in enzymatic biosensors, the active site of the enzyme must be preserved after immobilization and satisfactory electrochemical communicability between the redox site and the electrode should be guaranteed.

Different electrochemical techniques can be used to characterize and evaluate biosensors: chrono(amperometry), chronopotentiometry, linear potential sweep (LPS), cyclic voltammetry (CV) (DC techniques), and electrochemical impedance spectroscopy (EIS) and AC voltammetries (AC techniques). For electrochemical characterization of electrode processes, CV and EIS are probably the most used electrochemical techniques. In general, when pre-treated surfaces and modified electrodes are characterized using these techniques, the reversibility, the electron charge transfer (*e.c.t.*) rate of a redox couple such as $Fe(CN)_6^{3-/4-}$, $Ru(NH_3)_6^{3+}$, etc., and the diffusion coefficient of the electroactive species are determined and compared with their standard behavior. If the electrochemical response is given by species in a stagnant solution containing an excess of supporting electrolyte, the response corresponds to a reversible *e.c.t.* controlled by diffusion of the electroactive species to or from the electrode surface. If the electrode is partially blocked with self-assembled monolayers (SAMs) or non-electroactive surface modifiers in the examined potential range, the electrochemical response may vary depending on the shape and size of the access to the active sites by the electroactive species from the solution. Generally, an electrochemical response resembling a less reversible system is observed. The reversibility of the system may decrease as the blocked fraction of surface area increases. It is easily detected by the increase of the difference between anodic and cathodic peak potentials and by the decrease of peak currents. For EIS studies, the decrease in the reversibility is observed by the increase in the modulus of electrochemical impedance and the decrease in the *e.c.t.* reaction rate [3]. However, if the electroactive species is attached or adsorbed on the electrode surface, the electrochemical response will depend on the distance between the redox center and the surface and accessibility of electrons to this redox center, on the position of the redox center into the molecule attached to the electrode surface, and the nature and state of the electrode surface. Generally, if a monolayer of the redox species is attached to the electrode surface and a reversible *e.c.t.* process takes place, the CV shows peak potential separation near zero. Similar response shows a film with several monolayers of the redox species and the electron exchange between the layers reversibly occurs [4].

Details on cyclic voltammetry and its applications are displayed in some textbooks [4-6]. Fundamentals and mathematical analysis of electrochemical impedance spectroscopy can be found in [7,8]. For some applications of CV and EIS to immunosensors characterization, the readers are referred to [9].

The main reasons for the large use of (chrono)amperometry are its simplicity in data collection due to the apparent facility in measuring the current related to the *e.c.t.* associated with the biosensor response. For example, if one compares the amperometric technique with the EIS [9], also used in biosensors characterization and monitoring, there is no doubt that the later is much more laborious; it is also true that EIS is a better technique to deeply investigate the global behavior of the system [3,10].

The electrochemical techniques are not able to identify the chemical nature of the products or reactants involved in certain electrochemical process, and then some non-electrochemical techniques complement and help us to understand the electrochemical processes. They can be associated to surface or bulk (solid, liquid or gas) analysis.

Amperometry is a voltammetric method in which two- or three-electrode cell configurations are used, and the potential applied between the WE and the auxiliary electrode (AE) results in a constant potential at the working *versus* a reference electrode (RE). If the current is measured as function of time we have the chronoamperometry technique. In this technique, a drastic and immediate change in the WE potential from an initial potential value E_i (where no faradaic reactions take place) to a final potential E_f (where the faradaic reaction of interest occurs) and the current is continuously measured. The analysis of the current-time (I-t) transients can be used to study many electrochemical processes as *e.c.t.* involving species in solution, new phase formation, adsorption, diffusion coefficient determination and so on. For chemical analysis, time-independent current is interesting to be obtained, which can be attained if the diffusion layer thickness is constant. It is possible to obtain by convection transport or using micro or mainly ultramicroelectrodes [11,12].

In order to develop an amperometric biosensor, special attention should be devoted to choose the WE, to conveniently prepare and modify its surface, and identify the electrochemical response related to the specific reaction involving any electroactive species present in the biosensor system, which may unequivocally indicate the presence of certain disease. The performance of the biosensor is strongly dependent on the *e.c.t.* reaction rate. The current generated at the WE measured using a two- or three-electrode cell configuration depends on the reaction rate. The steady-state current is proportional to the analyte concentration in the bulk, c_{bulk}. The area of RE (two-electrode configuration) or the AE (three-electrode configuration) needs to be at least 10 times wider than the WE, and then the reaction occurring at the AE is fast compared to that one occurring at the WE. To fulfill its role, the AE must be a good conductor of electricity and it must be placed in the cell in order to guarantee good distribution of electric field [13].

In order for facilitating the analyses of I-t curves and for getting the best sensitivity for the appropriate electrochemical reaction, the applied potential value can be chosen in such way that the surface concentration (c_{surf}) of the investigated species is zero. If c_{surf} is not zero the current will be lower and dependent on the potential and time. The corresponding equations and mathematical details can be found in [14-17].

Based on the comments presented before, some aspects about the transducer in amperometric biosensors should be considered:

- chemical nature of the working electrode, surface preparation and characterization;

- choosing the potential value of the working electrode;

- repeatability and sensitivity in (chrono)amperometry measurements.

1.1. Chemical nature of the working electrode, surface preparation and characterization

In this section, it will be presented different materials that have been used as transducers, mainly for amperometric immunosensors construction, electrode surface preparation and pre-treatments (when used), and electrochemical cell configurations.

Among these different materials, some can be mentioned, such as: gold, CD-trode, screen-printed electrodes, silver, mercury, graphite, glassy carbon, carbon nanotubes, gold nano-wires, gold nanoparticles, metallic oxide nanoparticles, carbon paste, boron-doped diamond and composites. These surfaces can be transformed with different modifiers to form SAMs, and composites which carry or incorporate the active components desired to construct the biosensor.

In aqueous medium, gold presents some advantages compared to platinum since it does not adsorb hydrogen and it has high overpotential for hydrogen-evolution reaction, which is appropriate to study cathodic processes. The real surface area can be determined measuring the charge involved in the reduction of the gold oxide layer formed at high overpotentials. This area can be very different from the geometrical one. In the case of carbon paste electrode, the main advantages are ease of preparation, versatility in the chemical modification and its rapid renewal. Glassy carbon electrode has low cost, high resistivity to chemical attack, very low permeability to gas, large potential window, it is easily polished and treated via potential scanning and it may improve the kinetic of some charge transfer reactions [4].

It is well known that the response of a solid electrode is strongly dependent on the surface preparation, i.e., the mechanical, chemical and electrochemical pre-treatment applied. Different from the liquid electrodes (Hg, Tl), the rate of *e.c.t.* at solid electrodes is extremely dependent on the surface condition. To a general procedure for surface preparation of solid electrodes, the readers may consult the literature [13]. Probably, the more critical consequence of this behavior of solid electrodes is the difficulty in renewing the surface in order to obtain reproducible electrochemical response. Also, preparation, characterization and control of the transducer surface play an important role in the following steps of the sensor construction, stability, quality of response and amount of SAM or other modifier component, and the success or fail of the developed device. These steps and properties are crucial for the immobilization of the biological molecules or other material on the transducer surface and subsequent interaction between the modified surface and the analyte, i.e., the final electrochemical response of the biosensor.

Massive or modified gold was also used to produce immunosensors. A gold electrode was repeatedly polished with 1.0 and 0.3 µm alumina slurry, successively sonicated in bi-distilled water and ethanol for 5 min, and dried in air [18]. Kheiri et al. [19] used similar procedure to pre-treat the gold electrode before modifying it with carbon nanotubes (CNTs) and other modifiers. The gold electrode was polished with 0.3 and 0.05 mm alumina powders in succession, thoroughly rinsed with double distilled water between each polishing step, successively sonicated with acetone and double distilled water, and dried at room temperature. Another strategy was adopted to clean and pre-treat the gold electrode surface to construct immunosensors [20]. Gold electrodes were first polished with aqueous alumina slurries of

25 and 1 μm, rinsed with MilliQ water, sonicated for 1 min, dried with argon, treated with cold piranha solution for 30 s, washed with Milli-Q water and argon dried. Afterwards, a preliminary electrochemical cleaning was performed by LPS between −0.2 and −1.8 V in 0.1 mol L⁻¹ KOH, followed by CV in 2 mol L⁻¹ H₂SO₄ at 0.2 V s⁻¹ for 30 cycles or until stable CVs were recorded.

Gold electrodes array, consisting of 16 gold working electrodes where each WE was placed between an Ag pseudo-reference and a gold AE, were used to prepare amperometric immunosensors for tumor detecting [21]. A thin film of gold or platinum was modified with CNTs to construct an amperometric immunosensor for rheumatoid arthritis [22]. To construct amperometric immunosensors for detection of Chagas disease, transducers were prepared by sputtering gold on Si and Si₃N₄ in argon atmosphere [23]. The silicon-gold slices were annealed at 1000 °C for 5 s, cooled at air atmosphere and room temperature, vigorously washed with distilled water and dried with purified compressed air.

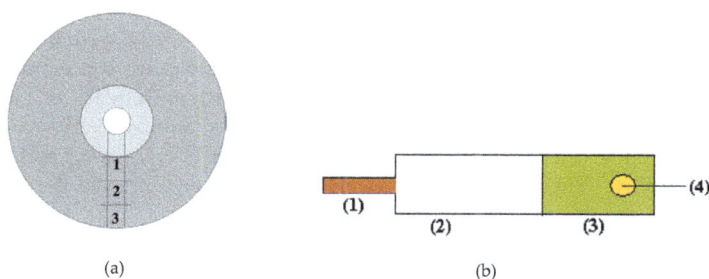

(a)   (b)

Figure 1. Regions of CD-Rs: (1) inner, (2) central and (3) out border. (B) CDtrode: (1) electric contact of copper, (2) PTFE tape to fix the electric contact, (3) 1KFA25 Kapton tape® applied on the surface to delimit the area of the electrode, and (4) area of the working electrode [29].

Gold-based substrates produced by sputtering can be substituted, with advantages, by metallic substrates obtained from recordable compact discs (CD-Rs) [24]. These devices present comparable electrochemical performance to commercial gold electrodes, they are easily constructed and versatile, of low cost to be used and discarded in cases of fouling, surface oxidation, irreversible adsorption, and so on, and are user-friendly because electrode polishing is not necessary [25,26]. In general, the gold CD-R has a gold film thickness of 50-100 nm and it can be also used as WE (CDtrode). As-received CD-R pieces may be treated with 69-70% HNO₃ for 5-10 min to remove the polymeric layers, cleaned with 95-98% sulfuric acid and abundantly washed with ethanol and/or water. Recently, Foguel et al. (2011) [27,28] developed an amperometric immunosensor for Chagas disease using CDtrode prepared by the procedure described above. It was observed that the voltammetric response of CDtrode depends on the procedure applied to remove the protective polymeric-based layer, the subsequent chemical or electrochemical treatments, trade of CD-R and also sometimes the region of the CD-R. Foguel [29] also investigated in more detail the use of different CD-R trades, nominated as AA, BB and CC, and different regions (out border, center and inner) of

CD-Rs (Fig. 1a). The polymeric layers covering the gold surface were removed by different procedures: (a) careful manual removal with tweezers, vigorous washed with distilled water and dried with purified compressed air; (b) the procedure described by Lowinsohn et al. [26]; (c) the procedure described in (b) and the area of the electrode limited by a mask of toner. Figure 1b illustrates the final setup of the electrode.

Surface roughness, CD track height and thickness, and the distance between CD tracks (cavities thickness) were measured by AFM. For AA CD-R the measured parameters were almost invariable in the inner, center and out border regions of the CD-R; BB CD-R presented almost the same surface roughness and CD track height in all regions, high difference in track thickness, the inner border tracks are thicker and the distance between them is higher; the inner and border parts of the CC CD-R showed similar tracks height, thickness and roughness values, but varied the distance between CD tracks among the different regions of the CD-R, and different values for all parameters in the center compared with the other regions. These results indicated that the AA CD-R is the only one that showed a more homogeneous gold surface and, therefore, it should present the best electrochemical behavior. FE-SEM analysis showed differences in the CDtrodes surface: CD tracks were better defined when the polymeric layers were manually removed and flatter when concentrated HNO_3 was used. Unmodified electrode surfaces were initially characterized by CV of 1×10^{-3} mol L^{-1} $Fe(CN)_6^{4-}$ in 0.5 mol L^{-1} H_2SO_4 aqueous solution (higher *e.c.t.* rate) at 50 mV s^{-1} without or with an application of 10 cycles from +0.2 to +1.5 V / Ag|AgCl|KCl$_{sat.}$ in 0.5 mol L^{-1} H_2SO_4 solution at 50 mV s^{-1}. When the polymeric layers were manually removed the I-E profiles were bad-defined and this procedure was abandoned.

Figure 2 shows cyclic voltammograms (CVs) recorded for unmodified CDtrode, constructed from BB CD-R, in 0.1 mol L^{-1} phosphate buffer (PB) solution at pH 7.0 containing 1×10^{-3} mol L^{-1} $Fe(CN)_6^{3-/4-}$ at 50 mV s^{-1}: (A) after removal the protective layers from the gold surface using the procedure (b); (B) after applying the procedure (b) followed by 10 cycles from +0.2 to +1.5 V / Ag|AgCl|KCl$_{sat.}$ in 0.5 mol L^{-1} H_2SO_4 solution at 100 mV s^{-1} and 10 cycles from −0.4 to +0.7 V / Ag|AgCl|KCl$_{sat.}$ in 1.0×10^{-3} mol L^{-1} $Fe(CN)_6^{3-/4-}$ + 0.1 mol L^{-1} PB solution at pH 7, at 50 mV s^{-1}.

It is clear that the I-E profile described in Fig. 2B resembles the response of a reversible charge transfer process, while the I-E profile in Fig. 2A suggests a non-reversible charge transfer process. Many factors can be involved in this electrochemical response. All of them are related to the surface nature of the solid electrode: the presence of protective material residues and other dirt, contaminants, oxides generated during the acid attack, defects and heterogeneities on the surface present in the original material or caused by the chemical attack. In this case, the mechanical procedure which is applied in many solid electrodes is not applicable. The chemical etching recommended to gold by using "piranha" or strong alkaline solution may also damage the delicate surface mainly at stressed regions of gold deposit. Therefore, the chemical etching is not recommended for CDtrodes. The adsorbed species can be removed and the electrode surface activated by potential cycling between the potentials of H_2 and O_2 evolution reactions. This process makes the surface reproducible and repeatable, and may improve the reversibility of the electrode process, as observed in Figure 2.

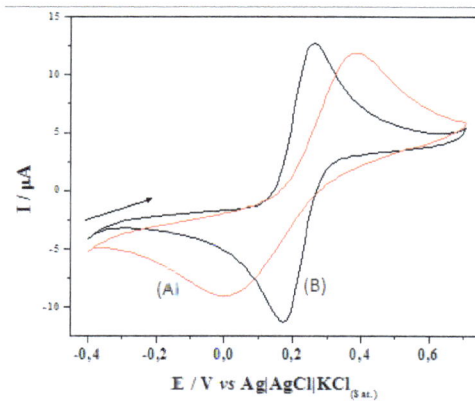

Figure 2. CVs of 1.0 x 10^{-3} mol L^{-1} Fe(CN)$_6$$^{3-/4-}$ in 0.1 mol L^{-1} PB solution at pH 7.0, 50 mV s^{-1} on gold CDtrode in which the protective layers were removed by procedure ɔ (A) and *b* followed by 10 cycles in 0.5 mol L^{-1} H$_2$SO$_4$ solution from +0.2 to +1.5 V / Ag|AgCl|KCl$_{sat}$ at 100 mV s^{-1} and 10 cycles in 1.0 x 10^{-3} mol L^{-1} Fe(CN)$_6$$^{3-/4-}$ in 0.1 mol L^{-1} PB solution at pH 7.0 from −0.4 to +0.7 V / Ag|AgCl|KCl$_{sat}$ at 50 mv s^{-1} (B) [29].

In the case of screen-printed electrodes (SPEs), special care should be taken during handling to avoid irreversible damage. For instance, in recent studies [3,10], screen-printed gold-based electrodes were used as-received These SPEs are received in aluminum sealed package individually isolated from the atmosphere. The package of each electrode was opened just before using and avoiding surface contamination. Chemical etching is not recommended for SPE gold electrodes. Therefore, the SPEs were thoroughly washed with ethanol and Milli-Q water for further procedures. Similar procedure has been recommended in literature [30,31]. It was observed that some immobilization or electrochemical processes are not significantly influenced by surface pre-treatments [32], and, some cases, they are used as-produced or -received, without pre-treatment [33].

Carbon based materials (graphite, glassy carbon, carbon fibers, carbon-SPE, carbon-epoxy resin composites, nanotubes and boron-doped diamond) have been used in both unmodified and modified forms by incorporation of gold nanoparticles (GNP) or iron oxides nanoparticles (NPs) dispersed in a polystyrene polymer matrix to construct amperometric or other biosensors. Iron oxides NPs exhibit magnetic properties and are constituted by paramagnetic γ-Fe$_2$O$_3$ and Fe$_3$O$_4$ or modified with some specific groups or can be a core-shell structure, with a core (γ-Fe$_2$O$_3$) and shell (styrene-based copolymer). In a recent work, bare graphite electrodes were mechanically treated by wet polished on emery paper, thoroughly washed with distilled water and modified to construct an amperometric biosensor [34]. Glassy carbon was successively wet polished with 1.0, 0.3 and 0.05 mm alumina slurry until a mirror-like surface, and the surface was thoroughly rinsed between each polishing step with doubly distilled water. Afterwards, it was successively sonicated in 1:1 nitric acid, acetone and doubly distilled water, and allowed to dry at room temperature [35]. Carbon fiber electrodes are produced, mainly in connection with the preparation of high-strength composites by high-temperature pyrolysis of polymer textiles or via catalytic chemical vapor deposition [36]. A chitosan-modified carbon fiber electrode was used to develop a biosensor for dengue virus envelope protein detection [37]. The carbon fibers surfaces were sonicated in ultrasonic bath with 10% HNO_3 solution for 10 min, rinsed with distilled water and conveniently modified. Commercial available carbon SPE was treated by applying an anodic current of 25 μA for 2 min in 50 μL of 0.1 mol L^{-1} H_2SO_4 solution dropped on the SPE carbon electrodes and washed with 0.1 mol L^{-1} Tris buffer pH 7.2 [38].

Graphite powder may be used to prepare composites which can be modified by NPs and/or magneto NPs and used in amperometric sensor. Recently, graphite powder and epoxy resin were used by Pividori, et al. to develop a sandwich magneto immunoassay [39]. The modified magnetic NPs are captured by the magnetic field on the magneto electrode. Arrays of carbon-SPE electrodes were also used to construct immunosensors. The arrangement was washed with water to remove any adsorbed species and characterized by CV in 5.0 mmol L^{-1} $Fe(CN)_6^{3-}$ solutions [40].

The development, properties (good electrical conductivity, nanometer size, high aspect ratio and structure, electrochemical stability, high specific area and surface chemistry) and applications of CNTs, mainly in biosensors construction, were deeply discussed recently [41]. Both its high specific area, which allows the analyte to be accumulated on the surface, and the capability of increasing *e.c.t.* reaction rate increase the response signal and diminish the overpotential for some electrode reactions. However, the fundamental reasons for that are not still well-established. The electrochemical behavior of graphitic materials is in great part defined by edge defects and oxygen functionalities at the surface, and the properties of the CNTs are similar to the high oriented pyrolytic graphite (HOPG). Details of CNT growth, working electrode preparation, surface modification and its application to construct specific enzymatic biosensors and genosensors were described [41]. In general, the CNT electrodes are subjected to electrochemical treatments based on three different electrolytes: 0.1 mol L^{-1} HNO_3, 10 s at 1 V; 0.1 mol L^{-1} KCl, 60 s at 1.75 V and 1 mol L^{-1} NaOH, 60 s at 1 V / Ag|AgCl| KCl; the last one seems to be the best. The CNTs cleaning is based on oxidation of the amorphous carbon, and carboxylic moieties generation for further covalent functionalization. SPEs made of commercial or homemade carbon inks were modified with multiwall carbon nanotubes (MWCNTs) and Au NPs to construct immunosensors [42]. These electrodes are pre-treated by applying +1.5 V / Ag|AgCl|KCl$_{sat.}$ for 5 min in 0.1 mol L^{-1} NaOH solution, and chemical treated by 3:1 concentrated H_2SO_4 and HNO_3 solution. This modified surface is immersed in 0.5 mol L^{-1} H_2SO_4 solution containing 0.1 mmol L^{-1} $HAuCl_4$ and gold NPs which are deposited by applying 15 cycles from +1.0 to 0.0 V / Ag|AgCl|KCl$_{sat.}$ at 50 mV/s. The resulting surface is rinsed with deionized water and stored in 0.1 mol L^{-1} phosphate buffer saline (pH 7.0) before characterization. CNTs can be conveniently functionalized with amino groups, deposited GNP, generated an appropriate composite and applied on massive gold electrodes [19]. Composites of MWCNT-polystyrene were modified and also applied on gold or on platinum thin film, and used to construct an amperometric immunosensor for rheumatoid arthritis diagnosis [22].

1.2. Choosing the potential value of the working electrode

Several factors influence the choice of the best potential value to be applied to the working electrode in order to get the best sensitivity of the biosensor. Some criteria may be adopted: (a) all steps of the biosensor construction should be carefully characterized by electrochemical and non-electrochemical techniques; (b) the current peak or wave responsible for the biosensor response must be unequivocally determined; (c) the stability and repeatability of the system should be investigated by obtaining enough number of I-E curves or CVs for a series of biosensors prepared by the same methodology.

Theoretically, the potential to be applied should reduce to zero the surface concentration of active centers responsible for the amperometric biosensor response. At this potential current, is directly proportional to the analyte concentration and the effective electrode surface area. In practice, this potential value frequently corresponds to the peak potential of CV, which does not mean that the surface concentration is zero; it depends on the electrode process. As the current generated at this potential is the sum of all faradaic processes occurring, supposing that no significant charging current is present, the reaction of interest identification may not be easy. Getting satisfactory reproducibility and repeatability of the biosensor response may be a hard task, mainly if low currents are generated, which may require more sophisticated setup and/or more expensive instrumentation. Different studies have applied the peak potential obtained from the CVs or the peak current of CVs at a constant scan rate to evaluate the biosensor response.

1.3. Repeatability and sensitivity in (chrono)amperometry measurements

For surface-controlled electrode processes (adsorption, new phase formation, surface modifications and so on) the current-time curves recorded at constant potential are strongly dependent on the nature of the substrate, and the reproducibility is strictly related to the similarity between previous and renewed surfaces. At constant temperature and solution composition, the structure of the monoatomic layers at the renewed surface are not strictly similar to that recorded to the previous surface, which may leave to different I-E profiles. Therefore, the best practice is recording a great number of current transients for each investigated condition and using the average current value. In minor grade, the surface conditions also influence the current values even if electroactive species are in solution, due to changes in the surface roughness or adsorption of active or inactive species or the surface. The response of modified surfaces may also depend on the surface roughness, defects, heterogeneities, coating stability, impurities in the medium, etc.

Some techniques are more sensitive than others for specific properties of the system. For instance, EIS presents high sensitivity to any change on the electrode surface. Amperometric response for diffusion-controlled processes depends on c_{bulk}, diffusion coefficient, number of electrons/particle, applied potential and effective surface area, size and geometry of the working electrode, and it is inversely proportional to the square root of time. Therefore, the higher current is obtained at short measuring time and, in general, it exponentially decays, tending to a stationary value. The capacitive current contribution is higher at very short time, the faradaic current depends on the kinetic of the electrode process, and the total current reaches a stationary value for longer times. These two characteristics of the technique may result in lower sensitivity when compared to some other electrochemical techniques. The analytical current density can be increased by convection (flux, stirring or jet the analyte) during the electrolysis or using micro or ultramicroelectrodes. Decreasing the analyte concentration, the faradaic current decreases and approximates to the current background (charging current, surface oxidation or reduction processes, noise). Therefore, in classical polarography the charging current limits the detection from 5×10^{-6} to 1×10^{-5} mol L^{-1} interval. However, techniques with time dependences for capacitive and analytical currents favoring the analytical one (pulse polarography techniques) may offer lower limit of detection. All these pulse techniques are based on a sampled current potential-step (chronoamperometric) experiment [36].

Also, higher faradaic/capacitive currents ratio (lower limit of detection) can be obtained for redox processes which occur near the potential of zero charge of the working electrode. Therefore, if possible, the working electrode that must be chosen is the one with a potential of zero charge closest to the redox potential of the analyte.

In order to optimize the working conditions of the developed biosensor, other parameters and/or properties influencing its response should be investigated such as pH, operating potential, temperature, stability, repeatability, cut off, limit of detection and sensitivity.

Following biosensor for disease diagnosis based on antigen/antibody (immunosensor) or DNA/DNA (genosensor) or enzymatic catalytic reaction will be described.

2. Amperometric immunosensors

The immobilization of antigens or antibodies on the surface of electrochemical transducers led to the development of immunosensors for several substrates of interest in the biological, clinical and industrial areas [43-45]. Immunosensors combine the advantages of the electrode process and the high specificity of immunologic reactions [46]. The methods are very rapid, they have the advantage of requiring small sample volumes affording an increase in the number of analyzed samples, and enabling versatile transducers and different techniques for monitoring, thus lowering costs compared with conventional analytical methods.

The immunosensor is classified as optical, mass-sensitive or electrochemical according to the technique. The electrochemical immunosensor, according to the monitoring, is classified as amperometric, potenciometric, impedimetric and condutometric. As mentioned before, chrono-amperometric technique for the development of amperometric immunosensor compared with other electrochemical techniques, is simple, cheap, sensitive, its potential applied not affected sample and possibly portable measuring amperometric system.

Several amperometric immnunosensors have been developed for disease diagnosis as shown in Table 1.

Cavalcanti et al. [37] developed a chitosan modified fiber electrode for dengue virus envelope (DENV). Antibodies against DENV were covalently immobilized on the chitosan matrix after activation with sodium periodate. Amperometric response of the competitive immunoassays was generated by hydrogen peroxide with peroxidase conjugated to DENV and 2'-azino-bis-(-3-ethylbenzthiazoline-6-sulfonic acid) (ABTS) as mediator. The immunosensor showed a lower limit of detection for DENV (0.94 ng mL^{-1}) than previously described and a linear range from 1.0 to 175 ng mL^{-1}, in concentration levels clinically relevant for dengue virus diagnosis.

A novel amperometric immunosensor for the detection of the p24 antigen (p24Ag) from HIV-1 using gold nanoparticles (GNP), multiwalled carbon nanotubes (MWCNTs), and an acetone extracted propolis (AEP) film was developed by Kheiri et al. The GNP/CNT/AEP film provided a suitable surface for the immobilization of antibodies and prevented direct contact of the biomolecules with the substrate. Moreover, GNPs were synthesized in situ on the amino functionalized MWCNTs (MWCNTNH2) for antibody immobilization, which also improved the electrochemical signal of HRP-anti p24 Ab, thus enhancing the detection sensitivity of the reduction of H$_2$O$_2$ [19].

Two methods to diagnose hepatitis B [18,35] are described in Table 1 and both methods determine hepatitis B surface antigen based on gold nanoparticle. The method developed by Zhuo et al. [18] is based on the gold nanoparticles and horseradish peroxidase (HRP)-modified gold electrode for the determination of hepatitis B surface antigen (HBsAg). The system was optimized for a reliable determination of HBsAg in the range of 2.56-563.2 ng mL^{-1} with a limit of detection 0.85 ng mL^{-1}. Qiu et al. [35] also determined hepatitis B surface antigen using a glassy carbon electrode modified with an assembly of positively charged poly(allylamine)-branched ferrocene (PAA-Fc) and negatively charged gold nanoparticle. The concentration of the antigen can be quantified in the range 0.1 and 150 ng mL^{-1}, with a limit of detection 40 pg mL^{-1}.

González et al. [38] used screen-printed carbon electrodes to detect pneumolysin (PLY) in human urine. The voltammetric immunosensor is based on the electrochemical detection of indigo blue, produced by alkaline phosphatase (AP) when 3-indoxyl phosphatase (3-IP) is used as enzymatic substrate. It is prepared and evaluated for measuring this toxin in human urine samples. The single-use immunosensor is fabricated by deposition of biotinylated anti-PLY monoclonal antibodies onto pre-oxidised streptavidin coated screen-printed carbon electrodes (SPCEs). Rabbit polyclonal IgGs anti-PLY are used in combination with an anti-rabbit IgG alkaline phosphatase conjugate as detection antibodies.

The determination of the antigliadin antibodies from human serum samples is of vital importance for the diagnosis of an autoimmune disease such as celiac disease. Therefore, Rivera et al. determined antigliadin antibodies in real human serum using an electrochemical immunosensor with control over the orientation and packing of gliadin antigen molecules on the surface of gold electrodes. The orientation of the antigen on the surface has been achieved using a carboxylic ended bipodal alkanethiol that is covalently linked with amino groups of the antigen protein. Amperometric evaluation of the sensor with polyclonal antigliadin antibodies showed stable and reproducible low limits of detection (46 ng mL^{-1}; % RSD = 8.2, n = 5) [20].

Disease or infectious agent	Electrode / immobilization/Sample	Limit of detection
Dengue	Carbon fiber electrode / chitosan with antibody against dengue virus envelope protein / Hs [37]	0.94 ng mL^{-1}
Malaria falciparum	Graphite epoxy composite electrode / magnetic nanoparticle modified with monoclonal antibody against HRP2 / Hs [39]	0.36 ng mL^{-1}
	Screen-printed electrodes / multiwall carbon nanotubes and Au nanoparticles with rabbit anti-PfHRP-2 antibody / Hs [42]	8 ng mL^{-1}
HIV-1 p24 antigen	Multi-walled carbon nanotubes / Au nanoparticles with p 24 antibody / Hs [19]	0.0064 ng mL^{-1}
Hepatitis B	Glassy carbon electrode / assembly of positively charged poly(allylamine)-branched ferrocene (PAA-F$_c$) and hepatitis B surface antibody / Hs [35]	40 pg mL^{-1}
	Gold electrode / Au nanoparticles / HRP and hepatitis B surface antibody / Hs [18]	0.85 ng mL^{-1}
Pneumonia	Screen-printed carbon electrodes / biotinylated anti-pneumolysin monoclonal antibodies Hu [38]	0.12 ng mL^{-1}
Celiac disease	Gold electrode / gliadin antigen / Hs [20]	46 ng mL^{-1}
Urinary infection	Gold electrodes array / alkanethiolate SAM / monoclonal antibody anti-lactoferrin / Hu [47]	145 pg mL^{-1}
Tumor markers	Carbon screen-printed / capture antibody / Hs [40]	0.03-0.05 ng mL^{-1}
Colon cancer	Gold electrode arrays / anti-carcinoembryonic antibody / Hs [21]	0.2 ng mL^{-1}
Rheumatoid arthritis	Carbon nanotube composite electrodes / anti-citrullinated peptide antibody / Hs [22]	> 1:200 dilution Hs
Chagas disease	Au-SPE / antigenic protein (epimastigote membranes) / Hs [51]	−0.104 μA (cut-off)
	Au-SPE / antigenic Tc85 protein (trypomastigote membranes) / Hs [23]	−0.158 μA (cut-off)
	Au-CD-R transducer / antigenic Tc85 protein (trypomastigote membranes) / Hs [27]	−0.949 μA (cut-off)
	Gold electrode / anti-*Trypanosoma cruzi* G / Hs [52]	62 ng mL^{-1}

Human serum (Hs); Human urine (Hu)

Human serum (Hs); Human urine (Hu)

Table 1. Amperometric biosensors for diseases or infectious agents based on immunosensors.

Pan et al. [47] reported the development of an electrochemical immunosensor for direct detection of the urinary tract infection (UTI) biomarker lactoferrin from infected clinical samples. The electrode surfaces were coated with either a SAM of 11-mercaptoundecanoic acid (MUDA) or a mixed of MUDA and 6-mercapto-1-hexanol. A sandwich amperometric immunoassay was developed for detection of lactoferrin from urine, with a limit of detection 145 pg mL^{-1}.

Honglan et al. developed an electrochemical immunosensor array for the simultaneous detection of multiple tumor markers by incorporating electrochemically addressing immobilization and one signal antibody strategy. As a proof-of-principle, an eight-electrode array including six carbon screen-printed working electrodes was used as a base array for the analysis of two important tumor markers, carcinoembryonic antigen (CEA) and a-fetopro-tein (AFP) and a horseradish peroxidase-labeled antibody was used as a signal antibody. The result showed that the steady current density was directly proportional to the concen-tration of target CEA/AFP in the range from 0.10 to 50 ng mL^{-1} with a limit of detection 0.03 and 0.05 ng mL^{-1} for CEA and AFP, respectively [48].

Laboria et al. [21] reported on the development of an amperometric biosensor for detecting CEA in colon cancer detection based on the immobilization of anti-CEA monoclonal anti-body on a novel class of bipodal thiolated self-assembled monolayers containing reactive N-hydroxysuccinimide ester end groups The current variations showed a linear relationship with the concentration of CEA over the range of 0-200 ng mL^{-1} with a sensitivity of 3.8 nA mL ng^{-1} and a limit of detection 0.2 ng mL^{-1}, which is much below the commonly accepted concentration threshold (5 ng mL^{-1}) used in clinical diagnosis.

A simple amperometric immunosensor was constructed to be potentially used for the detec-tion of serum anticitrullinated peptide antibodies, which are specific for rheumatoid arthri-tis (RA) autoimmune disease. Sera of RA patients contain antibodies to different citrullinated peptides and proteins such as fibrin or filaggrin. Herein, a chimeric fibrin-filag-grin synthetic peptide was used as a recognition element anchored to the surface of a multi-walled carbon nanotube-polystyrene-based electrochemical transducer [22].

2.1. Amperometric immunosensors for malaria

Malaria is a serious tropical disease transmitted to humans via the female *Anopheles* mosqui-to and is caused by 4 species of protozoal parasites from the *Plasmodium* genus: *P. falciparum, P. vivax, P. ovale* and *P. malariae. P. falciparum* causes the most severe form of the disease and can be fatal if not correctly treated.

The *P. falciparum* parasite synthesizes several proteins containing large amount of amino acid histidine, which are commonly referred to as histidine-rich proteins (HRP). One of these, HRP2, with 34% histidine and 37% alanine shows a markedly high density among proteins [42].

In recent years, devices for the diagnosis of *P. falciparum* malaria based on HRP2 have signif-icantly gained importance. The abundance of the antigen and the resulting high sensitivity of the diagnostic devices combined with the simplicity of their application make them an obvious alternative in settings where microscopy is not available or not of sufficiently high quality standard [48].

Sharma et al. developed amperometric immunosensor for the detection of HRP2 in the sera of humans with *P. falciparum* malaria. For this purpose, disposable screen-printed electrodes were modified with multiwall carbon nanotubes and Au nanoparticles. Nano-Au/MWCNT/SPEs yielded the highest-level immunosensing performance among the electrodes, with a limit of detection 8 ng mL^{-1} [42].

Castilho et al. [39] used, for the first time, magneto immunoassay-based strategies for the detection of *P. falciparum* histidine-rich protein 2 related to malaria using magnetic micro-

nanoparticles. The immunological reaction for the protein PfHRP-2 was successfully performed in a sandwich assay on magnetic micro- and nanoparticles by using a second
monoclonal antibody labeled with the enzyme horseradish peroxidase (HRP). Then the
modified magnetic particles were easily captured by a magneto sensor made of graphite-ep-
oxy composite (m-GEC) which was also used as the transducer for the electrochemical detection. The schematic representation for the detection of the *P. falciparum* antigen related to
malaria disease in human serum based on a sandwich assay performed on magnetic beads
or nanoparticles modified with a IgM monoclonal antibody (anti-HRP2-MB and anti-HRP2-
MNP, respectively) and using a second IgG monoclonal antibody labeled with the enzyme
horseradish peroxidase (anti-HRP2-HRP) electrochemical signal is showed in the Figure 3.

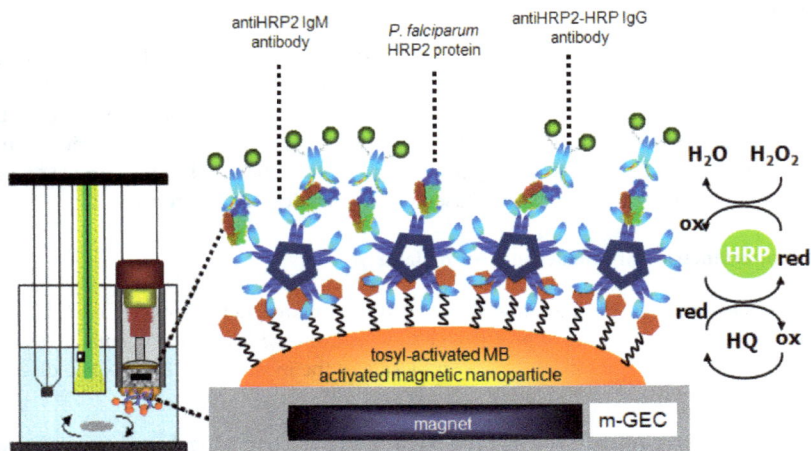

Figure 3. Schematic representation of the experimental details for the electrochemical magneto immunosensor [39].

The electrochemical signal was determined by polarizing the m-GEC electrode at a working
potential of −0.100 V / Ag|AgCl. The electrochemical signal was based on the enzymatic ac-
tivity of the HRP after the addition of hydrogen peroxide as the substrate and hydroquinone
as a mediator. The electrochemical magneto immunosensor coupled with magnetic nanoparticles have shown a limit of detection 0.36 ng mL^{-1} [39].

2.2. Amperometric immunosensors for Chagas disease

Chagas disease, also known as American trypanosomiasis, is a neglected tropical disease
caused by the hemoflagellate *Trypanosoma cruzi* (*T. cruzi*). An estimated 10-15 million people
are infected worldwide, mostly in Latin America where Chagas disease is endemic. More
than 25 million people are at risk of the disease. There is no vaccine for Chagas disease;
therefore, vector control and diagnostic tests are effective methods of preventing Chagas

disease. Blood screening is necessary to prevent infection through transfusion and organ transplantation [49].

The detection of antigen in the blood sera could be useful just for the acute phase of Chagas disease. Detection of anti-*T. cruzi* antibodies in the serologic investigation is the method of choice for the etiological diagnosis of Chagas disease in the chronic phase, considering the specificity and sensitivity of the tests used in the clinical analysis routine. Traditional in clinical practice are the following serological tests using *T. cruzi* antigens: indirect hemaglutination, indirect immunofluorescence and enzyme-linked immunosorbent assay (ELISA) [50].

The methodology for clinical diagnosis must be sensitive and with high reproducibility and repeatability. Different analytical methodologies were developed and amperometric immunosensors were constructed and applied for diagnosis of various diseases stages.

Antigenic proteins (Ag) of *T. cruzi* epimastigote membranes were used for construction of an amperometric immunosensor for serological diagnosis. Proteins with molecular mass ranging from 30 to 100 kDa were immobilized on gold surface of screen-printed electrode treated with self- assembled monoyers (SAMs) of cysteamine (CYS) and glutaraldehyde (GA). Antibodies (Ab) present in the serum of patients with Chagas disease were captured by the immobilized antigens and the affinity interaction was monitored by chronoamperometry at a potential of -400 mV / Ag\|AgCl\|KCl$_{sat.}$ using peroxidase-labeled IgG (Ac*) conjugate and hydrogen peroxide, iodide substrate. Figure 4 shows a scheme of the reactions involved in the steps of SAMs formation, antigen *T. cruzi* immobilization on GA-CYS SAMs and immunoassays. The incubation time to allow maximum antigen-antibody and antibody-peroxidase-labeled IgG interactions was 20 min with a reactivity threshold at -0.104 µA [51]. Another amperometric immunosensor was developed using a specific glycoprotein of the trypomastigote surface (Tc85). The purified recombinant antigen also was immobilized on cysteamine and glutaraldehyde self-assembled monolayers. The affinity reaction was monitored directly using amperometry through a secondary antibody tagged to peroxidase at -400 mV / Ag\|AgCl\|KCl$_{sat.}$ [23]. In both amperometric immunosensors, peroxidase enzyme catalyses the I_2 formation in the presence of hydrogen peroxide and potassium iodide, and the reduction current intensity was measured at a given potential with screen-printed electrodes. The immunosensor was applied to sera of chagasic patients and patients having different systemic diseases with a reactivity threshold at -0.158 µA. Amperometric immunosensor also was developed for determination of Chagas disease through a gold based electrode obtained from a recordable compact disc (CD-R transducer) modified with 4-(methylmercapto)benzaldehyde for the immobilization of Tc85 protein of the *T. cruzi*. The immunoassays were carried out using positive and negative sera from Chagas disease patients and immunoglobulin conjugated with peroxidase enzyme. The immunosensor presented -0.949 µA as cut-off value and was applied in sera samples [27]. It is important to note that the cut-off value obtained for each immunosensor is different because the transducer modifications are not the same.

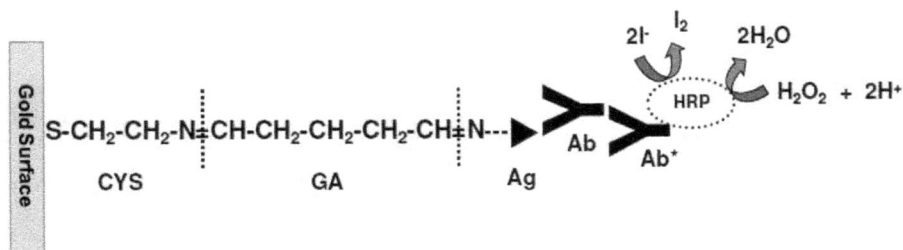

Figure 4. Scheme of the immobilization of antigenic protein on gold modified with SAMs and immunoassays.

Recently, Belluzo et al. applied strategy orientation recombinant proteins to develop amperometric biosensors to diagnose Chagas disease. The gold electrode was modified with thiol and activated the thiolated surface with carbodiimide which allow the subsequent reaction with the amine moieties of the protein Lys residues. The immunoassay involved serum sample anti-*T. cruzi* (analyte), peroxidase-conjugated anti-human immunoglobulin G and with 62 ng mL^{-1} limit of detection [52].

3. DNA based biosensors

Electrochemical biosensors that use DNA, also called genosensors, can be used for analysis and determination of base sequences of DNA to diseases diagnose. DNA molecule has structural features that allow its immobilization on electrode surfaces as single or double helix [53]. Several electrode materials can be modified with DNA, and DNA biosensors can be used for hybridization studies in order for disease diagnosis, mutation detection [54] and also for DNA damage [55] analysis and for detection of antioxidant capacity of many compounds [56]. In this part of the chapter, the focus is on amperometric biosensors for hybridization studies.

DNA hybridization technology has been applied in biosensor systems for diagnosis and it can be considered rapid, with simplicity of execution and lower cost. Hybridization process involves the formation of the DNA duplex by annealing two complementary single strands. The single-stranded DNA (ss-DNA) modified electrode identifies the complementary sequence of nucleic acid in the sample solution leading to the formation of a hybrid double-stranded (ds-DNA). This identification is effective and specific even in the presence of non-complementary sequences [57]. The stability of the hybridization depends on the nucleotide sequences of both strands. A perfect match in the sequence of nucleotides produces very stable ds-DNA, whereas one or more base mismatches impart increasing instability that can lead to weak hybridization of strands [58].

The ability to immobilize the probe DNA in a predictable manner while maintaining their affinity for complementary DNA is an important aspect of genosensors development. The appropriate immobilization is strictly dependent on the characteristics of the transducer, since each of the different immobilization strategies can lead to the proper orientation of bio-molecules, allowing to control the probes conformational freedom, making them accessible for interaction with target DNA and providing minimal steric hindrance. Random DNA attachment to the electrode surface can result in chemical modifications of genetic material basic components, which consequently may cause the decrease in the specificity of layer recognition.

The hybridization event can be direct or indirectly monitored [57,59,60]. Direct detection or label-free detection involves the measurement of changes in electrochemical signals related to the electroactivity of DNA bases, most commonly guanine oxidation. After the hybridization, the steric conformation of the DNA molecule protects the guanine oxidation, causing an electrochemical signal decrease, since the oxidation sites of the base are in the internal parts ds-DNA molecule [61]. Although this method is simple and sensitive, the direct oxidation of DNA requires relatively high potential. Other disadvantage is that such measurement of the decreased anodic signal of the immobilized probe cannot be used for detecting targets containing guanine bases. An alternative is the use of inosine-substituted probes. Guanines in the probe sequence are substituted by inosine residues (pairing with cytosines) and the appearance of a guanine signal upon hybridization with the target enables a new detection method for DNA hybridization [62].

Indirect hybridization detection protocol can be based on the incorporation of electroactive indicators. These compounds, usually cationic metal complexes or organic compounds, have different affinities for the double-stranded DNA (formed after the hybridization process) when compared with single-stranded DNA, preferentially binding with ds-DNA in the groove, by intercalation or electrostatic interaction. Due to variation of redox indicator concentrations near the electrode surface, the resulting current signal indicates the hybridization event. An example of this kind of biosensor is described by Gao & Tansil [63]. After hybridization, a threading intercalator called PIND-Ru was introduced into the biosensor. PIND-Ru selectively intercalated with double-stranded DNA (ds-DNA) and became immobilized on the biosensor surface. The redox moieties of the interacted PIND-Ru showed excellent catalytic activity towards oxidation of amines observed by amperometry at 0.65 V / Ag|AgCl. The current was proportional to the target DNA concentration and a limit of detection 1.5 pmol L^{-1} was determined.

The use of enzymes has shown a good sensitivity for indirect electrochemical hybridization detection. The target DNA sequence is previously labeled with a redox active enzyme which catalyses a redox reaction and further generates an electrochemical change [64]. An electrochemical genosensor array for the individual and simultaneous detection of two high-risk human papillomavirus (HPV) DNA sequences using horseradish peroxidase enzyme (HRP) labeled DNA probes was developed by Civit et al. [65,66]. Using polymerase chain reaction (PCR) products of three specific high-risk HPV sequences, HPV 16, 18 and 45, it was possible to detect DNA in picomolar range. A high specificity of the sensor array was observed with negligible hybridization signal with the non-specific target.

A DNA sensor for West Nile Virus (WNV) was developed by Ionescu et al. [67]. In this work, aminated DNA probe was immobilized on the electrode, followed by hybridization of the WNV complementary DNA target and an additional hybridization process with a complementary biotinylated WNV DNA, resulting in an extremely sensitive detection limit (1 fg mL^{-1}) of WNV DNA target.

Genosensors based on enzyme label have also been applied for diagnosis of some kind of cancer, for example, acute promyelocytic leukemia. Lin et al. [68] employed oligonucleotide derivative that hybridizes with very high affinity to perfectly complementary targets. Hybridization event was monitored by the HRP. The biosensor was applied in PCR amplicon from the fusion gene, which plays an important role in leukemogenesis. Another DNA biosensor for detection of promyelocytic leukemia/retinoic acid receptor alpha fusion gene is described by Wang et al. [69]. This biosensor, based on a 'sandwich' sensing mode, involves a pair of capture probe immobilized at electrode surface and biotinyl reporter probe as an affinity tag for streptavidin-horseradish peroxidase. It allowed detecting the complementary DNA standard concentration range from 0.05 to 5.0 nmol L^{-1}. A large number of studies describe the use of enzymes to monitor amperometrically DNA or RNA hybridization in order to analyze other diseases or infectious agents and some of them are included in Table 2.

As described above, there are many works about DNA biosensor for disease detection or diagnosis purposes. In our research group, we have been working in the development of genosensors for hepatitis C virus (HCV) detection. According to World Health Organization (WHO), hepatitis C affects about 170 million people worldwide and more than 350,000 people die from hepatitis C-related liver diseases each year. Since it rarely causes specific symptoms, hepatitis C is one of the most serious public health problems [70]. In general, the goal of a detection strategy is the simplification of the analytical methodology to a practical level, with a minimum demand of operator skills. In this way, HCV biosensors have become an alternative for diagnosis.

In the first work, we studied a piezoelectric biosensor [71]. Gold electrodes from quartz crystal microbalance were modified with oligonucleotides for detection of hepatitis C virus in serum. Avidin or streptavidin were immobilized and used for attachment of biotinylated DNA probes from four different sequences. The piezoelectric biosensors were used to monitor the DNA resulting from samples from HCV contaminated patients and the results compared with the standard RT-PCR procedure (test kit Roche Amplicor®). The samples characterized as positive in the Amplicor test were able to hybridize with at least one of the four probes immobilized on the piezosensor. However, some of the samples appearing as negative in the Amplicor assay also provided hybridization with some of the immobilized probes. This inconsistency might be explained by different sequences of probes used in the piezosensor assay and in the Amplicor assay (sequence unknown). These results are considered preliminary as not all parameters affecting the hybridization reaction were optimized and the effect of temperature on the double strand formation and stability of hybridized complex on the surface of piezosensor is critical. In our case, all measurements were carried out at room temperature (25 °C), thus allowing for hybridization and duplex formation probably even in the case of only a partial matching between the probe and the amplicon.

Disease or infectious agent	Electrode / immobilization	Sample	Limit of detection	Reference
Colorectal Cancer	Gold / SAM	Synthetic oligonucleotides	5.85 pmol L^{-1}	[72]
Celiac Disease	Gold electrode / SAM	Synthetic oligonucleotides	0.01 nmol L^{-1}	[73]
Pseudomonas aeruginosa	Gold / SAM	Total RNA isolated from *P. aeruginosa*	0.012 pg µL^{-1}	[74]
Uropathogenic bacteria	Gold array / SAM	16S rRNA from bacterial lysis	0.3 fmol L^{-1}	[75]
	Gold array / SAM	16S rRNA from bacterial lysis	0.5 ng µL^{-1} for *E. coli* total RNA	[76]
	Biosensor array	16S rRNA from bacterial lysis	10^4 cfu mL^{-1}	[77]
	Gold SPE / SAM	16S rRNA from bacterial lysis	---	[78]
Escherichia coli	Fe$_2$O$_3$@Au core/shell nanoparticle / SAM	*E. coli* genomic DNA	0.01 pmol L^{-1}	[79]
	Screen-printed electrodes- magnetic beads / STA-biotin	PCR products	0.01 cfu mL^{-1}	[80]
	Gold electrode array / STA-biotin	rRNA from *E. coli*	1000 cells without PCR	[81]
Staphylococcus aureus	Graphite-epoxy electrodes / adsorption onto a nylon membrane	Synthetic oligonucleotides	---	[82]
Enterobacteriaceae family	Gold screen-printed electrodes - magnetic beads / Tetrathiafulvalene	PCR products	5.7 fmol	[83]
Streptococcus pneumoniae	Gold electrode and magnetic beads / STA-biotin	PCR products	1.1 nmol L^{-1}	[84]

SAM: self-assembled monolayer; STA: streptavidin; PCR: polymerase chain reaction; rRNA: Ribosomal ribonucleic acid; cfu: colony-forming unit.

Table 2. Amperometric biosensors for diseases or infectious agents based on DNA or RNA hybridization.

A selective and sensitive label free electrochemical detection method of DNA hybridization for HCV was proposed in cooperation with Dr. M. Josowicz's research group [85]. DNA probes of specific sequence HCV type-1 were immobilized on polypyrrole films deposited on Pt microelectrodes. The monitoring of the hybridization with the complementary DNA was based on electrostatic modulation of the ion-exchange kinetics of the polypyrrole film and it allowed the detection of HCV-1 with a limit of detection 1.82×10^{-21} mol L^{-1}. With this biosensor, HCV-1 DNA detection did not show unspecific interactions in the presence of mismatched sequences from different HCV genotypes as 2a/c, 2b, and 3.

An advantage of the construction of DNA biosensors is the use of disposable electrodes. These electrodes have a low construction cost, good reproducibility of the area, the possibility of large scale production, and the absence of surface inactivation. Different disposable electrodes as recordable gold CD-R and pencil graphite electrodes (PGE) have being used.

Using PGE, we developed a disposable HCV genossensor with thin films siloxane-poly(propylene oxide) hybrids prepared by sol-gel method and deposited on the electrode surface by dip-coating process [86]. The streptavidin (STA) was encapsulated in the films and biotinylated 18-mer DNA probes for hepatitis C virus (genotypes 1, 2a/c, 2b and 3) were immobilized through STA, since strong interaction occurs between the avidin (or streptavidin) and biotin. The complementary DNA was hybridized to the target-specific oligonucleotide probe immobilized and followed by avidin-peroxidase labeling. Hybridization event was detected by amperometrically monitoring the enzymatic response at $-0.45V$ / Ag|AgCl using H_2O_2 as enzyme substrate and KI as electron mediator. Negative and positive controls and positive samples of sera patients were analyzed and the HCV 1, 2a/c, 2b and 3 oligonucleotide probes immobilized on PGE were able to distinguish positive and negative sera samples.

Chemometric studies were applied to the development of another biosensor for hepatitis C virus using PGE [87]. Fractional factorial and factorial with center point design were applied in order to simultaneously evaluate the variables of interest that have significant influence on the biosensor response. MINITAB software generated level combinations for all factors used in the assays. Then the sensor current was measured by controlled potential amperometric technique for each of these level combinations. This strategy had several advantages, such as a reduced number of experimental runs, more information obtained and biosensor delineation, in which the biosensor response permitted the optimal experimental conditions to be determined. It was possible to optimized concentration and incubation time for all biomolecules studied with this biosensor using the developed methodology. We also demonstrated the applicability of full factorial and fractional factorial designs to the immobilization of DNA molecules at a gold electrode built using a recordable compact disc (CDtrode) [88].

For DNA immobilization on electrode surfaces, the optimization of many parameters is necessary, such as: biomolecules concentration and incubation time. In this way, the biosensor for HCV, illustrated in Figure 5, was developed using chemometric experiments applied to steps 4-6 (Figure 5). The evaluated variables were the degree of dilution and incubation time of DNA probes for HCV-1, dilution and incubation time of complementary DNA, and concentration and incubation time of conjugate avidin-HRP, which was the label for hybridization accompanied by amperometry measurements. After establishment of all optimized parameters for biomolecule immobilization, the amperometric genosensor was applied to HCV-1 DNA detection in different HCV-infected patients, which had been previously analyzed by the standard qualitative Amplicor hepatitis C Virus Test. The results showed that the current intensities for the positive samples were higher than those for the negative samples. The factorial design procedure enables identification of critical parameters, while knowledge of the chemistry involved enables further refinement of the technique, where necessary. Full and fractional factorial design methods were employed for the optimization of a biosensor for hepatitis C diagnosis, and could be extended to other types of DNA-based biosensors.

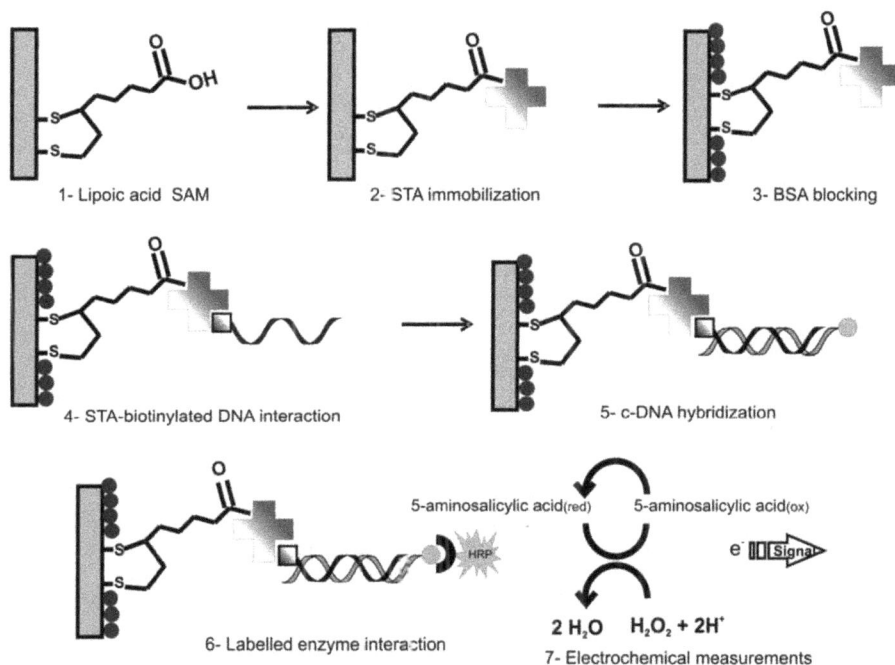

Figure 5. Scheme of DNA biosensor construction with gold CDtrodes [88].

According to the literature, biosensors rank fourth among the techniques used for the detection and classification of pathogens, behind the polymerase chain reaction (PCR), culture and colony counting and ELISA methods [89]. The reason for that is DNA biosensors offer several advantages, such as the ability to analyze complex fluids, high sensitivity, compatibility with compact instrumentation technology and portability, becoming a good alternative for application in clinical chemical analysis.

4. Enzyme based biosensor

Enzymes play a critical role in the metabolic activities of all living organisms and are widely applied in biotechnology. Abnormality of the enzyme metabolism systems leads to a number of metabolic diseases [90]. Diseases associated with components of the enzyme metabolism or with the enzyme activities are broadly applied in clinical examinations as special markers as some examples displayed on Table 3.

Disease	Enzyme	Electrode / immobilization	Limit of detection	Reference
Diabetes mellitus	glucose oxidase	Gold nanocomposite/poly(pyrrole propylic acid)	50 mmol L^{-1}	[91]
		Graphene/nafion Film	30 mmol L^{-1}	[92]
Uremia	urease	Rhodium nanoparticles/acrylonitrile copolymer membrane	500 mmol L^{-1}	[93]
		Platinum and graphite composite/ urease covered with dialysis membrane	---	[94]
Heart failure, Respiratory insufficiency, Metabolic Disorders	lactate oxidase	Carbon screen-printed/mesoporous silica	18.3 µmol L^{-1}	[95]
		Carbon screen-printed/polysulfone-carbon nanotubes	1.5 mmol L^{-1} 3.46 µmol L^{-1}	[96]
Idiopathic urolithiasis, intestinal diseases	oxalate oxidase	Gold electrode/multi-walled carbon nanotube-gold nanoparticle composite	1 µmol L^{-1}	[97]
		Platinum/multi-walled carbon nanotubes-polyaniline composite film	3 µmol L^{-1}	[98]
Muscle damage	creatinine amidohydrolase	Platinum/multi-walled carbon nanotube-polyaniline composite film	0.1 µmol L^{-1}	[99]
		Platinum/PbO$_2$ layer-polyurethane membrane	0.8 µmol L^{-1}	[100]

Table 3. Amperometric biosensor for disease based on enzyme.

4.1. Biosensor for substrate determination

Cholesterol and its fatty acid ester are extremely important compounds for human beings since they are components of neural and brain cells and are precursors of other biological materials, such as bile acid and steroid hormones. However, high cholesterol accumulation in blood due to excessive ingestion results in fatal diseases, such as arteriosclerosis, cerebral thrombosis, myocardial infarction, coronary diseases and lipid metabolism dysfunction [101]. Brahim et al. [102] developed a rapid, two-step method for constructing cholesterol biosensors by entrapment of cholesterol oxidase within a composite poly(2-hydroxyethyl methacrylate) (p(HEMA))/polypyrrole (p(pyrrole)) membrane. The optimized cholesterol biosensor exhibited a linear response range from 500 µmol L^{-1} to 15 mmol L^{-1} and limit of detection 120 µmol L^{-1} toward cholesterol and was applied in the analysis of serum samples from hospitalized patients. A review on cholesterol biosensor is published by Arya [103].

Choline is used as a marker of cholinergic activity in brain tissue, especially in the field of clinic detection of neurodegenerative disorder diseases, such as Parkinson's and Alzheimer's diseases. Zhang et al. [104] presented an electrochemical approach for the detection of choline based on prussian blue (PB) modified iron phosphate nanostructures (PB-FePO$_4$), being the amperometric choline biosensor developed by immobilizing the enzyme choline oxidase on the PB-FePO$_4$ nanostructures and monitoring the formation of H$_2$O$_2$. The biosensor exhibited a low limit of detection (0.4 ± 0.05 μmol L^{-1}) and a wide linear range (2 μmol L^{-1} to 3.2 mmol L^{-1}). López et al. [105] designed a choline amperometric biosensor using as biological component choline oxidase entrapped in polyacrylamide microgels. The working electrode was prepared by holding the enzyme loaded microgels on a platinum electrode by a dialysis membrane. Under optimal conditions the biosensor presented high sensitivity for choline with limit of detection 8 μmol L^{-1}, and the response linear range from 20 μmol L^{-1} to 0.2 mmol L^{-1}. On the other hand, Lenigk et al. proposed methodology for the clinical purpose of evaluating anti-Alzheimer medicine based on the inhibition of acetylcholinesterase [106].

Phenylketonuria is a disease characterized by not metabolizing phenylalanine resulting in brain damage and mental retardation in children. A carbon paste electrode composed by paraffin oil, NAD$^+$, phenylalanine dehydrogenase, uricase and electron mediator was proposed [107] for aminoacid determination in urine sample. The reagentless biosensor presented a limit of detection 0.5 mmol L^{-1}.

Among biosensors for substrate determination, the most investigated and more successful on the commercial point of view is for glucose determination; probably because the diabetes mellitus is a world health problem, but also due to the stability of glucose oxidase (GOX).

The stability of enzymatic biosensors is important for the success of these devices as analytical instruments, and it is mainly dependent on the lifetime, or the rate of denaturation or inactivation of the immobilized enzyme [95]. Depending on the conditions of storage, temperature and method of immobilization, the enzyme can retain the activity from days to months [91-100], and is often one of the most important factors to take into account for the commercial viability of such device.

4.2. Biosensor for enzyme activities determination

Abnormal enzymes concentration can be related to diseases as shown.

Trypsin and trypsinogen levels are increased with pancreatitis disease like acute pancreatitis, cystic fibroses. Radioimmunoassay tests estimated 248 ± 94,9; 1100 ± 548 and 1399 ± 618 μg L^{-1} for healthy, chronic renal failure and acute pancreatitis, respectively. Ionescu et al. proposed a biosensor based on the suppression of GOX by steric hindrance due to a gelatin membrane and its reactivation by trypsin digestion of blocking membrane: the GOX was previously mixed with pyrrole and adsorbed onto platinum electrode after that the enzyme was entrapped into the polypyrrole film by electropolimerization at +0.8 V / Ag|AgCl| KCl$_{sat}$. LOD was 42 pmol L^{-1} and response time 10 min [108].

Aspartate aminotranferase is an enzyme to diagnose acute myocardial infarction [109]. A biosensor based on Os-HRP layer and a layer composed by hydroxyethylcellulose, microcrystalline cellulose, aspartic acid, cetoglutaric acid and pyridoxil onto the gold electrode was proposed by Guo, et al. [110]. The LOD was 10 U L^{-1}, shelf stability 2 months, response time 120 s.

Adenosine deaminase (ADA) level is a biomarker for liver disease. A printed Ir/C was modified by xanthine oxidase and purine nucleoside phosphorylase; through the H_2O_2 measurement at potential of +0.27 V / Ag|AgCl the ADA activities in blood sample were determined. Linear calibration curve from 0 to 36 U L^{-1} was obtained, which is suitable for discriminating a healthy individual from a person suffering of liver disease, 18 and 31.6 U L^{-1}, respectively [111].

Reviews on age-related disease [112], clinical chemistry [113], cancer clinical testing [114], technology of commercial glucose monitoring [115] and glucose biosensor based on carbon nanomaterials [116] have been recently published.

5. Concluding remarks

Two aspects are very important to consider in biosensor development: the biological component determines the selectivity while the transducer determines the sensitivity. To guarantee the maximum selectivity, the active center of a biological molecule must be chemically and/or physically accessible and as freer as possible of steric effects. The surface preparation and modification of the transducer need to be thought mainly to reach this goal. In this case, the affinity reaction between different molecules such as antigen/antibody or DNA/DNA or enzymatic catalytic reaction can be used for quantification of biological substances which are important for the medicine and clinical analysis. The tendency is to produce more and more sophisticated and specific surface transducers using surface engineering and nano-technological tools to get the best biosensor device. If this happens, health workers will believe more in this bioanalytical methodology and they may get benefits from it in the instant of giving to the patient an unequivocal diagnostic of disease.

Acknowledgment

The authors thank to FAPESP (Proc. 2008/08990-1, 2011/10707-9, 2008/07729-8, 2010/04663-6), CNPq (Proc. 305890/2010-7, 313307/2009-1), CAPES and PROPe-UNESP for financial support.

Author details

Antonio Aparecido Pupim Ferreira, Carolina Venturini Uliana, Michelle de Souza Castilho, Naira Canaverolo Pesquero, Marcos Vinicius Foguel, Glauco Pilon dos Santos, Cecílio Sadao Fugivara, Assis Vicente Benedetti and Hideko Yamanaka

Instituto de Química, UNESP - Univ Estadual Paulista, Brazil

References

[1] Thevenot D.R.; Toth K.; Durs⁻ R.A.; Wilson G.S. Electrochemical b⁻osensors: recommended definitions and clas⁻ifications. Pure and Applied Chemistry 1999; 12, 2333-2348.

[2] Cammann K. Bio-sensors based on ion-selective electrodes, Fresenius Zeitschrift Analytical Chemistry 1977; 287, 1-9.

[3] Ferreira A.A.P.; Fugivara C.S.; Barrozo S.; Suegama P.H.; Yamanaka H.; Benedetti A.V.; Electrochemical and spectroscopic characterization of screen-printed gold based electrodes modified with self-assembled monolayers and Tc85 protein. Journal of Electroanalytical Chemistry 2009; 634, 111-122.

[4] Noel M.; Vasu K.I. Cyclic Votammetry and the Frontiers of Electrochemistry, Aspect Publications Ltd., 1990.

[5] Gosser Jr. D.K. Cylic Voltammetry - Simulation and Analysis of Reaction Mechanisms, VCH Publishers, 1993.

[6] Compton R.G., Banks C.E. Undestanding Voltammetry, Word Scientific ed., London, 2009.

[7] Barsoukov E., Macdonald J.R. Impedance spectroscopy theory, experiment, and applications, John Wiley & Sons, USA, 2005.

[8] Orazem M. E.; Tribollet B. Electrochemical impedance spectroscopy John Wiley & Sons, Inc., Hoboken, N. J., 2008.

[9] Ferreira A.A.P.; Fugivara C.S.; Yamanaka H.; Benedetti A.V. Preparation and characterization of immunosensors for disease diagnosis. In: Serra A.P. (ed). Biosensors for Health, Environment and Biosecurity. Rijeka: In Tech; 2011. p.183-214.

[10] Ferreira A.A.P.; Alves M.J.M.; Barrozo S.; Yamanaka H.; Benedetti A.V. Optimization of incubation time of protein Tc85 in the construction of biosensor: Is the EIS a good tool? Journal of Electroanalytical Chemistry 2010; 643, 1-8.

[11] Bagotsky V.S., Fundamentals of Electrochemistry, 2nd edition, John Wiley & Sons. Inc., NJ, 2006.

[12] Wightman R.M.; Wipf D.O. In: Bard A.J. (ed.). Electroanalytical Chemistry, vol. 15, Marcel Dekker, New York, 1989.

[13] Westbroek P.; Priniotakis G.; Kiekens P. Analytical electrochemistry in textiles, CRC Press, NY, 2005.

[14] Bard A.J.; Faulkner L.R. Electrochemical Methods – Fundamentals and Applications, 2nd edition, John Wiley & Sons. Inc., NY, 2001.

[15] Macdonald D.D. Transient Techniques in Electrochemistry, 1st edition, NY, 1977.

[16] Fleishmann M.; Pons S.; Robson D.; Schmidt P.P. Ultramicroelectrodes. Datatech Science Morganton, N. C., 1987.

[17] Wightman R.M.; Wipf D.O. Voltammetry at Ultramicroelectrodes In: Bard A.J. (ed.). Electroanalytical Chemistry,15., New York: Marcel, 1989.

[18] Zhuo, Y.; Yuan, R.; Chai, Y.; Zhang, Y.; Li, X. I.; Zhu, Q.; Wang, N. An amperometric immunosensor based on immobilization of hepatitis B surface antibody on gold electrode modified gold nanoparticles and horseradish peroxidase. Analytica Chimica Acta 2005; 548, 205-210.

[19] Kheiri, F.; Sabzi, R.E.; Jannatdoust, E.; Shojaeefar, E.; Sedghi, H. A novel amperometric immunosensor based on acetone-extracted propolis for the detection of the HIV-1 p24 antigen. Biosensors and Bioelectronics 2011; 26, 4457-4463.

[20] Rosales-Rivera, L.C.; Acero-Sánchez, J.L.; Lozano-Sánchez, P.; Katakis, I.; O'Sullivan, C.K. Electrochemical immunosensor detection of antigliadin antibodies from real human serum. Biosensors and Bioelectronics 2011; 26, 4471-4476.

[21] Laboria, N.; Fragoso, A.; Kemmner, W.; Latta, D.; Nilsson, O.; Botero, M. L.; Drese, K.; O'Sullivan, C. K. Carcinoembryonic antigen in colon cancer samples based on monolayers of dentritic bipodal scaffolds. Analytical Chemistry 2012; 82, 1712-1719.

[22] Villa, M. G.; Jiménez-Jorquera, C.; Haro, I.; Gomara, M. J.; Sanmartí, R.; Fernández-Sánchez, C.; Mendoza, E. Carbon nanotube composite peptide-based biosensors as putative diagnostic tools for rheumatoid arthritis. Biosensors and Bioelectronics 2011; 27, 113-118.

[23] Ferreira, A.A.P.; Colli, W.; Alves, M.J.M.; Oliveira, D.R.; Costa, P.I.; Güell, A.G.; Sanz, F.; Benedetti, A.V.; Yamanaka, H. Investigation of the interaction between Tc85-11 protein and antibody anti-T. cruzi by AFM and amperometric measurements. Electrochimica Acta 2006; 51, 5046-5052.

[24] Angnes L.; Richter E.M.; Augelli M.A.; Kume G.H. Gold electrodes from recordable CDs. Analytical Chemistry 2000; 72, 5503-5506.

[25] Richter E.M.; Augelli M.A.; Kume G.H.; Mioshi R.N.; Angnes L. Gold electrodes from recordable CDs for mercury quantification by flow injection analysis. Fresenius Journal Analytical Chemistry 2000; 366, 444-448.

[26] Lowinsohn D.; Richter M.A.; Angnes L.; Bertotti M. Disposable gold electrodes with reproducible area using recordable CDs and toner masks. Electroanalysis 2006; 18, 89-94.

[27] Foguel, M.V.; Santos, G.P.; Ferreira, A.A.P.; Yamanaka, H.; Benedetti, A.V. Amperometric immunosensor for Chagas'disease using gold CD-R transducer. Electroanalysis 2011; 23, 2555-2561.

[28] Foguel M.V.; Uliana C.V.; Tomaz P.R.U.; Marques P.R.B.O.; Yamanaka H.; Ferreira A. A. P. Avaliação da limpeza de CDtrodo construídos a partir de CD de ouro gravável/fita adesiva de galvanoplastia. Eclética Química 2009; 34, 59-66.

[29] Foguel M.V. Master Dissertation, Instituto de Química-UNESP, 2010.

[30] Kaláb T.; Skládal P. Disposable amperometric immunosensor for 2,4-dichlorophenoxyacetic acid. Analytical Chimica Acta 1995; 304, 361-368.

[31] Navrátilová I.; Skaládal P. The immunosensors for measurement of 2,4-dichlorophenoxyacetic acid based on electrochemical impedance spectroscopy. Bioelectrochemistry 2004; 62, 11-18.

[32] Carpini G.; Lucarelli F.; Marrazza G.; Mascini M. Oligonucleotide-modified screen-printed gold electrodes for enzyme-amplified sensing of nucleic acids. Biosensors and Bioelectronics 2004; 20, 167-175.

[33] García-González R.; Fernández-Abedul M.T.; Pernía A.; Costa-García A. Electrochemical characterization of different screen-printed gold electrodes. Electrochimica Acta 2008; 53, 3242-3249.

[34] Bóka B.; Adányi N.; Virág D.; Sebela M.; Kiss A. Spoilage detection with biogenic amine biosensors, comparison of different enzyme electrodes. Electroanalysis 2012; 24, 181-186.

[35] Qiu, J.D.; Huang, H.; Liang, R.P. Biocompatible and label-free amperometric immunosensor for hepatitis B surface antigen using a sensing film composed of poly (allylamine)-branched ferrocene and gold nanoparticles. Microchimica Acta 2011; 174, 97-105.

[36] Wang J. Analytical Electrochemistry, 2nd ed., Wiley VCH, NY, 2000.

[37] Cavalcanti, I.T.; Silva, B.V.M.; Peres, N.G.; Moura, P.; Sotomayor, M.I.F. G.; Dutra, R. F. (2012). A disposable chitosan-modified carbon fiber electrode for dengue virus envelope protein detection. Talanta 2012; 91, 41-46.

[38] González, M.D., García, M.B.G., García, A.C. Detection of pneumolysin in human urine using an immunosensor on screen-printed carbon electrodes. Sensors and Actuators B 2006; 113, 1005-1011.

[39] Castilho, M.S.; Laube, T.; Yamanaka, H.; Alegret, S.; Pividori, M.I. Magneto immunoassays for plasmodium falciparum histidine-rich protein 2 related to malaria based on magnetic nanoparticles. Analytical Chemistry 2011; 83, 5570-5577.

[40] Honglan, Q., Chen, L., Qingyun, M , Qiang, G., Chengxiao, Z. Sensitive electrochemical immunosensor array for the simultaneous detection of multiple tumor markers. Analyst 2012; 137, 393-399.

[41] Pacios M.; Martín-Fernández I.; Villa R.; Godignon P.; Del Valle M.; Bartrolí J.; Esplandiu M.J. Carbon nanotubes as suitable electrochemical platforms for metallopro-

tein sensors and genosensors. In: Naraghi M. (ed). Carbon Nanotubes – Growth and Applications, Rijeka: InTech; 2011. p. 299-324.

[42] Sharma, M.K.; Rao, V.K.; Agarwal, G.S.; Rai, G.P.; Gopalan, N.; Prakash, S.; Sharma, S.K.; Vijayaraghavan, R. Highly sensitive amperometric immunosensor for detection of Plasmodium falciparum histidine-rich protein 2 in serum of humans with malaria: comparison with a commercial kit. Journal of Clinical Microbiology 2008; 46, 3759-3765.

[43] Luppa, P.B.; Sokoll, L.J.; Chan, D.W. Immunosensors – principles and applications to clinical chemistry. Clinica Chimica Acta 2001; 314, 1-26.

[44] Mello, L.D., Kubota, L.T. Review of the use of biosensors as analytical tools in the food and drink industries. Food Chemistry 2002; 77, 237-256.

[45] Kim, J.H.; Yeo, W.H.; Shu. Z.; Soelberg, S.D.; Inoue, S.; Kalyanasundaram, D.; Ludwig. J.; Furlong. C.E.; Riley, J.J.; Weigel, K.M.; Cangelosi, G.A.; Oh, K.; Lee, K.H.; Gao, D.; Chung, J.H. (2012). Immunosensor towards low-cost, rapid diagnosis of tuberculosis. Lab on a Chip. 2012; 12, 1437-40.

[46] Ramírez, N.B.; Salgado, A.M.; Valdman, B. The evolution and developments of immunosensors for health and environmental monitoring: problems and perspectives. Brazilian Journal of Chemical Engineering 2009; 26, 227-249.

[47] Pan, A.N., Y.; Sonn, G.A; Sin, M.L.Y.; Mach, K.E.; Shih, M-C.; Gau, V.; Wong, P.K.; Liao, J.C. Electrochemical immunosensor detection of urinary lactoferrin in clinical samples for urinary tract infection diagnosis. Biosensors and Bioelectronics 2010; 26, 649-654.

[48] Noedl, H.; Yingyuen, K.; Laoboonchai, A.; Fukuda, M.; Sirichaisinthop, J.; Miller, R.S. (2006). Sensitivity and specificity of an antigen detection ELISA for malaria diagnosis. American Journal of Tropical Medicine and Hygiene 2006; 75, 1205-1208.

[49] World Health Organization (WHO). Chagas disease (American trypanosomiasis). WHO Fact Sheet 2010; 340.

[50] Brener, Z.; Andrade, Z. A.; Barral-Netto, M. Trypanosoma cruzi e doença de Chagas. Guanabara Koogan, Rio de Janeiro, 2000.

[51] Ferreira, A.A.P.; Colli, W.; Costa, P.I.; Yamanaka, H. Immunosensor for the diagnosis of Chagas' disease. Biosensors and Bioelectronics 2005; 21, 175-181.

[52] Belluzo, M.S; Ribone, M.E.; Camussone, C.; Marcipar, I.S.; Lagier, C. M. Favorably orienting recombinant proteins to develop amperometric biosensors to diagnose Chagas' disease. Analytical Chemistry 2011; 408, 86-94.

[53] Yang, M.; Msgovern, M.V.; Thompson, M. Genosensor technology and the detention of interfacial nucleic acid chemistry. Analytica Chimica Acta 1997; 346, 259-275.

[54] Garcia, T.; Fernandez-Barrena, M.G.; Revenga-Parra, M.; Nunez, A.; Casero, E.; Parie-nte, F.; Prieto, J.; Lorenzo, E. Disposable sensors for rapid screening of mutated genes. Analytical and Bioanalytical Chemistry 2010; 39, 1385-139.

[55] Fojta, M. Detecting DNA damage with electrodes. Perspectives in Bioanalysis 2005; 1, 385-431.

[56] Prieto-Simón, B.; Cortina, M.; Campás, M.; Calas-Blanchard, C. Electrochemical bio-sensors as a tool for antioxidant capacity assessment. Sensors and Actuators B 2008; 129, 459-466.

[57] Teles, F.R.R.; Fonseca, L.P. Trends in DNA biosensors. Talanta 2008; 77, 606-623.

[58] Fojta, M. Electrochemical sensors for DNA interactions and damage. Electroanalysis 2002; 14, 1494-1500.

[59] Labuda, J.; Brett, A.M.O.; Evtugyn, G.; Fojta, M.; Mascini, M.; Ozsoz, M.; Palchetti, I.; Paleček, E.; Wang, J. Electrochemical nucleic acid-based biosensors: concepts, terms, and methodology (IUPAC Technical Report). Pure and Applied Chemistry 2010; 82, 1161-1187.

[60] Belluzo, M.S.; Ribone, M.E.; Lagier, C.M. Assembling amperometric biosensors for clinical diagnostics. Sensors 2008; 8, 1366-1399.

[61] Kerman, K.; Kobayashi, M.; Tamiya, E. Recent trends in electrochemical DNA biosen-sor technology. Measurement Science and Technology 2004; 15, R1.

[62] Wang, J.; Rivas, G.; Fernandes, J.R.; Paz, J.L.L.; Jiang, M.; Waymire, R. Indicator-free electrochemical DNA hybridization biosensor. Analytica Chimica Acta 1998; 375, 197-203.

[63] Gao, Z.; Tansil, N. A DNA biosensor based on the electrocatalytic oxidation of amine by a threading intercalator. Analytica Chimica Acta 2009; 636, 77-82.

[64] Sassolas, A.; Leca-Bouvier, B.D.; Blum, L.J. DNA Biosensors and Microarrays. Chemi-cal Reviews 2008; 108, 109-139.

[65] Civit, L.; Fragoso, A.; O'Sullivan, C.K. Electrochemical biosensor for the multiplexed detection of human papillomavirus genes. Biosensors and Bioelectronics 2012; 26, 1684-1687.

[66] Civit, L.; Fragoso, F.; Hölters, S.; Dürst, M.; O'Sullivan, C.K. Electrochemical geno-sensor array for the simultaneous detection of multiple high-risk human papilloma virus sequences in clinical samples. Analytica Chimica Acta 2012; 715, 93-98.

[67] Ionescu, R. E.; Herrmann, S.; Cosnier, S.; Marks, R. S. A polypyrrole cDNA electrode for the amperometric detection of the West Nile Virus. Electrochemistry Communi-cations 2006; 8, 1741-1748.

[68] Lin, L.; Liu, Q.; Wang, L.; Liu, A.; Weng, S.; Lei, Y.; Chen, W.; Lin, X.; Chen, Y. Enzyme-amplified electrochemical biosensor for detection of PML–RARαfusion gene
based on hairpin LNA probe. Biosensors and Bioelectronics 2011; 28, 277-283.

[69] Wang, K.; Sun, Z.; Feng, M.; Liu, A.; Yang, S.; Chen, Y.; Lin, X. Design of a sandwich
mode amperometric biosensor for detection of PML/RARαfusion gene using locked
nucleic acids on gold electrode. Biosensors and Bioelectronics 2011; 26, 2870-2876.

[70] World Health Organization. WHO: Hepatitis C. <http://www.who.int/mediacentre/
factsheets/fs164/en/>. (accessed 4 June 2012).

[71] Skládal, P.; Riccardi, C.S.; Yamanaka, H.; Costa, P.I. Piezoelectric biosensors for real-
time monitoring of hybridization and detection of Hepatitis C virus. Journal of Viro-
logical Methods 2004; 117, 145-151.

[72] Wang, Z.; Yang, Y.; Leng, K.; Li, J.; Zheng, F.; Shen, G.; Yu, R. A sequence-selective
electrochemical DNA biosensor based on HRP-labeled probe for colorectal cancer
DNA detection. Analytical Letters 2008; 41 24-35.

[73] Ortiz, M.; Torréns, M.; Canela, N.; Fragoso, A.; O'Sullivan, C.K. Supramolecular con-
finement of polymeric electron transfer mediator on gold surface for picomolar detection of DNA. Soft Matter 2011; 7, 10925-10930.

[74] Liu, C.; Zeng, G. M.; Tang, L.; Zhang, Y.; Li, Y. P.; Liu, Y. Y.; Li, Z.; Wu, M.S.; Luo, J.
Electrochemical detection of Pseudomonas aeruginosa 16S rRNA using a biosensor
based on immobilized stem-loop structured probe. Enzyme and Microbial Technology 2011; 49 266- 71.

[75] Liao, J.C.; Mastali, M.; Gau, V.; Suchard, M.A.; Møller, A.K.; Bruckner, D.A.; Babbitt,
J. T.; LI, Y.; Gornbein, J.; Landaw, E. M.; Mccabe, E. R. B.; Churchill, B. M.; Haake, D.
A. Use of electrochemical DNA biosensors for rapid molecular identification of uro-
pathogens in clinical urine specimens. Journal of Clinical Microbiology 2006; 44,
561-570.

[76] Elsholz, B.; Worl, R.; Blohm, L.; Albers, J.; Feucht, H.; Grunwald, T.; Jurgen, B.;
Schweder, T.; Hintsche, R. Automated detection and quantitation of bacterial RNA
by using electrical microarrays. Analytical Chemistry 2006; 78, 4794-4802.

[77] Mach, K. E.; Du, C. B.; Phull, H.; Haake, D. A.; Shih, M. C.; Baron, E. J.; Liao, J. C.
Multiplex pathogen identification for polymicrobial urinary tract infections using biosensor technology: A Prospective Clinical Study. The Journal of Urology 2009; 182,
2735-2741.

[78] Liao, J.C.; Mastali, M.; Li, Y.; Gau, V.; Suchard, M. A.; Babbitt, J.; Gornbein, J.; Landaw, E.M.; Mccabe, E.R.B.; Churchill, B.M.; Haake, D.A. Development of an advanced
electrochemical DNA biosensor for bacterial pathogen detection. Journal of Molecular Diagnostics 2007; 9, 158-168.

[79] Li, K.; Lai, Y.; Zhang, W.; Jin, L. Fe2O3@Au core/shell nanoparticle-based electrochemical DNA biosensor for Escherichia coli detection. Talanta 2011; 84 607-613.

[80] Loaiza, O.A.; Campuzano, S.; Pedrero, M.; García P.; Pingarrón, J.M. Ultrasensitive detection of coliforms by means of direct asymmetric PCR combined with disposable magnetic amperometric genosensors. Analyst 2009; 134, 34-37.

[81] Gau, J.J.; Lan, E.H.; Dunn, B.; Ho, C.M.; Woo, J.C.S. A MEMS based amperometric detector for E. Coli bacteria using self-assembled monolayers. Biosensors and Bioelectronics 2001; 16, 745-755.

[82] Pividori, M.I.; Merkoçi, A.; Alegret, S. Dot-blot amperometric genosensor for detecting a novel determinant of b-lactamase resistance in Staphylococcus aureus. Analyst 2001; 126, 1551-1557.

[83] Loaiza, O.A.; Campuzano, S.; Pedrero, M.; Pividori, M.I.; García, P.; Pingarrón, J.M. Disposable magnetic DNA sensors for the determination at the attomolar level of a specific Enterobacteriaceae family gene. Analytical Chemistry 2008; 80, 8239-8245.

[84] Campuzano, S.; Pedrero, M.; García, J.L.; García, E.; García, P.; Pingarrón, J.M. Development of amperometric magnetogenosensors coupled to asymmetric PCR for the specific detection of Streptococcus pneumonia. Analytical and Bioanalytical Chemistry 2011; 399, 2413-2420.

[85] Riccardi, C.S.; Kranz, C.; Kowalik, J.; Yamanaka, H.; Mizaikoff, B.; Josowicz, M. Label-free DNA detection of Hepatitis C virus based on modified conducting polypyrrole films at microelectrodes and atomic force microscopy tip-integrated electrodes. Analytical Chemistry 2008; 80, 237-245.

[86] Riccardi, C.S.; Dahmouche, K.; Santilli, C.V.; Costa, P.I.; Yamanaka, H. Immobilization of streptavidin in sol-gel films: application on the diagnosis of Hepatitis C virus. Talanta 2006; 70, 637-643.

[87] Uliana, C.V.; Riccardi, C.S.; Tognolli, J.O.; Yamanaka, H. Optimization of an amperometric biosensor for detection of Hepatitis C virus using fractional factorial designs. Journal of the Brazilian Chemical Society 2008; 19, 782-787.

[88] Uliana, C.V.; Tognolli, J.O.; Yamanaka, H. Application of factorial design experiments to the development of a disposable amperometric DNA biosensor. Electroanalysis 2011; 23, 2607-2615.

[89] Lazcka, O.; Campo, F.J; Muñoz, F.X. Pathogen detection: A perspective of traditional methods and biosensors. Biosensors and Bioelectronics 2007; 2, 1205-1217.

[90] Raja, M.M.M.; Raja, A.; Imran, M.M.; Santha, A.M.I.; Devasena, K. Enzymes application in diagnostic prospects. Biotechnology 2011; 10, 51-59.

[91] Senel, M.; Nergiz, C. Novel amperometric glucose biosensor based on covalent immobilization of glucose oxidase on poly(pyrrole propylic acid)/Au nanocomposite. Current Applied Physics 2012; 12, 1118-1124.

[92] Zhang, Y.; Fan, Y.; Wang, S.; Tan, Y.; Shen, X.; Shi, Z. Facile Fabrication of a gra-
      phene-based electrochemical biosensor for glucose detection. Chinese Journal of
      Chemistry 2012, 30, 1163-1167.

[93] Gabrovska, K.; Ivanov, J.; Vasileva, I.; Dimova, N.; Godjevargova, T. Immobilization
      of urease on nanostructured polymer membrane and preparation of urea ampero-
      metric biosensor. International Journal of Biological Macromolecules 2011; 48,
      620-626.

[94] Pizzariello, A.; Stredansky, M.; Stradanska, S.; Miertus, S. Urea biosensor based on
      amperometric pH-sensing with hematein as a pH-sensitive redox mediator. Talanta
      2001; 54, 763-772.

[95] Shimomura, T.; Sumiya, T.; Ono, M.; Ito, T.; Hanaoka, T. Amperometric L-lactate bio-
      sensor based on screen-printed carbon electrode containing cobalt phthalocyanine,
      coated with lactate oxidase-mesoporous silica conjugate layer. Analytica Chimica Ac-
      ta 2012; 714, 114-120.

[96] Perez, S.; Sanchez, S.; Fábregas, E. Enzymatic strategies to construct L-lactate biosen-
      sors based on polysulfone/carbon nanotubes membranes. Electroanalysis 2012; 24,
      967-974.

[97] Pundir, C.S.; Chauhan, N.; Rajneesh; Verma, M.; Ravi. A novel amperometric biosen-
      sor for oxalate determination using multi-walled carbon nanotube-gold nanoparticle
      composite. Sensors and Actuators, B 2011; 155, 796-803.

[98] Yadav, S.; Devi, R.; Kumari, S.; Yadav, S.; Pundir, C.S. An amperometric oxalate bio-
      sensor based on sorghum oxalate oxidase bound carboxylated multiwalled carbon
      nanotubes-polyaniline composite film. Journal of Biotechnology 2011; 151, 212-217.

[99] Yadav, S.; Kumar, A.; Pundir, C.S. Amperometric creatinine biosensor based on co-
      valently coimmobilized enzymes onto carboxylated multiwalled carbon nanotubes/
      polyaniline composite film. Analytical Biochemistry 2011; 419, 277-283.

[100] Shin, J.H.; Choi, Y.S.; Lee, H.J.; Choi, S.H.; Ha,J.; Yoon, I.J.; Nam, H.; Cha, G.S.A pla-
      nar amperometric creatinine biosensor employing an insoluble oxidizing agent for
      removing redox-active interferences. Analytical Chemistry 2011; 73, 5965-5971.

[101] Li, X-R.; Xu, J-J.; Chen, H-Y. Potassium-doped carbon nanotubes toward the direct
      electrochemistry of cholesterol oxidase and its application in highly sensitive choles-
      terol biosensor. Electrochimica Acta 2011; 56, 9378-9385.

[102] Brahim, S.; Narinesingh, D.; Guiseppi-Elie, A. Amperometric determination of cho-
      lesterol in serum using a biosensor of cholesterol oxidase contained within a poly-
      pyrrole–hydrogel membrane. Analytica Chimica Acta 2001; 448, 27-36.

[103] Arya, S.A., Datta, M. Malhotra, B.D. Recent advances in cholesterol biosensor. Bio-
      sensors and Bioelectronics 2008; 23, 1083-1100.

[104] Zhang, H.; Yin, Y.; Wu, P.; Cai, C. Indirect electrocatalytic determination of choline by monitoring hydrogen peroxide at the choline oxidase-prussian blue modified iron phosphate nanostructures. Biosensors and Bioelectronics 2012; 31, 244-250.

[105] López, M. S-P.; Pérez, J.P.H.; Lopez-Cabarcos, E.; Lopez-Ruiz, B. Amperometric biosensors based on choline oxidase entrapped in polyacrylamide microgels. Electroanalysis 2007; 19, 370-378.

[106] Lenigk, R.;Lam, E.; Lai, A.; Wang, H.; Han, Y.; Carlier, P.; Rennenberg, R. Enzyme biosensor for studying therapeutics of Alzheimer's disease. Biosensors and Bioelectronics 2000; 15, 541-547.

[107] Weiss, D.J.; Dorris, M.; Loh, A.; Peterson, L. Dehydrogenase based reagentless biosensor for monitoring phenylketonuria. Biosensors and Bioelectronics 2007; 22, 2436-41.

[108] Ionescu, R.E.; Cosnier, S.; Marks, R.S. Protease amperometric sensor. Analytical Chemistry 2006; 78, 6327-31.

[109] Yamanaka, H., Laluce, C., Oliveira Neto, G. Assay of serum glutamic-oxaloacetic transaminase using malate dehydrogenase preparation obtained from S. aureofaciens. Analytical Letters 1995; 28, 2305-2316.

[110] Guo, Z.; Yang, Q.; Liu, H.; Liu, C.; Cai, X. The study of a disposable reagentless biosensor for fast test of aspartate aminotransferase. Electroanalysis 2008; 20, 1135-41.

[111] Bartling, B.; Li, L.; Liu, C. Detection of adenosine deaminase activity with a thick-film screen-printed Ir/C sensor to detect liver disease. Journal of the Electrochemical Society 2010; 157, 130-J134.

[112] He, Y.; Wu, Y.; Mishra, A.; Acha, V.; Andrews, T.; Hornsby, P.J. Biosensor Technology in aging research and age-related diseases. Ageing Research Reviews 2012; 11, 1-9.

[113] D'Orazio, P. Biosensors in clinical chemistry - 2011 update. Clinica Chimica Acta 2011; 412, 1749-1761.

[114] Rasooly, A.; Jacobson, J. Development of biosensors for cancer clinical testing. Biosensors and Bioelectronics 2006; 21, 1851-1858.

[115] Vashist, S.K.; Zheng, D.; Al-Rubeaan, K.; Luong, J.H.T.; Sheu, F. Technology behind commercial devices for blood glucose monitoring in diabetes management: A review. Analytica Chimica Acta 2011; 703, 124-136.

[116] Zhu, Z.; Garcia-Gancedo, L.; Flewitt, A.J.; Xie, H.; Francis Moussy, F.; Milne, W.I. A critical review of glucose biosensors based on carbon nanomaterials: carbon nanotubes and graphene. Sensors 2012; 12, 5996-6022.

Chapter 12

Functionalized GaN Based Transistors For Biosensing

Stephen J. Pearton, Fan Ren and Byung Hwan Chu

Additional information is available at the end of the chapter

1. Introduction

The US market size of chemical sensors is projected to increase 8.6% annually to reach $6 billion in 2014. This growth will be sustained especially by high demand of biosensors for medical applications such as glucose monitoring, biomarker detection for infectious disease and cancer diagnosis. In addition, there will be strong demand in biodefense, environmental monitoring, food, and pharmaceutical industries. The biosensor market is forecast to reach $4.4 billion in 2014 in the US [1].

There is interest in developing sensors that could be used in point-of-care applications or on-field measurements. These sensors need to have high precision, compact size, fast response time and be sensitive to small amount of biological material. Field effect transistor structures (FETs) are promising for these applications. In a standard FET, the size and shape of the conductive channel between the source and drain terminals is controlled by an electrode that applies gate voltage. In place of the electrode, chemical characteristics of the active area could also play a crucial role in controlling the performance behaviors of the device. This property makes semiconductors ideal materials for sensors in many chemical and biochemical systems. Compared to methods such as protein assays [2], there is no need for optical components to translate the surface binding phenomena into a readable signal. Semiconductor properties including, surface current, potential, and impedance characteristics can be used to directly measure chemical or physical stimuli on the semiconductor surface [3-10]. Handheld, wireless-capable medical sensors are attractive for reducing financial and emotional cost of false-positive tests. The ability to make robust, inexpensive sensor arrays using semiconductor technology is one of the main driving forces behind this work. A typical semiconductor-based bio-sensor consists of a bio-receptor and transducer, which is generally a gateless field effect transistor structure. The bio-receptor is typically an enzyme or antibody layer attached to the gate of the FET.

To date, silicon-based sensors have been the main platform due to their low cost and maturity as well as the extensive experience base for chemical treatments on silicon oxide or glass. A drawback with Si sensors is that they generally cannot operate under harsh conditions of elevated temperature and /or pressure, and they are susceptible to degradation in chemically corrosive environments. This has led to interest in the use of GaN-based semiconductors as alternatives to silicon because of their greater chemical inertness in acidic and basic solutions, higher temperature of operation and ability to also emit and blue and ultraviolet light that can be used for fluorescence detection of particular bio-species. Other GaN applications already commercialized include blue/violet/white/UV Light Emitting Diodes (LEDs) used in traffic stoplights and full color displays, blue light lasers, which are employed in high density CD-ROM storage, and high resolution printers, high power microwave transistors with applications in new generations of radar and cell phone systems and low noise, radiation hard transistors for use in high temperature sensors and space-flight instrumentation.

In the area of detection of medical biomarkers, many different methods, including enzyme-linked immunsorbent assay (ELISA), particle-based flow cytometric assays, electrochemical measurements based on impedance and capacitance, electrical measurement of microcantilever resonant frequency change, and conductance measurement of semiconductor nanostructures. gas chromatography (GC), ion chromatography, high density peptide arrays, laser scanning quantitiative analysis, chemiluminescence, selected ion flow tube (SIFT), nanomechanical cantilevers, bead-based suspension microarrays, magnetic biosensors and mass spectrometry (MS) have been employed [1,2]. Depending on the sample condition, these methods may show variable results in terms of sensitivity for some applications and may not meet the requirements for a handheld biosensor. Most of the techniques mentioned earlier such as ELISA possesses a major limitation in that only one analyte is measured at a time. Particle-based assays allow for multiple detection by using multiple beads but the whole detection process is generally longer than 2 hours, which is not practical for in-office or bedside detection. Electrochemical devices have attracted attention due to their low cost and simplicity, but significant improvements in their sensitivities are still needed for use with clinical samples. Microcantilevers are capable of detecting concentrations of 10 pg/ml, but suffer from an undesirable resonant frequency change due to the viscosity of the medium and cantilever damping in the solution environment. In clinical settings, biomarkers for a particular disease state can be used to determine the presence of disease as well as its progress.

A promising FET sensing technology utilizes AlGaN/GaN high electron mobility transistors (HEMTs). HEMT structures have been developed for use in microwave power amplifiers due to their high two dimensional electron gas (2DEG) mobility and saturation velocity. The conducting 2DEG channel of AlGaN/GaN HEMTs is very close to the surface and extremely sensitive to adsorption of analytes. HEMT sensors can be used for detecting pH values, proteins, and DNA.The GaN materials system is attracting much interest for commercial applications of green, blue, and UV light emitting diodes (LEDs), laser diodes as well as high speed and high frequency power devices. Due to the wide-bandgap nature of the material, it is very thermally stable, and electronic devices can be operated at temperatures up to 500 °C. The GaN based materials are also chemically stable, and no known wet chemical etchant can etch these materials; this makes them very suitable for operation in chemically harsh environments. Due to the high electron mobility, GaN material based high electron mobility transistors (HEMTs) can operate at very high frequency with higher breakdown voltage, better thermal conductivity, and wider transmission bandwidths than Si or GaAs devices [3-10].

The high electron sheet carrier concentration of nitride HEMTs is induced by piezoelectric polarization of the strained AlGaN layer in the hetero-junction structure of the AlGaN/GaN HEMT and the spontaneous polarization is very large in wurtzite III-nitrides. This provides an increased sensitivity relative to simple Schottky diodes fabricated on GaN layers or FETs fabricated on the AlGaN/GaN HEMT structure. The gate region of the HEMT can be used to modulate the drain current in the FET mode or use as the electrode for the Schottky diode. A variety of gas, chemical and health-related sensors based on HEMT technology have been demonstrated with proper surface functionalization on the gate area of the HEMTs, including the detection of hydrogen, mercury ions, prostate specific antigen (PSA), DNA, and glucose.

In this chapter, we discuss progress in functionalization of these sensors for applications in detection of gases, pH measurement, biotoxins and other biologically important chemicals and the integration of these sensors into wireless packages for remote sensing capability.

2. Sensor Functionalization

One drawback of HEMT sensors is a lack of selectivity to different analytes due to the chemical inertness of the HEMT surface. This can be solved by surface modification with detecting receptors. Sensor devices should be usable with a variety of fluids having environmental and bodily origins, including saliva, urine, blood, and breath. For use with exhaled breath, the device may include a HEMT bonded on a thermo-electric cooling device, which assists in condensing exhaled breath samples. One of the key technical challenges in fabricating hybrid biosensors is the junction between biological macromolecules and the inorganic scaffolding material (metals and semiconductors) of the chip. For actual device applications, it is often necessary to selectively modify a surface at micro- and even nano-scale, sometimes with different surface chemistry at different locations. In order to enhance detection speed, especially at very low analyte concentration, the analyte should be delivered directly to the active sensing areas of the sensors. A common theme for bio/chem sensors is that their operation often incorporates moving fluids. For example, sensors must sample a stream of air or water to interact with the specific molecules they are designed to detect.

The general approach to detecting biological species using a semiconductor sensor involves functionalizing the surface (eg. the gate region of an ungated field effect transistor structure) with a layer or substance which will selectively bind the molecules of interest. In applications requiring less specific detection, the adsorption of reactive molecules will directly affect the surface charge and affect the near-surface conductivity. In their simplest form, the sensor consists of a semiconductor film patterned with surface electrodes and often heated to temperatures of a few hundred degrees Celsius to enhance dissociation of molecules on the exposed surface. Changes in resistance between the electrodes signal the adsorption of reactive molecules. It is desirable to be able to use the lowest operating temperature to maximize battery life in hand-held detection instruments.

Since GaN-based material systems are chemically stable, this should minimize degradation of adsorbed cells. The bond between Ga and N is ionic and proteins can easily attach to the GaN surface. This is one of the key factors for making a sensitive biosensor with a useful lifetime. HEMT sensors have been used for detecting gases, ions, pH values, proteins, and DNA temperature with good selectivity by the modification of the surface in the gate region of the HEMT. The 2DEG channel is connected to an Ohmic-type source and drain contacts. The source-drain current is modulated by a third contact, a Schottky-type gate, on the top of the 2DEG channel. For sensing applications, the third contact is affected by the sensing environment, i.e. the sensing targets changes the charges on the gate region and behave as a gate. When charged analytes accumulate on the gate area, these charges form a bias and alter the 2DEG resistance. This electrical detection technique is simple, fast, and convenient. The detecting signal from the gate is amplified through the drain-source current and makes this sensor very sensitive for sensor applications. The electric signal also can be easily quantified, recorded and transmitted, unlike fluorescence detection methods which need human inspection and are difficult to precisely quantify and transmit the data. Table 1 shows a summary of surface functional layers used with HEMT sensors for selective detection of various gases, toxins, cancers and biomarkers, heavy metals, pressure changes and marine pathogens.

3. pH Sensors

An important aspect of biosensing, particularly for health monitoring, is the need to monitor the concentration of several different species for calibration purposes or conditions such as pH. which can also be used to calibrate samples of breath condensate. The glucose oxidase enzyme (GOx) is commonly used in biosensors to detect levels of glucose for diabetics. However, the activity of GOx is highly dependent on the pH value of the solution. The pH value of a typical healthy person is between 7 and 8. This can vary significantly depending on the health condition of each individual, e.g. the pH value for patients with acute asthma was reported as low as 5.23 + 0.21 (n=22) as compared to 7.65 + 0.20 (n=19) for the control subjects. To achieve accurate glucose concentration measurement with immobilized GO_x, it is necessary to determine the pH value and glucose concentration with an integrated pH and glucose sensor.

The measurement of pH is needed in many applications, including medicine, biology, food and environmental science and oceanography. Solutions with a pH less than 7 are acidic and solutions with a pH greater than 7 are basic or alkaline. Ungated AlGaN/ GaN HEMTs exhibit large changes in current upon exposing the gate region to polar liquids [6]. The polar nature of electrolyte introduced to the surface produces a change of surface charge and hence surface potential at the semiconductor /liquid interface. The use of Sc_2O_3 gate dielectric in the gate region produces superior results to either a native oxide or UV ozone-induced oxide. The ungated HEMTs with Sc_2O_3 in the gate region exhibited a linear change in current between pH 3-10 of 37μA/pH. Figure 1 shows a scanning electron microscopy (SEM) image (top) and a cross-sectional schematic (bottom) of the completed device. The gate dimension of the device is 2×50 μm^2. The pH solution was applied using a syringe autopipette (2-20ul).

Detection	Mechanism	Surface Functionalization	Detection Limit
1. Gases			
H_2	Catalytic dissociation	Pd,Pt	10 ppm
CO_2	Absorption of water/ charge	Polyethylenimine/starch	1%
CO	Charge transfer	ZnO nanowires	50 ppm
O_2	Oxidation	InZnO	5%
2.Toxins			
Botulinum	Antibody	Thioglycolic acid/antibody	1 ng/ml
Anthrax Protective Antigen	Antibody	Thioglycolic acid/antibody	2µg/ml
3. Cancers			
Breast cancer	Antibody	Thioglycolic acid/c-erbB antibody	
Prostate Specific Antigen	Antibody	Carboxylate succimdyl ester/PSA antibody	10 pg/ml
4. Biomarkers			
DNA	Hybridization	3'-thiol-modified oligonuceotides	
Chloride ions	Anodization	Ag/AgCl electrodes, InN	10^{-8} M
Lactic acid	LOX immobilization	ZnO nanorods	167 nM
Glucose	GOX immobilization	ZnO nanorods	0.5 nM
Proteins	Conjugation/hybridization	Aminoprpoylsilane/biotin	
pH	Absorption of polar molecules	Sc_2O_3 , ZnO	±0.01
KIM-1	Antibody	KIM-1 Antibody	1 ng/ml
Traumatic Brain Injury	Antibody	TBI Antibody	1 µg/ml
5. Heavy Metals			
Hg+ with Na, Pb, Mg ions	Chelation	Thioglycolic acid/Au	1 nM
6. Marine pathogens/diseases			
Perkinsus marinus	Antibody	Thioglycolic acid/anti-P marinus antibody	
Vitellogenin	Antibody	Thioglycolic acid/anti-vitellogenin antibodies	1% serum of 4 µg/ml

Table 1. Summary of surface functional layers used with HEMT sensors.

The pH solution made by the titration method using HCl, NaOH and distilled water. The electrode was a conventional Acumet standard Ag/AgCl electrode. Figure 2 shows the current at a bias of 0.25V as a function of time from HEMTs with Sc_2O_3 in the gate region exposed for 150s to a series of solutions whose pH was varied from 3-10. The current is increased upon exposure to these polar liquids as the pH is decreased. The change in current was 37 µA/pH. The HEMTs show stable operation with a resolution of ~0.1 pH over the entire pH range.

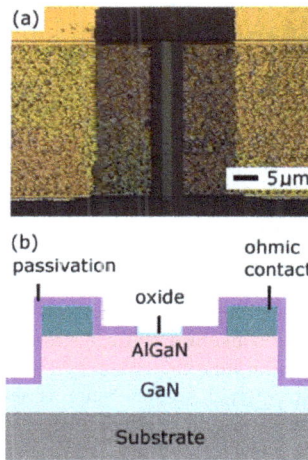

Figure 1. SEM and schematic of gateless HEMT.

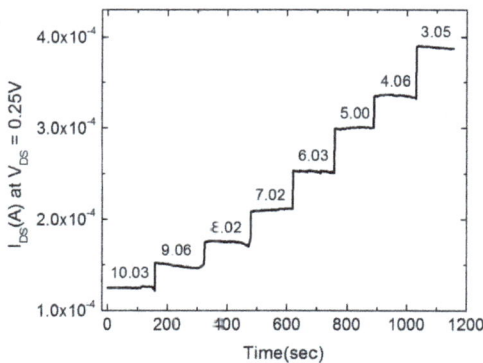

Figure 2. Change in current of HEMT with pH from 3-10.

4. Clinically Relevant Gas Detection

(i) O_2 Sensing

The current technology for O_2 measurement, referred to as oximetry, is small and convenient to use [11,12]. However, the O_2 measurement technology does not provide a complete measure of respiratory sufficiency. A patient suffering from hypoventilation (poor gas exchange in the lungs) given 100% oxygen can have excellent blood oxygen levels while still suffering from respiratory acidosis due to excessive CO_2. The O_2 measurement is also not a complete measure of circulatory sufficiency. If there is insufficient blood flow or insufficient hemoglobin in the blood (anemia), tissues can suffer hypoxia despite high oxygen saturation in the blood that does arrive. The current oxide-based O_2 sensors can operate at very high temperatures, such as the commercialized solid electrolyte ZrO_2 (700ºC) or the semiconductor metal oxides such as TiO_2, Nb_2O_5, $SrTiO_3$, and CeO_2 (>400ºC). However, it remains important to develop a low operation temperature and high sensitivity O_2 sensor to build a small, portable and low cost O_2 sensor system for biomedical applications.

Oxide-based materials are widely used and studied for oxygen sensing because of their low cost and good reliability. The commercialized solid electrolyte ZrO_2 has been widely used in automobiles for oxygen sensing in combustion processes. The electrolyte metal oxide oxygen sensor usually uses a reference gas and operates at high temperature (700˚C) [12]. The semiconductor metal oxides mentioned above do not need the reference gas, but they still need to be operated at a considerably high temperature (>400˚C) in order to reach high sensitivity, which means a high power consumption for heating up the sensors [13-15]. For biomedical applications, such as monitoring oxygen in the breath for a lung transplant patient, a portable and low power consumption O_2 sensor system is needed. Therefore, it is crucial to develop a low operating temperature and high sensitivity O_2 sensor for those applications.

The conductivity mechanism of most metal oxides based semiconductors results from electron hopping from intrinsic defects in the oxide film and these defects are related to the oxygen vacancies generated during oxide growth [13-15]. Typically, the higher the concentration of oxygen vacancies in the oxide film, the more conductive is the film. InZnO (IZO) films have been used in fabricating thin film transistors and the conductivity of the IZO is also found to depend on the oxygen partial pressure during the oxide growth. The IZO is a good candidate for O_2 sensing applications.

The schematic of the oxygen sensor based on oxide-functionalized HEMTs is shown at the top of Figure 3. The bottom part of the figure shows the device had a strong response when it was tested at 120˚C in pure nitrogen and pure oxygen alternately at a bias voltage of 3V. When the device was exposed to oxygen, the drain-source current decreased, whereas when the device was exposed to nitrogen, the current increased. The IZO film provides a high oxygen vacancy concentration, which makes the film readily sense oxygen and create a potential on the gate area of the AlGaN/GaN HEMT. A sharp drain-source current change demonstrates the combination of the advantage of the high electron mobility of the HEMT and the high oxygen vacancy concentration of the IZO film. Because of these advantages, this oxygen sensor can operate with a high sensitivity at a relatively low temperature compared to many oxide-based oxygen sensors which operate from 400˚C to 700˚C. The combination of IZO films and the AlGaN / GaN HEMT allows realization of low operation temperature and low power consumption oxygen sensor.

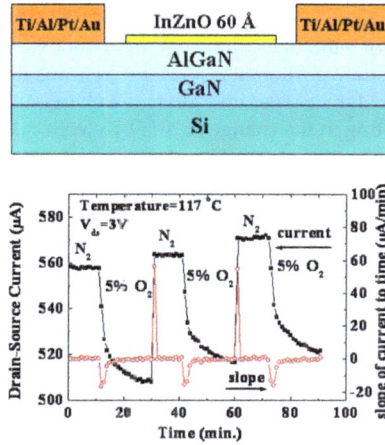

Figure 3. Schematic of AlGaN/GaN HEMT based C2 sensor (top) and drain current of IZO functionalized HEMT sensor measured at fixed source-drain during the exposure to different O2 concentration ambients. The drain bias voltage was 0.5 V and measurements were conducted at 117 oC.

(ii) CO_2 Sensing

The detection of carbon dioxide (CO_2) gas is important for global warming, biological and health-related applications such as indoor air quality control, process control in fermentation, and in the measurement of CO_2 concentrations in patients' exhaled breath with lung and stomach diseases. In medical applications, it can be critical to monitor the CO_2 and O_2 concentrations in the circulatory systems for patients with lung diseases in the hospital. The current technology for CO_2 measurement typically uses IR instruments, which can be very expensive and bulky [16-22].

The most common approach for CO_2 detection is based on non-dispersive infrared (NDIR) sensors, which are the simplest of the spectroscopic sensors. The best detection limits for the NDIR sensors are currently in the range of 20-10,000 ppm. The key components of the NDIR approach are an infrared (IR) source, a light tube, an interference filter, and an infrared (IR) detector. In operation, gas enters the light tube. Radiation from the IR light source passes through the gas in the light tube to impinge on the IR detector. The interference filter is positioned in the optical path in front of the IR detector such that the IR detector receives the radiation wavelength that is strongly absorbed by the gas whose concentration is determined while filtering out the unwanted wavelengths. The IR detector produces an electrical signal that represents the intensity of the radiation impinging upon it. It is generally considered that the NDIR technology is limited by power consumption and size.

In recent years, monomers or polymers containing amino-groups, such as tetrakis (hydroxyethyl)ethylenediamine, tetraethylene-pentamine and polyethyleneimine (PEI) have been used for CO_2 sensors to overcome the power consumption and size issues found in the NDIR approach [21-24]. Most of the monomers or polymers are utilized as coatings of surface acoustic wave transducers. The polymers are capable of adsorbing CO_2 and facilitating a carbamate reaction. PEI has also been used as a coating on carbon nanotubes for CO_2 sensing by measuring the conductivity of nanotubes upon exposing to the CO_2 gas. For example, CO_2 adsorbed by a PEI coated nanotube portion of a NTFET (nanotube field effect transistor) sensor lowers the total pH of the polymer layer and alters the charge transfer to the semiconducting nanotube channel, resulting in the change of NTFET electronic characteristics [25-28].

The HEMT-based device relies on the interaction between CO_2 and amino group-containing compounds. Addition of starch into the PEI enhances the absorption of the water molecules into the PEI/starch thin film. The reaction mechanism is expected to be that primary amine groups, $-NH_2$, on the PEI main chain react with CO_2 and water forming $-NH_3^+$ ions and the CO_2 molecule became OCOOH- ions. Thus, the charges, or the polarity, on the PEI main chain were changed. The electrons in the two-dimensional electron gas (2DEG) channel of the AlGaN/GaN HEMT are induced by piezoelectric and spontaneous polarization effects. The PEI/starch was coated on the gate region of the HEMT. The charges of the PEI changed through the reactions between $-NH_2$ and CO_2 as well as water molecules are then transduced into a change in the concentration of the 2DEG in the AlGaN/GaN HEMTs.

PEI/starch functionalized HEMT sensors were found to be capable of measuring different CO_2 concentration at temperatures as low as 108 C and a fixed source-drain bias voltage of 0.5 V. The current increased with the introduction of CO_2 gas. This was due to the net positive charges increased on the gate area, thus inducing electrons in the 2DEG channel. The response to CO_2 gas had a wide dynamic range from 0.9% to 50%. Higher CO_2 concentrations were not tested because there is little interest in these for medical related applications. The response times were on the order of 100 seconds. The drain current changes were linearly proportional to the CO_2 concentration for all the tested temperatures and higher sensitivity for the higher testing temperatures. There was a noticeable change of the sensitivity from the sensors tested at 61 C to those tested at 108 C. The sensors exhibited reversible and reproducible characteristics [29].

5. Biotoxin Sensors

Reliable detection of biological agents in the field and in real time is challenging. The objective of this application is to develop and test a wireless sensing technology for detecting logicalb toxins. A significant issue is the absence of a definite diagnostic method and the difficulty in differential diagnosis from other pathogens that would slow the response in case of a terror attack. Our aim is to develop reliable, inexpensive, highly sensitive, handheld sensors with response times on the order of a few seconds, which can be used in the field for detecting biological toxins.

The current methods for toxin sensing in the field are generally not suited for field deployment and there is a need for new technologies. The current methods are impractical because such tests can only be carried out at centralized locations, and are too slow to be of practical value in the field. These still tend to be the methods of choice in current detection of toxins, eg. the standard test for botulinum toxin detection is the 'mouse assay', which relies on the death of mice as an indicator of toxin presence [30].

(i) Botulinum

Antibody-functionalized Au-gated AlGaN/GaN HEMTs show great sensitivity for detecting botulinum toxin. The botulinum toxin was specifically recognized through botulinum antibody, anchored to the gate area, as shown in Figure 4. We investigated a range of concentrations from 0.1 ng/ml to 100 ng/ml. The source and drain current from the HEMT were measured before and after the sensor was exposed to 100 ng/ml of botulinum toxin at a constant drain bias voltage of 500 mV. Figure 5 (top) shows a real time botulinum toxin detection in PBS buffer solution using the source and drain current change with constant bias of 500 mV. No current change can be seen with the addition of buffer solution around 100 seconds, showing the specificity and stability of the device. In clear contrast, the current change showed a rapid response in less than 5 seconds when target 1 ng/ml botulinum toxin was added to the surface. The abrupt current change due to the exposure of botulinum toxin in a buffer solution was stabilized after the botulinum toxin thoroughly diffused into the buffer solution. Different concentrations (from 0.1 ng/ml to 100 ng/ml) of the exposed target botulinum toxin in a buffer solution were detected. The sensor saturates above 10ng/ml of the toxin. The limit of detection of this device was below 1 ng/ml of botulinum toxin in PBS buffer solution. The source-drain current change was nonlinearly proportional to botulinum toxin concentration, as shown in Figure 5 (bottom). Figure 6 shows a real time test of botulinum toxin at different toxin concentrations. This result demonstrates the real-time capabilities of the chip [31]. These tests are typical for our sensors, to demonstrate their quick response to different concentrations.

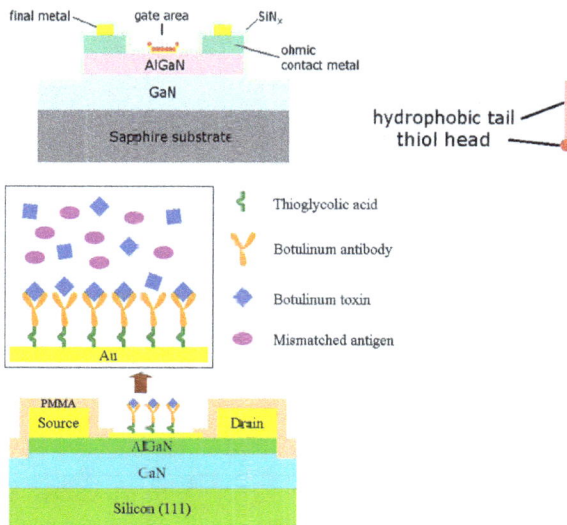

Figure 4. Schematic of AlGaN/GaN HEMT. The Au-coated gate area was functionalized with thioglycolic acid for botulinum detection.

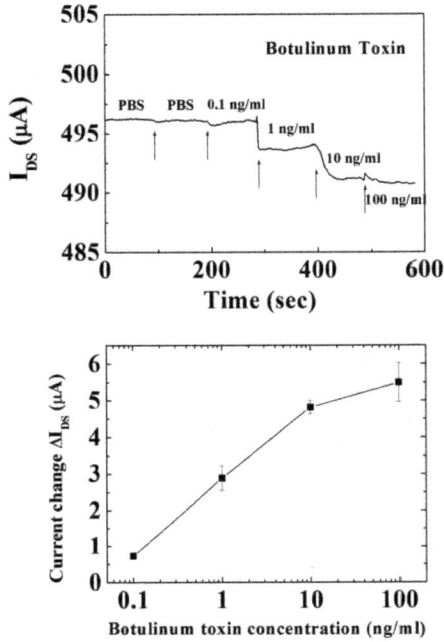

Figure 5. Drain current of an AlGaN/GaN HEMT versus time for botulinum toxir from 0.1 ng/ml to 100 ng/ml (top) and change of drain current versus different concentrations from 0.1 ng/ml to 100 ng/ml of botulinum toxin (bottom).

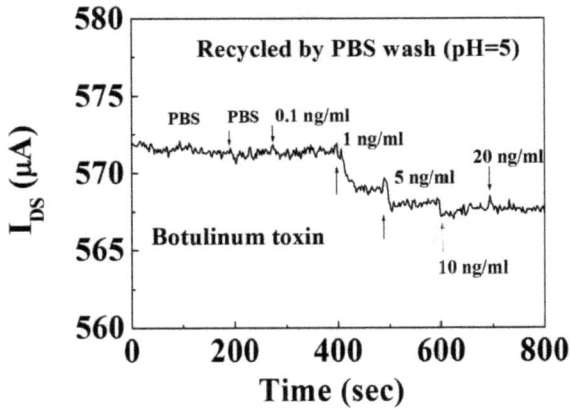

Figure 6. Real-time test from a used botulinum sensor washed with PBS in pH 5 to refresh the sensor.

6. Biomedical Applications

(i) Glucose

AlGaN/GaN HEMTs can be used for measurements of pH in exhaled breath condensate (EBC) and glucose, through integration of the pH and glucose sensor onto a single chip and with additional integration of the sensors into a portable, wireless package for remote monitoring applications [9,32]. The glucose was sensed by ZnO nanorod functionalized HEMTs with glucose oxidase enzyme localized on the nanorods. Figure 7 shows an optical microscope image of an integrated pH and glucose sensor chip and cross-sectional schematics of the completed pH and glucose device. The gate dimension of the pH sensor device and glucose sensors was $20 \times 50 \ \mu m^2$.

For the glucose detection, an array of 20-30 nm diameter and 2 μm tall ZnO nanorods were grown on the $20 \times 50 \ \mu m^2$ gate area. The lower right inset in Figure 7 shows closer view of the ZnO nanorod arrays grown on the gate area. The ZnO nanorod matrix provides a microenvironment for immobilizing negatively charged glucose oxidase (GO_x) while retaining its bioactivity, and passes charges produced during the GO_x and glucose interaction to the AlGaN/GaN HEMT. The GOx solution was prepared with concentration of 10 mg/mL in 10 mM phosphate buffer saline. After fabricating the device, 5 μl GO_x solution was precisely introduced to the surface of the HEMT using a pico-liter plotter. The sensor chip was kept at 4 °C in the solution for 48 hours for GO_x immobilization on the ZnO nanorod arrays followed by an extensively washing to remove the un-immobilized GO_x.

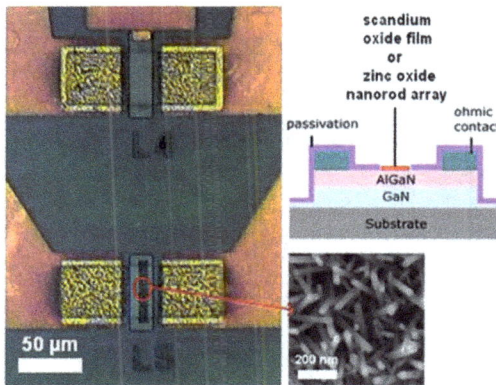

Figure 7. SEM image of an integrated pH and glucose sensor. The insets show a schematic cross-section of the pH sensor and also an SEM of the ZnO nanorods grown in the gate region of the glucose sensor.

To take the advantage of quick response (less than 1 sec) of the HEMT sensor, a real-time EBC collector is needed. The amount of the EBC required to cover the HEMT sensing area is very small. To condense 3 μl of water vapor, only ~ 7 J of energy need to be removed for each tidal breath, which can be easily achieved with a thermal electric module, a Peltier device. The AlGaN/GaN HEMT sensor is directly mounted on the top of the Peltier unit, which can be cooled to precise temperatures by applying known voltages and currents to the unit. During our measurements, the hotter plate of the Peltier unit was kept at 21°C, and the colder plate was kept at 7 °C by applying bias of 0.7 V at 0.2 A. The sensor takes less than 2 sec to reach thermal equilibrium with the Peltier unit. This allows the exhaled breath to immediately condense on the gate region of the HEMT sensor.

The HEMT sensors were not sensitive to switching of N_2 gas, but responded to applications of exhaled breath pulse inputs from a human test subject, The principal component of the EBC is water vapor, which represents nearly all of the volume (>99%) of the fluid collected in the EBC. The measured current change of the exhale breath condensate shows that the pH values are within the range between pH 7 and 8. This range is the typical pH range of human blood. The sensors do not respond to glucose unless the enzyme is present, as shown in Figure 8. Although measuring the glucose in the EBC is a noninvasive and convenient method for the diabetic application, the activity of the immobilized GO_x is highly dependent on the pH value of the solution. The GOx activity can be reduced to 80% for pH = 5 to 6. If the pH value of the glucose solution is larger than 8, the activity drops off very quickly [15]. When the glucose sensor was used in a pH controlled environment, the drain current stayed fairly constant. The human pH value can vary significantly depending on the health condition. Since we cannot control the pH value of the EBC samples, we needed to measure the pH value while determining the glucose concentration in the EBC. With the fast response time and low volume of the EBC required for HEMT based sensor, a handheld and real-time glucose sensing technology can be realized.

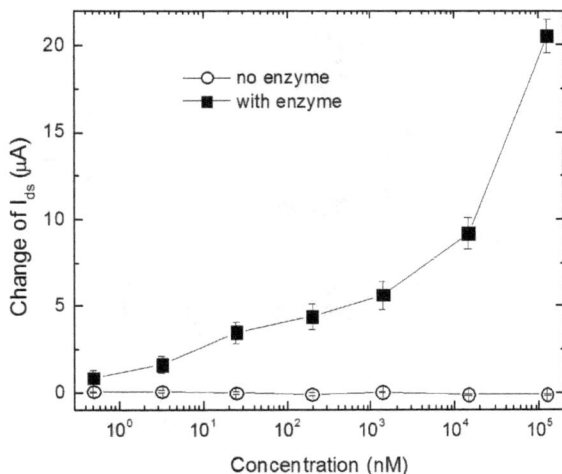

Figure 8. Change in drain-source current in HEMT glucose sensors with and without localized enzyme.

(ii) Prostate Cancer Detection

Prostate cancer is the second most common cause of cancer death among men in the United States and 1 in 6 men will be diagnosed with prostate cancer during his lifetime [3,33,34]. The most commonly used serum marker for diagnosis of prostate cancer is prostate specific antigen (PSA). Prostate cancer can often be found early by testing the amount of (PSA in the patient's blood. It can also be detected on a digital rectal exam (DRE). Most men have PSA levels under 4 nanograms per milliliter of blood. When prostate cancer develops, the PSA level usually goes up above 4 nanograms per milliliter; however, about 15% of men with a PSA below 4 will have prostate cancer on biopsy. Generally PSA testing approaches are costly, time-consuming and need sample transportation.

Antibody functionalized Au-gated AlGaN/GaN HEMTs shown schematically in Figure 9 were found to be effective for detecting PSA at low concentration levels. The PSA antibody was anchored to the gate area through the formation of carboxylate succinimdyl ester bonds with immobilized thioglycolic acid. The HEMT drain-source current showed a response time of less than 5 seconds when target PSA in a buffer at clinical concentrations was added to the antibody-immobilized surface. The devices could detect a range of concentrations from 1 µg/ml to 10 pg/ml. The lowest detectable concentration was two orders of magnitude lower than the cut-off value of PSA measurements for clinical detection of prostate cancer. Figure 10 shows the real time PSA detection in PBS buffer solution using the source and drain current change with constant bias of 0.5V[42]. No current change can be seen with the addition of buffer solution or nonspecific bovine serum albumin (BSA), but there was a rapid change when10 ng/ml PSA was added to the surface. The abrupt current change due to the exposure of PSA in a buffer solution could be stabilized after the PSA diffused into the buffer solution. The ultimate detection limit appears to be a few pg/ml [3].

(iii) Breast Cancer

The most effective and widely used diagnostic exam for breast cancer, the mammogram, is potentially harmful due to radiation exposure. Currently, the overwhelming majority of patients are screened for breast cancer by mammography [35-40]. This procedure involves a high cost to the patient and is invasive (radiation) which limits the frequency of screening. Breast cancer is currently the most common female malignancy in the world, representing 7% of the more than 7.6 million cancer-related deaths worldwide. More than one million mammograms are performed each year. According to the National Breast Cancer.

There is recent evidence to suggest that salivary testing for makers of breast cancer may be used in conjunction with mammography [36-40]. Saliva-based diagnostics for the protein c-erbB-2, have great prognostic potential. Soluble fragments of the c-erbB-2 oncoprotein and the cancer antigen 15-3 were found to be significantly higher in the saliva of women who had breast cancer than in those patients with benign tumors. These initial studies indicate that the saliva test is both sensitive and reliable and can be potentially useful in initial detection and follow-up screening for breast cancer. However, to fully realize the potential of salivary biomarkers, technologies are needed that will enable facile, sensitive, specific detection of breast cancer.

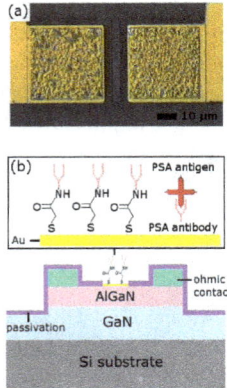

Figure 9. Schematic of HEMT sensor functionalized for PSA detection.

Antibody-functionalized Au-gated AlGaN/GaN high electron mobility transistors (HEMTs) show promise for detecting c-erbB-2 antigen [41]. The c-erbB-2 antigen was specifically recognized through c-erbB antibody, anchored to the gate area. We investigated a range of clinically relevant concentrations from 16.7 µg/ml to 0.25 µg/ml. The Au surface was functionalized with a specific bi-functional molecule, thioglycolic acid. We anchored a self-assembled monolayer of thioglycolic acid, $HSCH_2COOH$, an organic compound and containing both a thiol (mercaptan) and a carboxylic acid functional group, on the Au surface in the gate area through strong interaction between gold and the thiol-group of the thioglycolic acid. The device was incubated in a phosphate buffered saline (PBS) solution of 500 µg/ml c-erbB-2 monoclonal antibody for 18 hours before real time measurement of c-erbB-2 antigen.

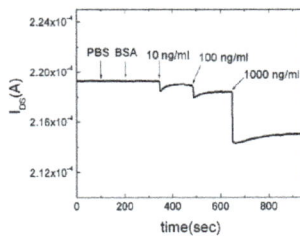

Figure 10. Drain current versus time for PSA detection when sequentially expcsed to PBS, BSA, and PSA.

Figure 11 (left) shows real time c-erbB-2 antigen detection in PBS buffer solution using the source and drain current change with constant bias of 500 mV. No current change can be seen with the addition of buffer solution around 50 seconds, showing the specificity and stability of the device. The current change showed a rapid response in less than 5 seconds when target 0.25 µg/ml c-erbB-2 antigen was added to the surface. The source-drain current change was nonlinearly proportional to c-erbB-2 antigen concentration, as shown in Figure 11 (right). Between each test, the device was rinsed with a wash buffer of pH 6.0 phosphate buffer solution to strip the antibody from the antigen. Clinically relevant concentrations of the c-erbB-2 antigen in the saliva and serum of normal patients are 4-6 µg/ml and 60-90 µg/ml respectively. For breast cancer patients, the c-erbB-2 antigen concentrations in the saliva and serum are 9-13 µg/ml and 140-210 µg/ml, respectively. Our detection limit suggests that HEMTs can be easily used for detection of clinically relevant concentrations of biomarkers.

(iv) Lactic Acid

Interest in developing improved methods for detecting lactic acid has been increasing due to its importance in clinical diagnostics, sports medicine, and food analysis. An accurate measurement of the concentration of lactate acid in blood is critical to patients that are in intensive care or undergoing surgical operations as abnormal concentrations may lead to shock, metabolic disorder, respiratory insufficiency, and heart failure. The concentration of lactate in human blood is typically 1~2 mmol/L at rest, but can rise to greater than 20mmol/L during various physiological and pathophysiological states, including intense exercise, shock (e.g., hypovolemia, congestive heart failure, septic shock), infections, respiratory insufficiency, and various metabolic disorders (e.g., inborn errors of metabolism such as congenital lactic acidosis). Since elevated concentrations of lactate may not only indicate the presence but also the severity of these clinically important disorders, accurate measurements of blood lactate concentration is critical to their proper management. Lactic acid concentrations can be used to monitor the physical condition of patients with chronic diseases such as heart failure, diabetes and/or chronic renal failure.

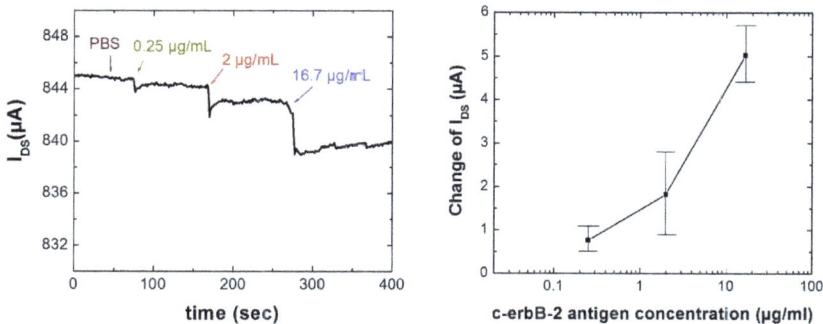

Figure 11. Drain current of an AlGaN/GaN HEMT over time for c-erbB-2 antigen from 0.25 µg/ml to 17 µg/ml (left) and change of drain current versus different concentrations from 0.25 µg/ml to 17 µg/ml of c-erbB-2 antigen.

A ZnO nanorod array, which was used to immobilize lactate oxidase oxidase (LOx), was selectively grown on the gate area using low temperature hydrothermal decomposition (Figure 12, top) [42,43]. The array of one-dimensional ZnO nanorods provided a large effective surface area with high surface-to-volume ratio and a favorable environment for the immobilization of LOx. The AlGaN/GaN HEMT drain-source current showed a rapid response when various concentrations of lactate acid solutions were introduced to the gate area of the HEMT sensor. The HEMT could detect lactic acid concentrations from 167 nM to 139 μM. Figure 12 (bottom) shows a real time detection of lactic acid by measuring the HEMT drain current sensor to solutions with different concentrations of lactic acid. The sensor was first exposed to 20 μl of 10 mM PBS and no current change could be detected with the addition of 10 μl of PBS at approximately 40 seconds, showing the specificity and stability of the device. A rapid increase in the drain current was observed when target lactic acid was introduced to the device surface. The sensor was continuously exposed to lactic acid concentrations from 167 nM to 139 μM.

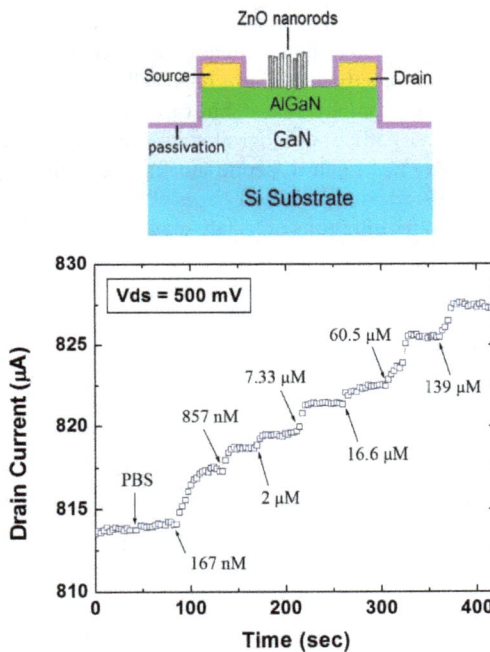

Figure 12. Schematic cross sectional view of the ZnO nanorod gated HEMT for actic acid detection (top) and plot of drain current versus time with successive exposure to lactic acid from 167 nM to 139 μM (bottom).

(v) Chloride Ion Detection

Chloride ions are also an essential counter-ion in our bodies [44-47]. Our kidneys balance the chloride in body fluids, such as serum, blood, urine, and exhaled breath condensate (EBC). Abnormal chloride ion concentration in serum may serve as an indicator for diseases such as renal diseases, adrenalism, and pneumonia. Chloride ion concentration can be a bio-marker for the level of pollen exposure in allergic asthma, chronic cough, and airway acidification related to respiratory disease. Also, the Cl⁻ concentration in EBC can be used as a reference for the other biomarkers in the EBC to estimate the dilution effect from the humidity in the ambient during the EBC collection.

Current analytical methods for measuring chloride ion include colorimetry, ion-selective electrode, x-ray fluorescence spectrometry, activation analysis, and ion chromatography. However, these methods involve high expertise levels and require expensive instruments that cannot be readily transported. The narrow gap semiconductor InN has positively charged surface donor states that function as fixed surface sites for the reversible anion coordination and it has been proposed as a useful material for sensing applications. InN thin film based potentiometric ion-selective sensors have been reported to detect Cl⁻ ions down to 1mM. In addition, HEMTs with a Ag/AgCl gate are found to exhibit significant changes in channel conductance upon exposing the gate region to various concentrations of chorine ion solutions. The Ag/AgCl gate electrode, prepared by potentiostatic anodization, changed electrical potential when it encountered chorine ions. The HEMT shown schematically in Figure 13 (top) source-drain current showed a clear dependence on the chorine concentration.

Figure 13 (bottom) shows the time dependence of Ag/AgCl HEMT drain current at a constant drain bias voltage of 500mV during exposure to solutions with different chlorine ion concentrations. The HEMT sensor was first exposed to DI water and no change of the drain current was detected with the addition of DI water at 100 seconds. There was a rapid response of HEMT drain current observed in less than 30 seconds when target of 10^{-8} M NaCl solution was switched to the surface at 175 sec. The limit of detection of this device was 10^{-8} M chlorine in DI-water. Between each test, the device was rinsed with DI water. These results suggest that our HEMT sensors are recyclable with simple DI water rinse.

Real time detection of chloride ion detection with AlGaN/GaN high electron mobility transistors (HEMTs) with an InN thin film in the gate region has also been demonstrated [45]. The sensor, shown schematically in Figure 14, exhibited significant changes in channel conductance upon exposure to various concentrations of NaCl solutions. The InN thin film provided fixed surface sites for reversible anion coordination. The sensor was tested over the range of 100nM to 100μM NaCl solutions. Figure 14 also shows the results of real time detection of Cl⁻ ions by measuring the HEMT drain current at a constant drain bias voltage of 500mV during exposure to solutions of different chloride ion concentrations. No change of the drain current was detected with the addition of DI water at 100 seconds. The small spike in the current is due to mechanical disturbance of the HEMT surface when the water was added. A rapid response of drain current was observed in less than 20 seconds when target of 100 nM NaCl solution was exposed to the surface at 200 seconds. The abrupt current change stabilized after the sodium chloride solution thoroughly diffused into water and reached a steady state. When the InN gate metal encountered chloride ion, the electrical potential of the gate was changed and resulted in the increase the pizeo-induced charge density in the HEMT channel. The measured drain current of the InN gated AlGaN/GaN HEMT in NaCl solutions was linearly proportional to the logarithm of chloride concentration, satisfying the Nernst equation. The pH value of the solutions did not affect the chloride ion concentration measurements.

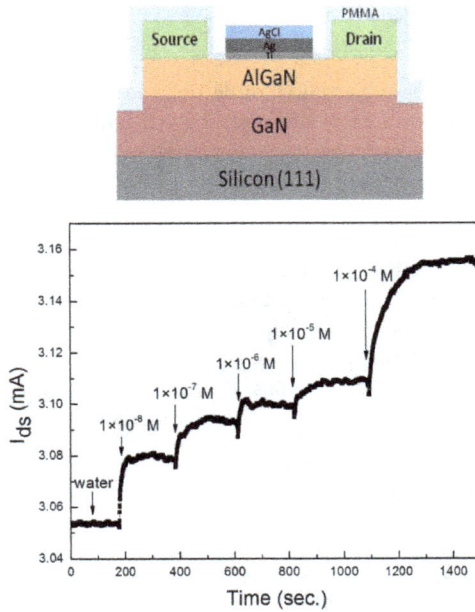

Figure 13. Schematic cross sectional view of a Ag/AgCl gated HEMT (top) and time dependent drain current of a Ag/AgCl gated AlGaN/GaN HEMT exposed to different concentrations of NaCl solutions (bottom).

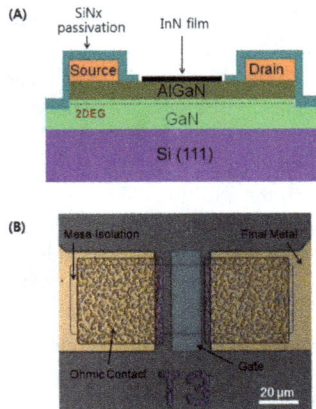

Figure 14. Schematic of an InN-gated HEMT sensor (top) and optical image of the gate region (bottom).

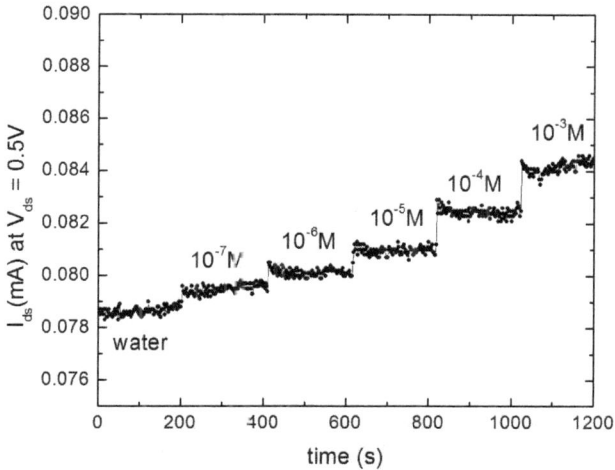

Figure 15. Source-drain current of InN-gated HEMT as different concentrations of Cl- ions were added.

(vi) Endocrine Disrupters

There have been many reports evaluating the adverse effects of endocrine disrupters (ED) on reproduction in wild animals, especially in aquatic environments [47-53]. A wide range of chemicals are considered EDs, including naturally occurring or improperly disposed estrogens and anthropogenic chemicals that were heavily used in the past. These chemicals promote feminization in wild life and also pose a threat to public health. Some reports suggest that ED can influence fetal development or act as a carcinogen. It is beneficial to develop tools that could accurately monitor the level of ED exposure.

Vitellogenin (Vtg) is a major egg yolk precursor protein used as a biomarker to indicate an organism's exposure to ED. The gene for this protein is expressed in the liver of oviparous animals under the control of estrogen. Male fish, under natural conditions, should have very low doses of Vtg since they do not produce eggs. However, if male fish are exposed to estrogen or to estrogen mimics in the environment, the Vtg gene is turned on. The dynamic range of this protein in normal male fish is 10-50 ng/ml in plasma and ~20 mg/ml in females producing eggs. There have been reports of finding as much as 100 mg/ml in some fish that were induced with estrogen. While the dynamic range is over 6 orders of magnitude, one normally finds 1~100 µg/ml in plasma in exposed males. Although Vtg from one species is limited in its application as a probe for another, some segments of Vtg are highly conserved among species, suggesting the possibility of developing antibodies with wide cross-reactivity. There have been few reports on the detection of analytes in a real solution such as serum. We have detected Vtg in both fish serum and phosphate buffer saline (PBS) using HEMT sensors with anti-vitellogenin antibodies attached to the gate region. A schematic of the sensor is shown in Figure 16.

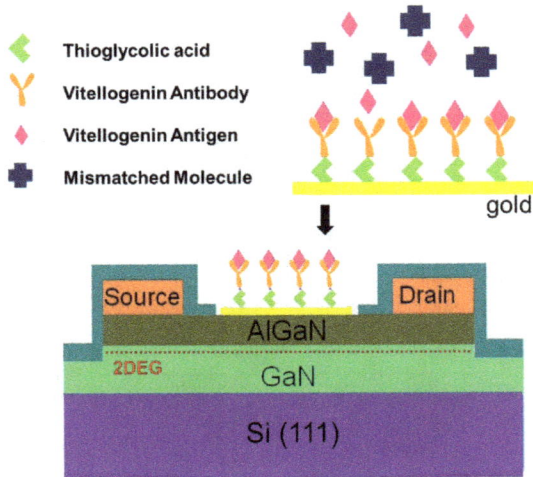

Figure 16. Cross sectional schematic of the vitellogenin-functionalized HEMT sensor.

Figure 17 shows the results of real time detection of Vtg. The drain current measurement began with 10 μL of PBS placing on the HEMT surface. Before introducing the Vtg solutions, an additional 1μL drop of PBS was added to the sensor. In comparison, a rapid response of HEMT drain current was observed in less than 10 seconds when the sensor was exposed to 5μg/mL of Vtg at 100 seconds. The abrupt current change stabilized after the VTG thoroughly diffused into the solution and reached a steady state. A larger signal change was observed when 10μg/mL of Vtg was added at 200 seconds. The sensor was exposed to higher Vtg concentrations of 50μg/mL and 100μg/mL sequentially for further real time test. The sensors were rinsed with 10mM PBS at pH=6 because antibodies have optimal reactivity at pH=7.4 and will release the antigen at a lower pH.

Perkinsus marinus (P. marinus), a protozoan pathogen of the oyster, is highly prevalent along the east coast of the United States. Perkinsus species (Perkinsozoa, Alveolata) are the causative agents of perkinsosis in a variety of mollusc species. Perkinsus species infections cause widespread mortality in both natural and farm-raised oyster populations, resulting in severe economic losses for the shellfish industry and detrimental effects on the environment. Currently, the standard diagnostic method for Perkinsus species infections has been fluid thioglycollate medium (FTM) assay detection. However, this method of detection requires several days. The polymerase chain reaction (PCR)-based technique is also used to diagnose Perkinsus, but it is quite expensive and time-consuming, and requires exquisite controls to assure specificity and accuracy. Clearly, such methods are slow and impractical in this age of global trade that requires rapid detection of such pathogens.

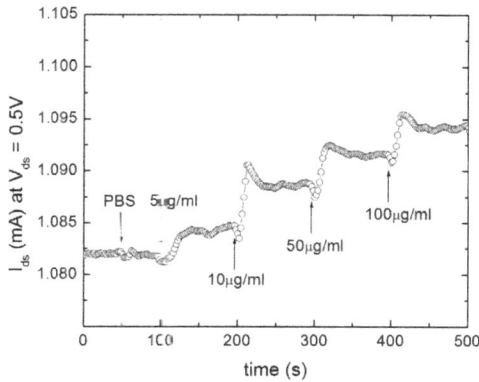

Figure 17. Real time source-drain current of sensors when introduced to 5, 10, 50, and 100 µg/mL of vitellogenin.

Figure 18 (top) shows a schematic device cross section with immobilized thioglycolic acid, and followed by anti-P. marinus antibody coating. Figure 18 (bottom) shows a picture of *Tridacna crocea*. The *Tridacna crocea* are extremely popular ornamental reef clams imported in huge numbers into the USA from the Indo-Pacific for the aquarium trade. *Tridacna crocea* are known to be vulnerable to P. marinus. This may pose a threat to domestic shellfish and negatively impact our desirability as trade partners. The infection status of the *Tridacna crocea* in this study was verified by histopatholcgy, FTM, and polymerase chain reaction assays.

Figure 19 shows real time P. marinus detection using source and drain current change with constant drain bias of 500 mV. No current change can be seen with the addition of buffer solution mixed showing the specificity and stability of the device. The current change showed a rapid response in less than 5 seconds when 2 ul of tank 2 water was added to the surface. Continuous 2 ul of the tank 2 water added into a buffer solution resulted in further decreases of drain current. Tank 2 housed sick and dying calms releasing P. marinus organisms into the water, which subsequently shed surface antigens readily detected by the sensors. Then, the sensor was washed with PBS (pH 6.5) and used to detect the P. marinus again. The recycled sensor still showed very good sensitivity as previously. These results demonstrated real-time P. marinus detection and reusability of the sensor.

7. Summary and Conclusions

We have summarized recent progress in AlGaN/GaN HEMT sensors. These devices take advantage of microelectronics, including high sensitivity, possibility of high-density integration, and mass manufacturability. The goal is to realize real-time, portable and inexpensive chemical and biological sensors and to use these as handheld exhaled breath, saliva, urine, or blood monitors with wireless capability. Frequent screening can catch the early development of diseases, reduce the suffering of the patients due to late diagnoses, and lower the medical cost.

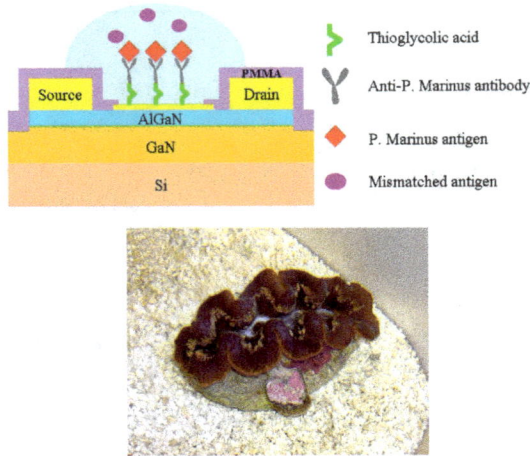

Figure 18. (top) Schematic of AlGaN/GaN HEMT sensor. The gate area was functionalized with anti-P. marinus antibody on thioglycolic acid. (bottom) A picture of the clam, which may carry the perkinsus.

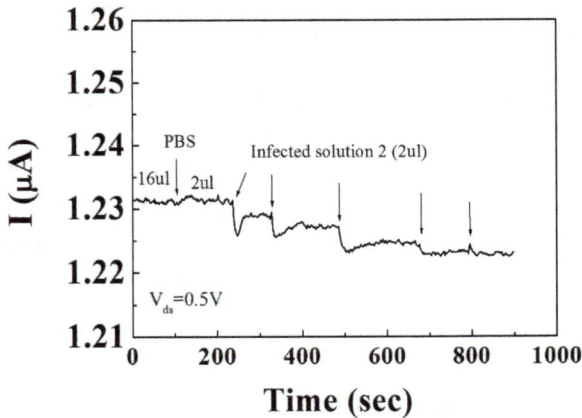

Figure 19. Real-time detection of P. marinus in an infected water from tank 2 before recycling the sensor with PBS wash (bottom).

There are many possible applications, including diabetes/glucose testing, hydrogen sensors, breast cancer testing, asthma testing, prostate testing and narcotics testing. The characteristics of these sensors include fast response (liquid phase-5 to 10 seconds and gas phase- millisecond), digital output signal, small device size (less than $100 \times 100 \ \mu m^2$) and chemical and thermal stability.

There are still some issues. First, the sensitivity needs to be improved to allow sensing in real body fluids, including blood and urine. Second, integrating multiple sensors on a single chip with automated fluid handling and algorithms is needed to analyze multiple detection signals. Third, a package that will result in a cheap product is needed. Fourth, the stability of surface functionalization layers in some cases is not conducive to long-term storage and this will limit the applicability of those sensors outside of clinics. There is a need for detection of multiple analytes simultaneously. However, there are many such approaches and acceptance from the clinical community is generally slow for regulatory concerns.

Acknowledgements

This work is supported by NSF. Collaborations with T. Lele, Y. Tseng, D. Dennis, W. Tan, B.P. Gila, W. Johnson and A. Dabiran are greatly appreciated.

Author details

Stephen J. Pearton[1*], Fan Ren[2] and Byung Hwan Chu[2]

*Address all correspondence to: spear@mse.ufl.edu

1 Department of Materials Science and Engineering, University of Florida, Gainesville, FL USA

2 Department of Chemical Engineering, University of Florida, Gainesville, FL USA

References

[1] Chemical Sensors (US industry forecasts for 2014 & 2019); Group, T. F., Ed.; http://www.the-infoshop.com/report/fd139268-us-chemical-sensor.html.

[2] Greenfield, R. A., Brown, B. R., Hutchins, J. B., Iandolo, J. J., Jackson, R., Slater, L., & Bronze, M. S. (2002). *The American J. Of The Medical Sciences*, 323(326), 326-332.

[3] Kang, B. S., Wang, H. T., Lele, T. P., Ren, F., Pearton, S. J., Johnson, J. W., Rajagopal, P., Roberts, J. C., Piner, E. L., & Linthicum, K. J. (2007). *Appl. Phys Lett.*, 91, 112106-112108.

[4] Wang, H. T., Kang, B. S., Chancellor, T. F., Jr., Lele, T. P., Tseng, Y., Ren, F., Pearton, S. J., Johnson, J. W., Rajagopal, P., Roberts, J. C., Piner, E. L., & Linthicum, K. J. (2007). *Appl. Phys. Lett.*, 91, 042114-042116.

[5]   Kang, B. S., Wang, H. T., Ren, F., Gila, B. P., Abernathy, C. R., Pearton, S. J., Dennis, D. M., Johnson, J. W., Rajagopal, P., Roberts, J. C., Piner, E. L., & Linthicum, K. J. (2008). *Electrochem. Solid-State Lett.*, 11, J19-J21.

[6]   Kang, B. S., Wang, H. T., Ren, F., Gila, B. P., Abernathy, C. R., Pearton, S. J., Johnson, J. W., Rajagopal, P., Roberts, J. C., Piner, E. L., & Linthicum, K. J. (2007). *Appl. Phys. Lett.*, 91, 012110-012112.

[7]   Kang, B. S., Pearton, S. J., Chen, J. J., Ren, F., Johnson, J. W., Therrien, R. J., Rajagopal, P., Roberts, J. C., Piner, E. L., & Linthicum, K. J. (2006). *Appl. Phys. Lett.*, 89, 122102-122104.

[8]   Kang, B. S., Ren, F., Wang, L., Lofton, C., Tan, W., Pearton, S. J., Dabiran, A., Osinsky, A., & Chow, P. P. (2005). *Appl. Phys. Lett.*, 87, 023508-023510.

[9]   Kang, B. S., Wang, H., Ren, F., Pearton, S. J., Morey, T., Dennis, D., Johnson, J., Rajagopal, P., Roberts, J. C., Piner, E. L., & Linthicum, K. J. (2007). *Appl. Phys. Lett.*, 91, 252103-252105.

[10] Pearton, S. J., Kang, B. S., Kim, S., Ren, F., Gila, B. P., Abernathy, C. R., Lin, J., & Chu, S. N. G. (2004). *J. Phys: Condensed Matter*, 16, R961-985.

[11] Chang, J. F., Kuo, H. H., Leu, I. C., & Hon, M. H. (1994). *Sensors and Actuators*, B84, 258-261.

[12] Logothetis, E. M. (1991). Automotive oxygen sensors. *in: N. Yamazoe (Ed.), Chemical Sensor Technology*, 3, Elsevier, Amsterdam.

[13] Xu, Y., Zhou, X., & Sorensen, O. T. (2000). *Sens. Actuators*, B65, 2-9.

[14] Castañeda, L. (2007). *Materials Science and Engineering*, B 139, 149-157.

[15] Gerblinger, J., Lohwasser, W., Lampe, U., & Meixner, H. (1995). *Sens. Actuators*, B26, 93-98.

[16] Sotter, E., Vilanova, X., Llobet, E., Vasiliev, A., & Correig, X. (2007). *Sens. Actuators*, B127, 567-572.

[17] Wang, Yu-Lin., Covert, L. N., Anderson, T. J., Lim, Wantae., Lin, J., Pearton, S. J., Norton, D. P., Zavada, J. M., & Ren, F. (2007). *Electrochemical and Solid-State Letters*, 11(3), H60-H62.

[18] Wang, Y.-L., Ren, F., Lim, W., Norton, D. P., Pearton, S. J., Kravchenko, I. I., & Zavada, J. M. (2007). *Appl. Phys. Lett.*, 90, 232103-232105.

[19] Wormhoudt, J. (1985). *Infrared Methods for Gaseous Measurements*, Marcel Dekker, New York.

[20] Manuccia, T. J., & Eden, J. G. (1985). *Infrared optical measurement of blood gas concentrations and fiber optical catheter*, U.S. Patent 4,509,522.

[21] Vasiliev, A., Moritz, W., Fillipov, V., Bartholomäus, L., Terentjev, A., & Gabusjan, T. (1998). *Sens. Actuators*, B49, 133-138.

[22] Savage, S. M., Konstantinov, A., Saroukan, A. M., & Harris, C. (2000). *Proc. ICSCRM'99*, 511-515.

[23] Mitzner, K. D., Sternhagen, J., & Galipeau, D. W. (2003). *Sensors and Actuators*, B9, 92-97.

[24] Wollenstein, J., Plaza, J. A., Cane, C., Min, Y., Botttner, H., & Tuller, H. L. (2003). *Sensors and Actuators*, B93, 350-356.

[25] Hu, Y., Zhou, X., Han, Q., Cao, Q , & Huang, Y. (2003). *Mat.Sci.Eng.*, B99, 41-46.

[26] Ling, Z., Leach, C., & Freer, R. (2001). *J.European Ceramic Society*, 21, 1977-1981.

[27] Rao, B. B. (2000). *Materials Chem.Phys.*, 64, 62-67.

[28] Mitra, P., Chatterjee, A. P., & Maiti, H. S. (1998). *Mater. Lett.*, 35, 33-38.

[29] Chang, C. Y., Kang, B. S., Wang, H. T., Ren, F., Wang, Y. L., Pearton, S. J., Dennis, D. M., Johnson, W., Rajagopal, P., Roberts, J. C., Piner, E. L., & Linthicum, K. J. (2008). *Appl.Phys. Lett.*, 92, 232102.

[30] Thomson, I. M., & Ankerst, D. P. (2007). *CMAJ*, 176, 1853-1857.

[31] Wang, Y. L., Chu, B. H., Chen, K. H., Chang, C. Y., Lele, T. P., Tseng, Y., Pearton, S. J., Ramage, J., Hooten, D., Dabiran, A., Chow, P. P., & Ren, F. (2008). *Appl. Phys. Lett.*, 93, 262101-262103.

[32] Kang, B. S., Wang, H. T., Ren, F., & Pearton, S. J. (2008). *J. Appl. Phys.*, 104, 031101-031103.

[33] American Cancer Society. (2007, 08 Nov.) *Detailed Guide: Prostate Cancer. What Are the Key Statistics About Prostate Cancer*, http://www.cancer.org/docroot/CRI/content/ CRI_2_4_1X_What_are_the_key_-statistics_for_prostate_cancer_36.asp?rnav=cri.

[34] Wang, J. (2006). *Biosens. Bioelectron.*, 21, 1887-1994.

[35] United States Department of Health and Human Services. (2007, 3 Nov.) *What is Breast Cancer?*, http://www.hhs.gov/breastcancer/whatis.html.

[36] Michaelson, J. S., Halpern, E., & Kopans, D. B. (1999). *Radiology*, 212(2), 551-558.

[37] Harrison, T., Bigler, L., Tucci, M., Pratt, L., Malamud, F., Thigpen, J. T., Streckfus, C., & Younger, H. (1998). *Spec. Care Dentist.* 18(3), 109-115.

[38] McIntyre, R., Bigler, L., Dellinger, T., Pfeifer, M., Mannery, T., & Streckfus, C. (1999). *Oral Surg. Oral Med. Oral Pathol. Oral Radiol. Endod.*, 88(6), 687-695.

[39] Streckfus, C., Bigelr, L., Dellinger, T., Pfeifer, M., Rose, A., & Thigpen, J. T. (1999). *Clin. Oral Investig.*, 3(3), 138-144.

[40] Streckfus, C., Bigler, L., Dellinger, T., Dai, X., Kingman, A., & Thigpen, J. T. (2000). *Clin Cancer Res.*, 6(6), 2363-2368.

[41] Chen, K. H., Kang, B. S., Wang, H. T., Lele, T. P., Ren, F., Wang, Y. L., Chang, C. Y., Pearton, S. J., Dennis, D. M., Johnson, J. W., Rajagopal, P., Roberts, C., Piner, E. L., & Linthicum, K. J. (2008). *Appl. Phys. Lett.*, 92, 192103.

[42] Chu, B. H., Kang, B. S., Ren, F., Chang, C. Y., Wang, Y. L., Pearton, S. J., Glushakov, A. V., Dennis, D. M., Johnson, J. W., Rajagopal, P., Roberts, J. C., Piner, E. L., & Linthicum, K. J. (2008). *Appl. Phys. Lett.*, 93, 042114-042116.

[43] Heo, Y. W., Norton, D. P., Tien, L. C., Kwon, Y., Kang, B. S., Ren, F., Pearton, S. J., & La Roche, J. R. (2004). *Mat. Sci. Eng.*, R47, 1-51.

[44] Hung, S. C., Hicks, B., Wang, Y. L., Pearton, S. J., Dennis, D. M., Ren, F., Johnson, J. W., Rajagopal, P., Roberts, J. C., Piner, E. L., Linthicum, K. J., & Chi, G. C. (2008). *Appl. Phys. Lett.*, 92, 193903-193905.

[45] Chu, Byung-Hwan, Lin, Hon-Way, Gwo, Shangjr, Wang, Yu-Lin, Pearton, S. J., Johnson, J. W., Rajagopal, P., Roberts, J. C., Piner, E. L., Linthicuni, K. J., & Ren, Fan. (2010). *Vac. Sci. Technol.*, B28, L5-7.

[46] Shekhar, H., Chathapuram, V., Hyun, S. H., Hong, S., & Cho, H. J. (2003). *IEEE Sensors*, 1, 67-71.

[47] Taylor, J., & Hong, S. (2000). *J. Lab. Clin. Med.*, 31, 563-570.

[48] Porte, C., Janer, G., Lorusso, L. C., Ortiz-Zarragoitia, M., Cajaraville, M. P., Fossi, M. C., & Canesi, L. (2006). *Comparative Biochemistry and Physiology, Part C*, 143, 303-309.

[49] Mosconi, G., Carnevali, O., Franzoni, M. F., Cottone, E., Lutz, I., Kloas, W., Yamamoto, K., Kikuyama, S., & Polzonetti-Magni, A. M. (2002). *Gen. Comparative Endocrinology*, 126, 125-130.

[50] Sumpter, J. P., & Jobling, S. (1995, Oct.) Environmental Health Perspectives. 103(7), *Estrogens in the Environment*, 173-178.

[51] Matozzo, V., Gagné, F., Marin, M. Gabriella., Ricciardi, F., & Blaise, C. (2008). *Environment International*, 34, 531-534.

[52] Watson, C. S., Bulayeva, N. N., Wozniak, A. L., & Alyea, R. A. (2007). *Steroids*, 72, 124-128.

[53] Garcia-Reyero, N., Barber, D. S., Gross, T. S., Johnson, K. G., Sep´ulveda, M. S., Szabo, N. J., & Denslow, N. D. (2006). *Aquatic Toxicology*, 78(358), 358-362.

[54] Wang, Y.-L., Chu, B. H., Chen, K. H., Chang, C. Y., Lele, T. P., Papadi, G., Coleman, J. K., Sheppard, B. J., Dungen, C. F., Pearton, S. J., Johnson, J. W., Rajagopal, P., Roberts, J. C., Piner, E. L., & Linthicum, K. J. (2009). *Appl. Phys. Lett.*, 94, 243901-243903.

Chapter 13

Amperometric Urea Biosensor Based Metallic Substrate Modified with a Nancomposite Film

Florina Brânzoi and Viorel Brânzoi

Additional information is available at the end of the chapter

1.Introduction

Urea is one of the final products of protein metabolism. Urea is an omnipresent compound present in blood and various organic fluids.

The urea concentration in the blood lies between 2.5-6.7 mM (15-40mg/dl) while pathophysiological range covers 30-150 mM (180-900 mg/dl). Theprimary function of the kidneys is to remove wastes from the body. These mayinclude the by-products of normal physiologic processes, drugs, and various toxins.When the kidneys malfunction, such substances begin to accumulate. Over time,progressive kidney failure can result in uremia [1-2]. On the other hand, urea can passes directly into the milk through diffusion.Therefore, milk is the second major biological sample for the study of urea concentration [3]. A periodic monitoring of urea in milk can be used to predict the state of animal's health and predict theprotein requirement in its diet [4]. Besides milk, presence of urea in agricultural land as apollutantdue to excessive use fertilizers is also widely known. Various methods were used for the determination of urea.Amongst these methods, detection through electrochemical mode is highly adopted and versatile. This method involves the use of electrochemical urea biosensor. In the development of electrochemical urea biosensors, immobilization of urease over modified electrodes is the key parameter which decides the sensitivity and reproductibility of the sensor.

Therefore, devices developed based on biocatalyst "urease" to analyze urea also known as urea biosensors are of vital importance [1-5]. For the fabrication of the urea biosensor, urease is immobilized over a substrate, which would can be a polymeric film electrodeposited on a metallic electrode. In a series of our previous papers, it was described the obtainment of different polymeric films, their mechanical and electrochemical characteristics as a function of chemical composition of synthesis solution, as a function of morphological structure and conditions for obtaining [6-11]. After urease was immobilized in polymer film, in this case in polyaniline film, the immobilized urease catalyzes the urea conversion into ammonium and bicarbonate ions based on enzyme substrate reaction. Many biosensors have been developed for the determination of urea in the biological samples namely spectrometry, potentiometry with application of pH sensitive electrode, conductometry, coulometry,amperometry and inductometry [12-24].

Urea determinationhas been performed regularly in the medical field to study the proper functioning ofthe kidney. It can say that, the urea biosensors mainly are used in the medical field and also, in the food industry. The food industry has the requirement of a sensing system to accurately analyze dairyproducts during their manufacture and quality control. Urea biosensor is a valuable tool for monitoring the urea content of adulterated milk. Urea can stress the environmentbecause it decomposes to ammonia, which is very toxic, and so it can pollute thestreams and rivers into which it drains. So urea biosensor can be an economical tool to monitor the concentration of urea to be between limits allowed. The commercial biosensors that are available, suffer the drawbacks of highcost, complicated construction, and require extra electrodes to compensate forelectrical interferences. Therefore, the development of cost effective and disposable biosensors for the detection of clinically important metabolites, such as urea, is a scientific matter of great importance.

In the present paper, a new polymeric film based on polyaniline was synthesized and employed as a new electron-mediating support material for fabricating urease-immobilized electrodes. For this purpose, the urease-immobilized electrodes were prepared using the electrodeposited polyaniline films of various thickness andapplication todetection ofurea by amperometric method.

2. Experimental

The electrochemical polymerizations were carried out using a conventional three electrodes system. A platinum electrode and a saturated calomel electrode (SCE) were used as counter and reference electrode, respectively. The reference electrode was placed in a separate cell and was connected to the electrolytic cell via a salt bridge that ends as a Luggin capillary in the electrolytic cell. This arrangement helps in reducing the ohmicresistance of the electrochemical system. The working electrode was made from a platinum disk. In this paper were used: aniline 99.5%, urease, urea, Ringer-Brown solution,sulphuric acid 98% which were purchased from -Sigma-Aldrich or Fluka and all were of analytical grade.Bidistilled water was used for all sample preparations. Cyclic voltammetry and electrochemical impedance spectroscopy were used to investigate the electrochemical characteristics of the obtained modified electrodes. Electrochemical experiments were carried out with an automated model VoltaLab 40 potentiostat / galvanostat with EIS dynamic (electrochemical impedance spectroscopy) controlled by a personal computer.

All the following potentials reported in this work are against the SCE. Scanning electron microscopy (SEM) was used to compare the microstructures of the deposited films.

3. Preparation of the modified electrodes

The Pt electrode was carefully polished with aqueous slurries of fine alumina powder 0.05 µm on a polishing cloth until a mirror finish was obtained. After 10 min sonication, the electrodes were immersed in concentrated H_2SO_4, followed by thorough rinsing with water and ethanol. The prepared electrodes were dried and used for modification immediately. The purepolyaniline(PANI) films were prepared from an aqueous solution of 0.2mol/L aniline + 0.25 mol/L H_2SO_4 by cyclic voltammetry on the potential range of -250 mV to +900mV with a scan rate of 10mV/s and for a number of 20 cycles, when a thick film of PANI was electrodeposited. In this way, it was obtained PANI/platinum substrate modified electrode. Enzymatic electrode type PANI/Urease/Pt was obtained by dripping method. Aqueous urease solution of 0.76 mg/mL was prepared using 38 mg urease (65.7 u.a. /mg) and 50 mL bidistilled water in order to get the final enzyme layer casting solution 10µL of the enzyme-layer casting solution was pipetted out onto the PANI/platinum substrate modified electrode surface and allowed to dry. We shall assume that the urease solution sipped into the porous PANI layer. Hence, we shall refer to the resulting enzyme biosensor by the full shorthand: PANI/UreaseDrop/Pt substrate (where Drop means that enzymatic electrode was obtained by dripping method). The biosensor thus made was always kept in the working buffer (in our case, the Ringer - Brown solution) at 4^0C when not in use, and rinsed with deionized water between experiments.

Then, we tried to obtain an enzymatic electrode type PANI/Urease/Pt by electropolymerization from a synthetic solution containing: 0.2 mol/L aniline + 0.25 mol/L H_2SO_4 + 0.76 g/L urease, by cyclic voltammetry on the potential scanning range of -250 to +900 mV at a scan rate of 10mV/s and for a cycles number of 20. Several attempts have been made but there was no co-deposition, consequently the enzymatic electrode was not obtained. New attempts were made and finally, the enzymatic electrode was obtained by co-deposition. But,in this case was obtained first a thinpolyaniline film by electropolymerization from a synthetic solution containing: 0.2 mol/L aniline + 0.25 mol/L H_2SO_4, by cyclic voltammetry on the potential scanning range of -250 to +900 mV at a scan rate of 10mV/s and for a number of 10 cycles. The obtained modified electrode PANI/Pt was rinsed with bidistiled water and then it was immersed in another synthetic solution containing 0.2 mol/L aniline + 0.1 mol/L H_2SO_4 + 0.76 g/L urease. Using the cyclic voltammetry on the potential range of -250 to +900 at a scan rate of 10 mV/s and for 10 cycles, the enzymatic electrode PANI/Urease-COD/Pt was successfully obtained.In this case we shall refer to the resulting enzyme biosensor by the full shorthand PANI/UreaseCOD/Pt,where COD is co-deposition.

4. Results and discussions

Amperometry is most commonly used technique for biosensors based on conductingpolymers. Devices based on amperometry measure the change in current as a consequence of specific chemical reactions which take place at biotransducer electrode surface under none-

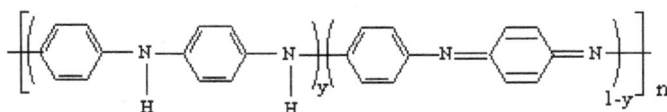

Scheme 1.

quilibrium condition. The principle of the amperometry is based on the efficiency of the electron transfer between the biomolecule and underlying electrode surface in presence of electron mediator or conducting polymer. Moreover, amperometric biosensors are not only limited to the redox enzyme but also related to the biocatalyst reaction and interaction of the reaction product with conducting polymer to induce change in current [15-16]. Urea biosensor is the typical example of biocatalyticamperometric biosensor where ammonium ion which is a product of biocatalytic reaction interacts with polymer to induce a change in conductivity of the polymer. As mentioned before, in urea biosensors the enzyme immobilized to the electrode surface catalyzes the hydrolysis of the urea, in an overall reaction leading to the formation of ammonium, bicarbonate and hydroxide ions as shown below:

$$Urea + 3H_2O \xrightarrow{\text{Urease}} 2NH_4^+ + HCO_3^- + OH^- \tag{1}$$

The ionic products of the above reaction change the electronic properties of the biosensor electrode (modified with conducting polymers), which can be observed by various electrochemical techniques, in this case was used amperometric technique. Devices based on amperometry measure the change in current as a consequence of specific chemical reactions which take place at biotransducer electrode surface under nonequilibrium condition.

For the beginning, the enzymatic electrode type PANI/UreaseDrop/Ptwas obtained by dripping method. In figure 1 are shown the polymerization cyclovoltammograms of aniline for obtainment of PANI/Pt modified electrode. Hence, these voltammogramswere recordedduring the growth of PANI film. As we can see from figure 1, at the cyclic potential sweep on the range -250 mV up to +900 mV, on the cyclic voltammograms appear three anodic oxidation peaks while, at the reverse potential sweep on the cathodic branch appears also three reduction peaks. This behaviour can be explained in the following mode: it is well known that polyaniline can exist in three different oxidation states such as leucoemeraldine (fully reduced form), emeraldine(partially oxidized form) and pernigraniline (fully oxidized form) as shown in the following scheme:

A very important characteristic of polyaniline consists in the fact that its structural units contain two different entities with different ponderables. Taking into account this property we can write thus: when y = 1, we have leucoemeraldine base, when y = 0 we obtained pernigraniline base and when y = 0.5 an intermediate state between leucoemeraldine and pernigraniline is obtained which is called emeraldine base. These forms of polyaniline are dependent on the applied potential. At the increasing anodic potential sweep the oxidation forms of polyaniline are obtained and on the anodic branch of the cyclovoltammogram appear the oxidation peaks while at the reverse potential sweep the reduction processes take place, on the cathodic branch of the cyclovoltammograms appear the reduction peaks. The three polyaniline oxidation forms correspond to the three anodic oxidation peaks while, the three polyaniline (PANI) reduction forms correspond to the three reduction peaks from the cathodic branch of cyclovoltammograms [6-11].

On the platinum surface of the working electrode a polyaniline thick film was electrodeposited and for the obtainment of the enzymatic electrode, the urease was immobilized in the PANI thick film by the dripping method (see chapter intitulated: Preparation of the modified electrodes).

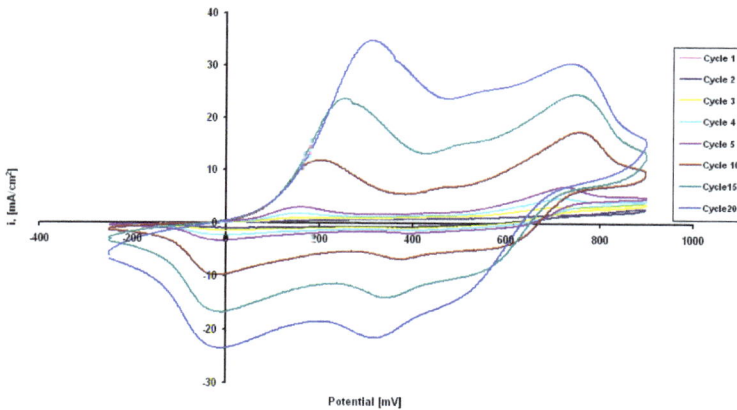

Figure 1. The obtainment of a PANI thick film by cyclic voltammetry from a synthesis solution of 0.2M aniline + 0.25 M H$_2$SO$_4$ using a potential scan rate of 10mV/s, on the potential range of -250 mV up to +900 mV and for 20 cycles.

Further, the obtained enzymatic electrode PANI/UreaseDrop/Pt was rinsed with bidistiled water and then immersed in a Ringer – Brown solution. Using the cronoamperometry method were registered the response currents at the addition of the urea samples. Hence, the testing of the obtained biosensor was carried out by amperometry method at the constant potential, in our case, at the open circuit potential.

In figure 2 is shown the variation diagram of response current at the successive additions of the urea samples.

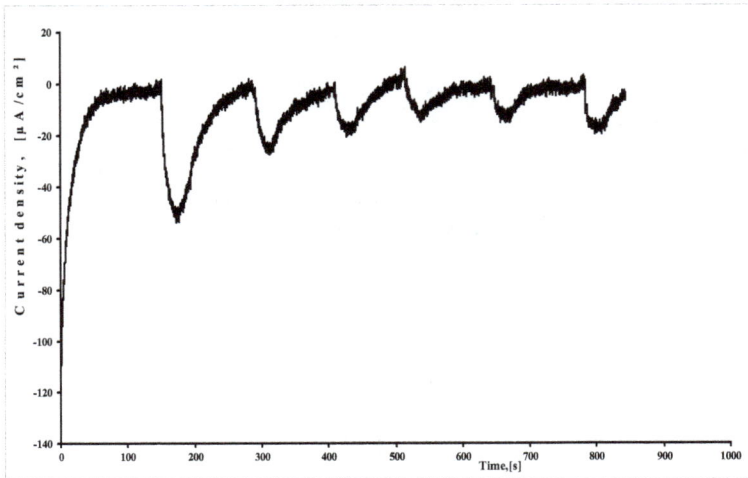

Figure 2. Response value of the peak current versus time at successive additions of the urea samples of 0.05mM (1 ml urea solution 0.05M).

Bellow, in figures 3 are presented the calibration curves of urea biosensor obtained by dripping method on different concentration ranges.

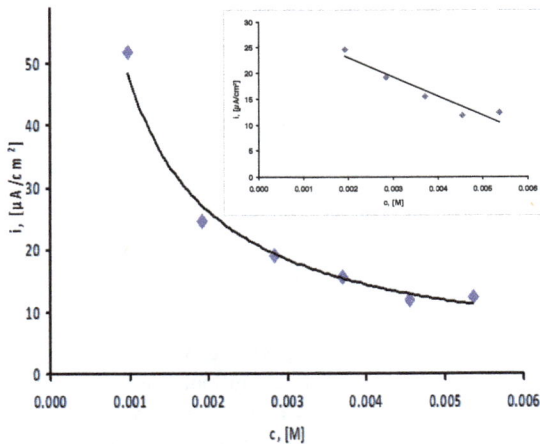

Figure 3. Calibration curve of urea biosensor type PANI/UreaseDrop/Pt namely,value of response current versus concentration of urea additions. Inset shows linear range of variation of response current value versus concentration of urea additions for urea biosensor type PANI/UreaseDrop/Pt.

Analyzing the obtained results (see figures 2 and 3), it can be observed that, these results are not satisfactory and for this reason it was fabricated a new urea biosensor namely PANI/ UreaseCod/Pt so how it was described in chapter intitulated: Preparation of the modified electrodes. As we shown, initially it was obtained a thin film of PANI (see figure 4) and then the urea biosensor of type PANI/UreaseCod/Pt (see figure 5). For this reason, the obtained modified electrode type PANI thin film/Pt was immersed in a synthesis solution of 0.2 mol/L aniline + 0.1 mol/L H2SO4 + 0.76 g/L urease, and then the electrode potential was scaned on a scanning range of -250 to +900 mV at a scan rate of 10mV/s and for a number of 10 cycles.

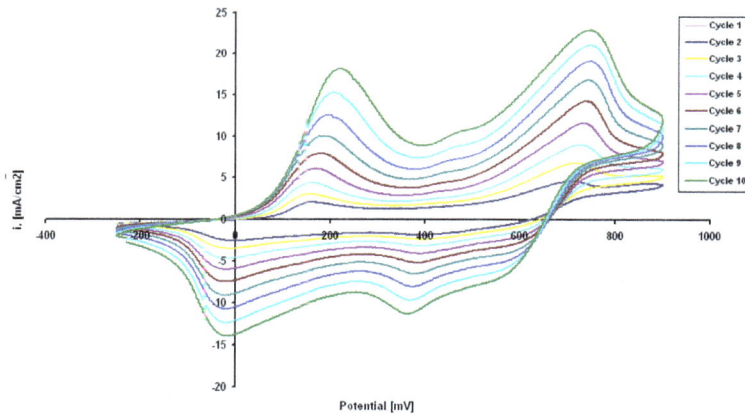

Figure 4. The obtainment of a PANI thin film by cyclic voltammetry from a synthesis solution of 0.2M aniline + 0.25 M H_2SO_4 using a potential scan rate of 10mV/s, on the potential range of-250 mV up to +900 mV and for 10 cycles.

Analyzing in comparison to figure 5 the figure 1,it can be observed that,the shape of cyclo-voltammograms from figure 5 differ very much comparative to shape those from figure 1.This fact points out that,the redox processes are very different in the two electrdeposited films, respectively,in the thick PANI film (figure 1) and in the PANI/Urease co-electrodeposited film (figure 5). In the same time, this fact proves that the composition and morphological structure of the two electrodeposited films are different and that means that the urease was entrapped in polyaniline matrix.Thus,the urease enzymatic electrode type PANI/UreaseCOD/Ptwas formed.

The testing of the obtained biosensor was carried out by amperometry method at the constant potential, in our case,at the open circuit potential, in a Ringer-Brown solution and the response currents were registered at the addition of the urea samples and after different times.

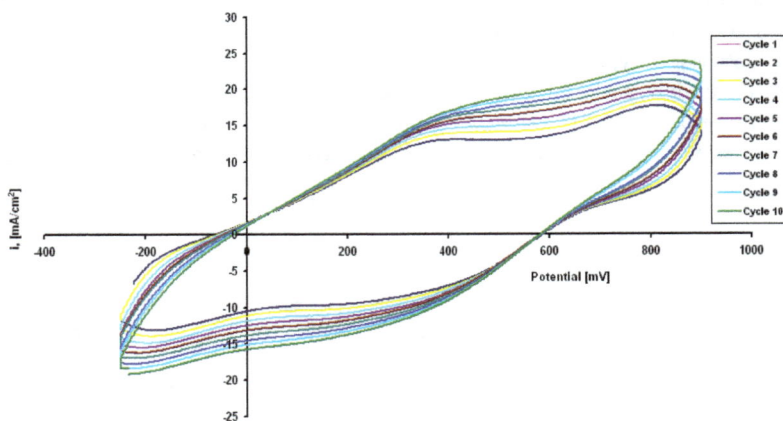

Figure 5. The recorded cyclovoltammograms at the urea biosensor obtainment by electropolymerization from a synthesis solution of 0.2 mol/L aniline + 0.1 mol/L H_2SO_4 + 0.76 g/L urease, on the potential scanning range of -250 to +900 mV at a scan rate of 10mV/s and for a number of 10 cycles.

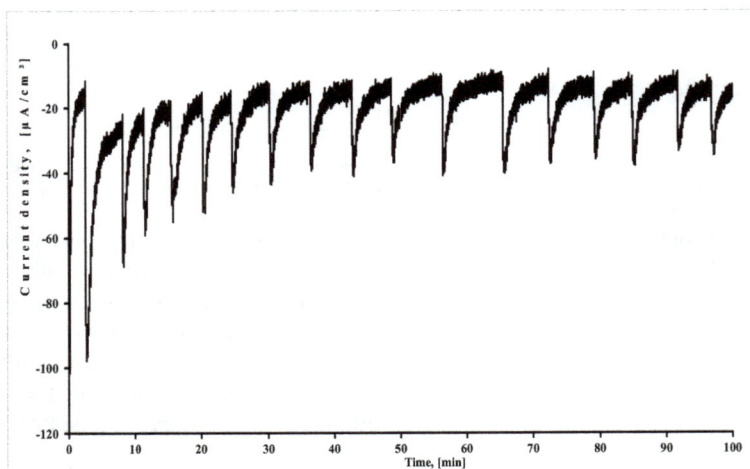

Figure 6. The response value of the peak current versus time at the successive additions of urea samples of 0.05 mM (1 ml solution urea 0.05M), for urea biosensor type PANI/UreaseCod/Pt.

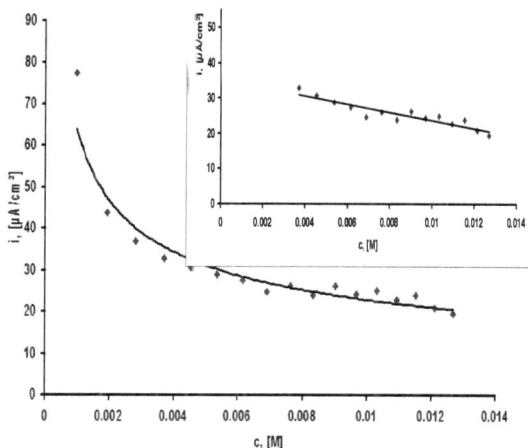

Figure 7. Calibration curve of urea biosensor of type PANI/UreaseCod/Pt which shows the variation of the response current versus urea concentration. Inset shows linear response range of urea biosensor type:PANI/UreaseCod/Pt

Further,the influenceoftestingtime on the value of the response currentwasstudied. Infigure 8 is shown the variation diagram of the response current after 24 hours. As, it can be observed, the activity of urea biosensor is still high enough and this fact proves that the sensor type: PANI/UreaseCod/Pt is much better than the sensor type: PANI/UreaseDrop/Pt. This fact points out that, obtainment of the enzymatic electrode by co-electrodeposition lead to results much more good and to a biosensor much more stable and sensitive, see in comparison the figures 2, 6 and 8.

Enzymatic electrode	Response current peak					Response current peak (after 24 hours)				
i_1, [μA/cm²]	i_2, [μA/cm²]	i_3, [μA/cm²]	i_4, [μA/cm²]	i_5, [μA/cm²]	i_1, [μA/cm²]	i_2, [μA/cm²]	i_3, [μA/cm²]	i_4, [μA/cm²]	i_5, [μA/cm²]	
PANI/ UreaseDrop/ Pt 51.72	24.61	19.19	15.61	10.93	5.24	3.27	2.16	1.45	0.67	
PANI/ UreaseCod/P t 77.44	43.68	36.65	32.72	30.57	24.97	23.89	22.74	21.97	19.89	

Table 1. The values of response currents for urea biosensors synthesised in different modes,only for the firstfive samplesof ureaadded and different times

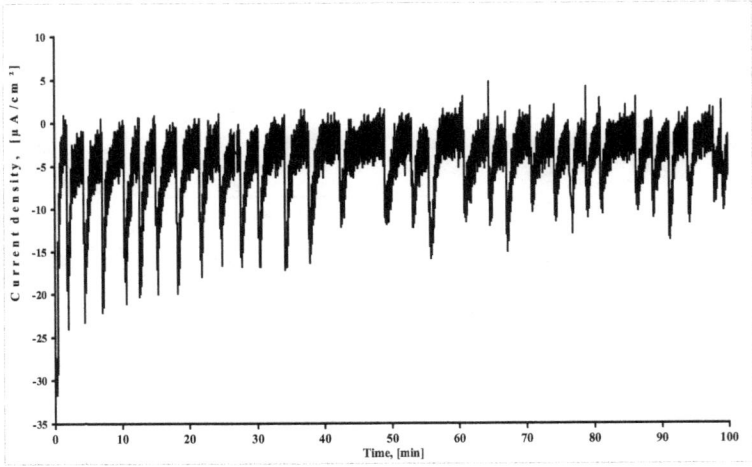

Figure 8. Response value of peak current versus time for urea biosensor type PANI/UreaseCod/Ptto successive additions of urea samples of 0.05mM (1mL urea solution of 0.05 M) after 24 hours.

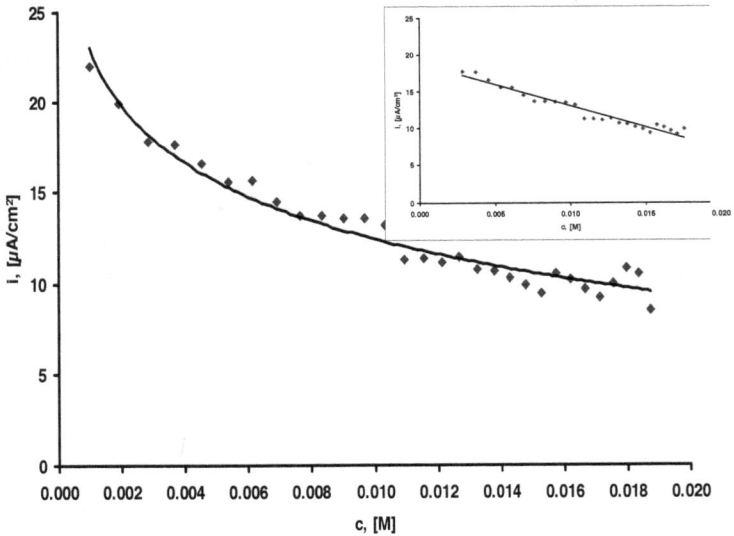

Figure 9. Calibration curve of urea biosensor of type PANI/UreaseCod/Pt which shows the variation of the response current versus urea concentration.Inset shows the response linear range for urea biosensor type:PANI/UreaseCod/Pt

Further, in table 1 are given the values of response current at the succesive additions of urea samples, immediately after biosensor preparation and after 24 hours.

Comparative analysis of current response values in Table 1 and Figures 2, 6 and 8, it can be observed that, in the case of urea biosensor type PANI/UreaseCod/Pt the response currents are much higher than the response currents of the biosensor typePANI/UreaseDrop/Pt, this means that sensitivity is much better than for biosensor type: PANI / UreaseDrop / Pt. Analyzing the values of response current after 24 hours from table 1, it can be observed that, for the biosensor typePANI/UreaseDrop/Pt, the values of the response currents have decreased very much in comparison to values of response current for biosensor type PANI/UreaseCod/Pt. This facts points out that the stability of biosensor type, PANI/UreaseCod/Pt is much higher than for biosensor type PANI / UreaseDrop /Pt.For this reason, is advisable to obtainthe urea biosensors by co-electrodeposition of polyaniline and urease enzyme. Hence, the obtainment of urea biosensor type PANI/UreaseCod/Pt by co-electrodeposition leads to a much more strong immobilization of urease into polymeric matrix and this fact means that, the stability and sensitivity of fabricated biosensor is much more higher than the biosensor typePANI/UreaseDrop/Pt obtained by dripping method.

The results of analysis carried out on milk samples contaminated with urea were in good concordance with experimental results given above in the paper.

Figure 10. SEM images of different magnitudes for polyaniline electrodeposited film

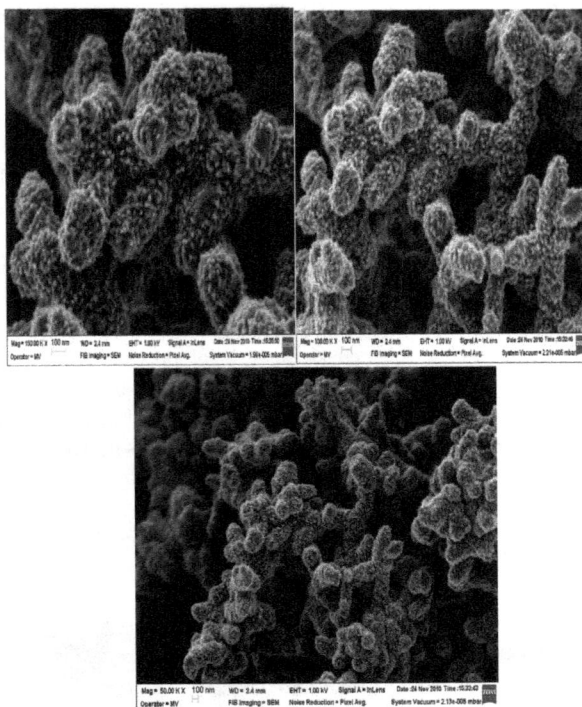

Figure 11. SEM images of different magnitudes for PANI/Urease co-electrodeposited film.

For a better understanding of these electrodeposited films behaviourwas given SEM images more of different magnitudes – see figures 10 -11.

Comparative analysis of the SEM images from figures 10 and 11,points out that, morphological structure of the two films differ greatly of and also, one can see how the biomolecules of urease are entrapped into polyaniline matrix.In this way it can be explained the different behaviour of the two biosensors namely PANI/UreaseDrop/Pt and PANI/UreaseCod/Pt in the same conditions. Also, in the same mode the difference between the shapes of the cyclovoltammograms of the two electrodeposited films can be explained.

5. Conclusions

The result presented indicates that the electrochemical behaviour of polymer film electrodes is strongly dependent on the actual morphology of the polymer matrix;

The morphology of polymer matrix is an essential factor for the processes occurring in these films and in the design of electrodes for practical purposes;

The rate of charge transport which is of vital importance in the use of these systems for electrocatalytic purposes can be influenced by the temperature, the nature of the supporting electrolyte and its concentration;

The stability of the surface layer may be influenced by applying the proper conditions of film preparation, as well as high electrolyte concentration;

The obtained modified electrodes are stable, highly permeable for ions but at the same time fulfill the conditions needed for fast electron transfer.

Author details

Florina Brânzoi[1] and Viorel Brânzoi[2]

*Address all correspondence to: fbrinzoi@chimfiz.icf.ro.; iv_branzoi@chim.upb.ro

1 Institute of Physical Chemistry, Bucharest, Romania

2 Department of Applied Physical Chemistry and Electrochemistry, University Politehnica of Bucharest, , Romania

References

[1] E. Karakus, S. Pekyardımc andE. Kılıc, Artificial Cells, Blood Substitutes, and Biotechnology. 2005, 33, 329

[2] C. Eggenstein, M. Borchardt, C. Diekmann, B. Grundig, C. Dumschat, K. Cammann, M. Knoll and F. Spener,. Bioelectron. 1999, 14, 33.

[3] M. J. R. Leema, S.M. V.Fernandes and A. O. S. Rangel,J. Agric Food Chem, 200452, 6887

[4] G. Hof, M. D. Vervoorn, J. Lenaers and S. Tamminga,.J. Dairy Sci. 1997, 80, 3333

[5] A. V. Rebriiev and N. F. Starodu ̌, Electroanalysis 2004,16, 1891

[6] V. Branzoi, Florina Branzoiand L. Pilan, Surface and Interface Analysis, 2010. 42, 1266-1273,

[7] A.Prună, V. Branzoi and F. Branzoi, Rev Roum Chim, 2010, 55, 293-298.

[8] V. Branzoi,F. Branzoi and A. Pruna, Rev. Roum. Chim., 2011, 56, 73-84.

[9] A.Prună, V.Branzoi and Florina. Branzoi, Journal Applied Electrochemistry, 2011, 41, 77-81.

[10] V. Branzoi, A. Prună, and F. Branzoi, *Molecular Crystal and Liquid Crystals*, 2008, *485*, 853-861

[11] V.Branzoi,F. Branzoi and L. Pilan, *Materials Chemistry and Physics*, 2009,*118*, 197-203.

[12] D.Liu, K. Ge, K. Cheng, L. Nie andS.Yao, *Anal. Chim. Acta*, 1995,*307*, 61.

[13] M. Mascini, *Sens. Actuators* B 1995, 29 121.

[14] X. Xie, A. A. Suleiman and G. G. Guilbault, *Talanta* 1991, *38*, 1197.

[15] S. B. Adeloju, S. J. Shaw and G. G. Wallace, *Anal. Chim. Acta,* 1993, 281, 611

[16] S. B. Adeloju, S. J. Shaw and G. G. Wallace, *Anal. Chim. Acta*, 1993, 281, 621.

[17] M. F. Esteve and S. Alegret, J. Chem. Educ. 1994, 71, A67.

[18] D. Liu, M. E. Meyerhoff, H. D. Goldberg and R. Brown, *Anal. Chim. Acta*, 1993,*274*, 37.

[19] K. Chen, D. Liu, L.Nie and S. Yao, *Talanta*, 1994, 41, 2195.

[20] A. S. Jdanova, S. Poyard, A. P. Soldatkin, N. Jaffrezic-Renault and C. Martelet, *Anal. Chim. Acta* 1996, *321* 35.

[21] H. Sangodkar,S. Sukeerthi and R. S. Srinivasa, *Anal. Chem.* 1996,*68*, 779.

[22] R. E. Adams and P. W. Carr, *Anal. Chem.* 1978, *50*, 944.

[23] S. B. Adelju, S. J. Shaw and G. G. Wallace, *Anal. Chimica Acta*, 1996, 323, 107

[24] W. O. Ho, S. Krause, C. J. McNeil, J. A. Pritchard, R. D. Armstrong, D. Athey and K.Rawson, *Anal. Chem.* 1999,*71*, 1940

Chapter 14

Macromolecular Biosensors Based on Proteins Involved in Bile Stones Formation

José Manuel Bravo-Arredondo,
María Josefina Robles-Aguila,
María Liliana Marín-García,
Bernardo A. Frontana-Uribe, Abel Moreno and
María Eugenia Mendoza

Additional information is available at the end of the chapter

1. Introduction

Biomineralization refers to the process of obtaining biominerals in living organisms, and can be both a pathological and a non-pathological process [1]. This term was coined at the beginning of the eighties of the past century. and opened a new way of inspiring materials for biomedical applications [2].

Nowadays, the understanding of many diseases, like gallbladder stones related to pathological aspects of biominerals' formation in living organisms, is based on the knowledge of the chemical recognition between biological macromolecules, calcium salts and cholesterol molecules [3-5]. This kind of chemical interaction has played an important role in the development of biologically inspired biosensors. A biosensor is an analytical device consisting of two elements in spatial proximity: (1) a biological recognition element able to interact specifically with a target; (2) a transducer able to convert the recognition event into a measurable signal [6].

According to the mechanism of biological signaling used, biosensors are classified into five major types, one of them is the biomimetic one; in this sense, a biomimetic biosensor is an artificial or synthetic sensor that mimics the function of a natural biosensor [7]. Some examples of these types of sensors are the quercetin-modified wax-impregnated graphite electrode (Qu/WGE) for the purpose of detecting uric acid (UA) in the presence of ascorbic acid

(AA) [8]; a cell-based biosensor platform of neuron silicon interface with acid-sensing taste receptor cells cultured on light addressable potentiometric sensor (LAPS) [9]; a Langmuir-Blodgett film of tyrosinase incorporated in a lipidic layer with lutetium bisphthalocyanine as an electron mediator for the voltammetric detection of phenol derivatives [10]; a mixed self-assembled monolayers (SAMs) functionalized with specific olfactory receptors (ODR-10) constructed on the sensitive area of surface acoustic wave (SAW) chip [11]. Recently, a new type of chemical biosensor based on intramineral proteins of eggshells for carbonate ions detection has been published elsewhere [12].

The precipitation of the cholesterol and the calcium salts are commonly found in the bladder (bile), whereas the calcium carbonate, which is the major component, is normally found in the pancreas, at alkaline pH [13]. This formation usually follows the principles of crystal growth and the simple chemical solubility rules. However, it has been recently observed that the formation of gallstones or minerals of biogenic origin (grown in these biological vesicles) leads the growth process in extreme conditions, even devoid of water in some cases. It is not clear whether some of the genes (mostly in mammals) could be involved in most of the pathological processes, however, it has been proved the ethnical influence on the medical diagnosis [14]. Recent publications have shown that there are genes already identified (mainly in marine organisms) whose role in activating biomineralization processes has been tested by molecular biology techniques [15]. Nowadays, the knowledge about the genes involved in the formation of certain skeleton in marine spicules is, in general clearly identified. However, our understanding of the role of any of the proteins in biomineralization is scarce, so is our understanding of the role of the matrix proteins as well as protein-protein, and protein-mineral interaction [16].

Concerning the crystallization of cholesterol in human bile, there are some proteins from the serum involved in this matter, the immunoglobulines IgM, IgA, IgG as well as the proteins α_1-Acid Glycoprotein (AAG, usually called Orosomucoid), Phospholipase C and Aminopeptidase N. According to a recent publication, only three proteins from this list showed a potent enhancement and a promoting effect on the crystallization of cholesterol: IgM, IgA and AAG protein [17]. Particularly, orosomucoid a protein of 42 kDa, is one of the most abundant proteins of serum proteins with no well-known physiological function. However, a number of biological activities have been described for orosomucoid, such as promotion of collagen fibril formation, inhibition of platelet aggregation, inhibition of heparin accelerated antithrombin III mediated activation of thrombin and factor Xa, binding of Δ-4 Ketosteroids, sequestration of a glycosaminoglycan cofactor in the lipoprotein lipase reaction and the preferential binding of more than 60 cationic therapeutic drugs [18]. The orosomucoid protein has an unusual pI of around 3 and an extraordinary carbohydrate content of approximately 40% (w/w), comprising five N-linked glycans [19]. However, although the importance of this orosomucoid in human physiology seems to be high, there are no more structural data from the native glycoprotein obtained at high resolution. The orosomucoid was firstly crystallized in 1984 [18], and since then no more structural data of that native glycoprotein at high resolution have been published. The high-resolution crystallographic structure was not available for a long time. However, the three-dimensional structure has been recently available at 1.8Å resolution of a recombinant AAG protein crystallized in a tetragonal $P4_12_12$ space group. This recombinant human AAG produced as monomeric, yet unglycosylated protein in E. coli was obtained via secretion into the bacterial periplasm, where formation of its two-disulphide bonds was facilitated in the oxidizing environment [20]. The former crystals of the native AAG protein were obtained in a hexagonal P622 or $P6_222$ or the enantiomorph $P6_422$ space group in the presence of chlorpromazine at 18 °C [18]. Interestedly, the co-crystallization of the tetragonal AAG by soaking method in the presence of the well-known ligands (phenothiazine tranquilizers) like chlorpromazine, bromazepam or diazepam, failed [20]. This is due to the unglycosylated 3D structure. It seems that the carbohydrates play an important role into the 3D structure-function of the native glycoprotein when making the chemical recognition between this AAG protein and some common drugs.

In this chapter we evaluate, from the electro-analytical point of view, the plausible role into the chemical interaction and chemical recognition of sodium carbonate, bilirubin, cholesterol with α_1-Acid Glycoprotein (AAG) usually found in biogenic minerals in bile. Additionally, we show the effect of different gel media on the crystallographic habits of synthetically grown crystals of cholesterol. These crystals were characterized by X-ray powder diffraction. Finally, based on our results, we propose a new design of a biologically inspired biosensor.

2. Experimental

Cyclic Voltammetry. All the electroanalytical assays to investigate the analyte-protein interaction for example, sodium carbonate, cholesterol, and bilirubin with the α_1-Acid Glycoprotein were performed using a Potentiostat/Galvanostat PG580 from UNISCAN Instruments (UK). The potential was ranging from 0 to 1.80 Volts versus a Saturated Calomel Electrode (as a reference electrode). The velocity for this electrochemical analysis was 50 mVs^{-1}.

Figure 1. Electrochemical cell used for cyclic voltammetry experiments.

Figure 1 shows the electrochemical cell used for this determination. Three electrodes were used along the experiments. The working electrode (5 mm diameter) was a gold microelectrode Au10 from Autolab Electrochemical Instruments (USA). The auxiliary electrode was a platinum wire of 0.5 mm in diameter. The gold electrode was polished using a diamond paste of 0.25 microns in particle size. The gold electrode was sonicated several times in order to remove the impurities. The auxiliary electrode was cleaned with a proper fine texture sandpaper to eliminate oxides on the surface. As an inert electrolyte 0.5M Potassium Chloride (Strem Chemicals, Inc. Newburyport, Code 7447-40-7) was used with a purity of 99.999%. The highly pure α_1-Acid Glycoprotein (AAG protein) was purchased from Sigma (Code G9885) without further purification. The electroanalytical plots, as those shown in Figure 2, were obtained taking into account the gold electrode saturation for each scanning cycle as well as removing the background from the oxidation of gold at 1.3V. This value corresponds to a maximum peak into the anodic response of the voltammogram (right-hand side of the plot current versus potential) at each analyte concentration. This AAG protein (50 µg) was absorbed on the gold electrode until the water was evaporated. For each analyte five different concentrations were done as well as duplicates for each single concentration. The electrochemical experiment is carried out until the analyte is put into contact with the protein and after an equilibrium time of 2 minutes. Before running any of the experiments the oxygen was degased by bubbling pure nitrogen for 15 minutes. All dissolutions were prepared in a glass beaker of 5 mL with 0.1, 0.2, 0.3, 0.4, and 0.5 mM of each electroanalyte. The dissolutions were then transferred to the electrochemical cell. However, as bilirubin is not soluble in water, it had to be prepared in DMSO as follows: 2.92 mg were dissolved in 1 mL of DMSO-Water 80:20 and keeping the solution away from oxidation inside of an amber container. Cholesterol was prepared dissolving 1.93 mg in 1 mL of ethanol-water 80:20 and sealing it carefully to avoid any alcohol evaporation.

In order to investigate the chemical interaction between cholesterol and AAG protein, the cholesterol crystals were grown in different crystal growth media, solution, and two types of hydrogels in the Granada Crystallization Boxes via counter-diffusion methods [21]. In solution method, it was used an ethanol solution in the classic evaporation process; and in the gel method tetramethyl orthosilicate (TMOS) hydrogel was prepared by polycondesation reaction [22], and agarose hydrogel was prepared by heating and cooling method [23]. After obtaining the gel phase, the solution of cholesterol 0.5% w/v in ethanol was poured onto the top of each gel allowing cholesterol molecules to diffuse into the gel network for one week at 18 °C. After this week, the solution was removed and replaced with water to produce the crystal growth of cholesterol by reducing the solubility in water, changing the dielectric constant.

X-ray powder diffraction. Cholesterol crystals were characterized by using an Empyrean XRD system from PANalytical Instruments (Netherlands), into the following conditions: Cu K_α radiation, monochromator, time step = 0.004s, step size = 19.68, 2θ = 4 to 50 deg, at room temperature.

3. Results and discussion

The cyclic voltammetry for AAG protein showed an anodic response when interacting with carbonate, cholesterol and bilirubin (Figures 2a, 2b, and 2c). The protein alone did not show any electrochemical response in an aqueous solution of KCl 0.5M (curve not shown). This protein-substrate response was dependent of the concentration, and is characterized by an increment of current with the sequential increment of the analyte concentration. Only two of the three analytes (carbonate ions Figure 2a and bilirubin Figure 2c) investigated showed a strong protein interaction with them and generated a characteristic plot with increments of current. A small peak is observed at a potential value of 0.54 V, which corresponds to the oxygen reduction of the small amount of water electrolysis that occurs during the electro-chemical response at $E > 1.6$ V. The peak is more evident for the cholesterol (Figure 2b) that interacts in minor proportion with the protein. This behavior decreases the strength of the interaction, which is almost imperceptible for bilirubin (Figure 2c). This demonstrates that covered electrode with a stable protein-analyte layer generates a cleaner analytical signal avoiding the water electrolysis as a parasite reaction.

In order to have a clearer visualization of the effect, the electrochemical response of the protein was measured on the gold electrode without analyte (I_o). Then this curve was used to obtain a normalized plot I/I_o, which allowed standardizing all the experiments respect to the chemical interaction to the analyte (see Figure 3). The chemical interaction between the AAG protein and the specific analytes can be detected by either a big difference between I/I_0 or a big change value on the slope. When obtaining the normalized plots as shown in Figure 3, bilirubin showed the strongest chemical interaction, less intense followed by the carbonate ions. This behavior is particularly interesting since most of the preliminary investigations have claimed that AAG protein recognizes cholesterol molecules or works as a crystalliza-tion promoter in human bile, as that observed for different immunoglobulins like IgM, IgA, IgM or for phospholipase C and aminopeptidase N [17], but according to these results the interaction with bilirubin must not be discarded.

From these results it is clear that supersaturation of cholesterol in bile is a necessary condi-tion though not sufficient, for the formation of cholesterol gallstones [3]. Biliary proteins, which are capable of affecting the rate at which cholesterol crystallization occurs, are impor-tant factors in pathogenesis of cholesterol gallstone diseases [3]. The pathway is as follows: (1) AAG-bilirubin complex formation (2) cholesterol nucleation and (3), cholesterol crystal growth. The solubility of cholesterol is quite an interesting issue to take into account, when trying to investigate this chemical recognition in vivo experiments. Cholesterol in vitro is soluble in alcoholic solutions, but in vivo, cholesterol molecules should be solubilized by aqueous micellar solution or transported either by apolipoproteins or by any other biomole-cules in order to make it soluble. The cholesterol can be crystallized inside the bile, or be se-creted to the blood stream or sent directly to the stools [4]. This process based on the electrochemical observations, is in agreement with the enterohepatic circulation from the liver to the bile [11]. In this physiological process bilirubin needs the chemical interaction with cholesterol. Bilirubin is firstly conjugated with glucoronic acid in the liver by the enzyme glucuronyltransferase making it soluble in water, and perhaps with cholesterol too. Therefore, it makes sense that the first crystals of AAG protein were obtained by Schmidt in 1952 using ethanol as precipitating agent [24]. The α_1-Acid Glycoprotein is a cholesterol crystallization promoter conjugated with a complex bilirubin-cholesterol as shown in the electroanalytical results in Figure 2.

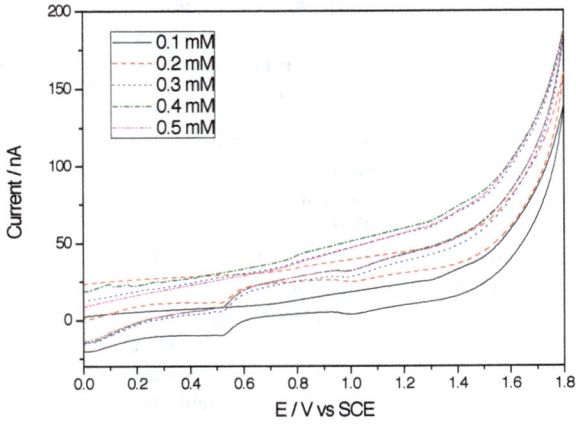

(a)

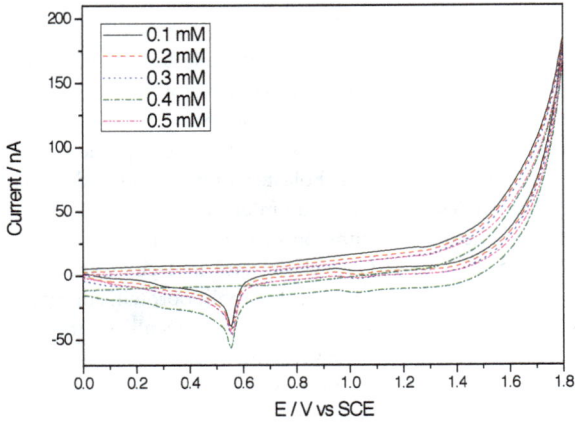

(b)

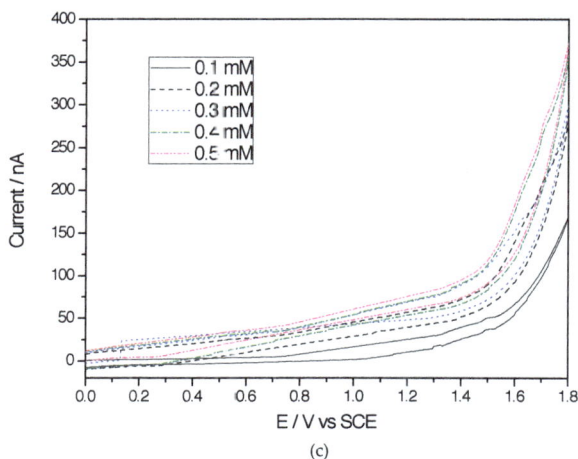

(c)

Figure 2. a. Cyclic voltammogram of AAG – KCl / Na$_2$CO$_3$ ranging the carbonates concentration from 0.1 to 0.5 mM scanning speed of 50 mVs^{-1}. b. Cyclic voltammogram of AAG – KCl / Cholesterol ranging the cholesterol concentration from 0.1 to 0.5 mM scanning speed of 50 mVs-1. c Cyclic voltammogram of AAG – KCl/Bilirubin ranging the bilirubin concentration from 0.1 to 0.5 mM scanning speed of 50 mVs-1

In humans, gallbladder sludge (gel-like media), a reversible pre-gallstone phase, consists of cholesterol crystals and bilirubin granules in a mesh of mocus called mucin. Mucin a glyco-protein commonly observed in bile, is the constituent of the core and non-cholesterol matrix of cholesterol gallstones. It is highly likely that gallbladder mucin is actively involved at a number of stages in cholesterol precipitation [25].

On the other hand, calcium salts are present in all pigment gallstones as compounds of one or more of the anions in bile: (i) carbonate; (ii) bilirubinate, and (iii) phosphate. In addition, since cholesterol stones have been found to contain pigment stone centers, perhaps bilirubin and AAG protein are complexed. We can postulate that the presence of calcium salts precip-itation in bile is a critical event in the initiation of cholesterol gallstones, so that the latter should be considered a two-stage process: (i) precipitation of calcium salts chemically bond-ed to AAG protein to form a macromolecular complex plus some components of the bile, for example mucin as a growth media, and then (ii) crystallization of cholesterol from its super-saturated state on this gel-like environment. It has been published that in vivo experiments cholesterol precipitation starts with thin filamentous structures, which evolve into helical and then tubular forms before breaking into characteristic flat cholesterol monohydrate plates [26].

From the crystal growth point of view, the crystal habit of cholesterol grown in vitro and in three different media (two of them hydrogels used to emulate mucin, a gel-like component usually found in bile) shows the vicissitudes that a crystal goes through when growing (Fig-ure 4). The crystallographic faces, shown on each gel media, describe the plausible shapes that cholesterol can display according to the transport and to the crystal growth processes, (the cholesterol keeps the same space crystallographic group). X-ray powder patterns of cholesterol lamellar-like crystals are shown in Figure 5. They have a good agreement with the pattern reported in the Powder Diffraction File (PDF, file 7-742), which corresponds to the triclinic phase, as well as those reported by several authors [26-28]. These cholesterol crystals should be grown in the presence of different combinations of AAG protein plus bilirubin, in order to check the protein as a promoter (nucleant) of cholesterol molecules.

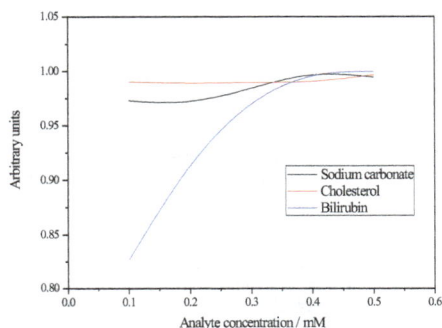

Figure 3. Normalized plot (I/I_o, see text for details) for anodic peaks corresponds to the maximum response of each voltammogram AAG-KCl, at different molar concentration for each analyte: sodium carbonate, cholesterol and bilirubin according to Figures 2a, 2b and 2c respectively.

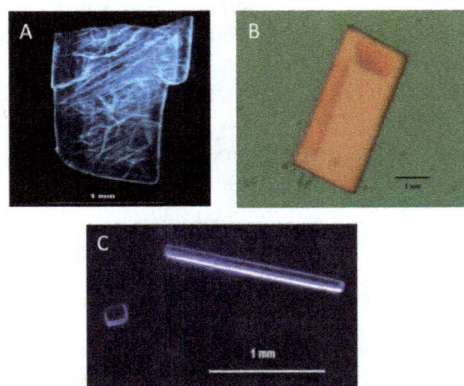

Figure 4. Optical images of Cholesterol crystals grown in: A) Solution Ethanol/Water, B) Tetramethyl orthosilicate hydrogel, C) Agarose hydrogel. The pictures (A) and (C) were taken between 90 degrees overcrossed-polarizers, while (B) picture corresponds to unovercrossed-polarizers. Different scale bars are included as inset on each picture.

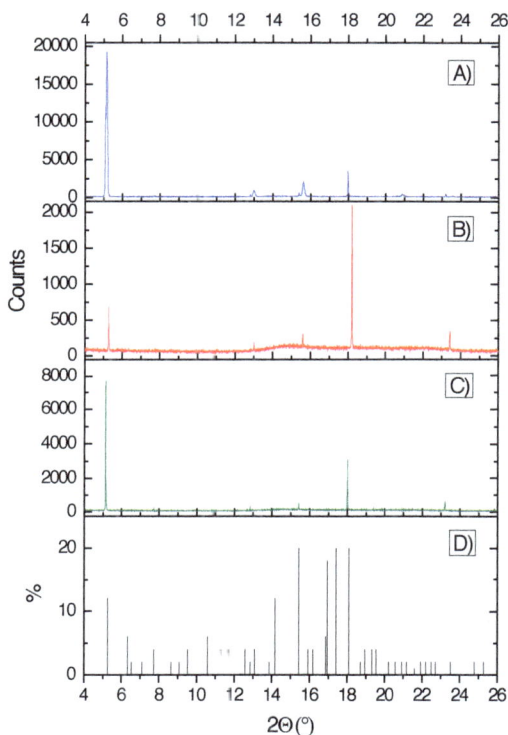

Figure 5. X-ray powder diffraction characterization plots of cholesterol crystals grown in three different media: A) alcoholic solution by evaporation method, B) tetramethyl orthosilicate hydrogel, C) agarose hydrogel and D) X-ray diffraction pattern (PDF file No. 7-742) of a cholesterol of the triclinic phase P1.

In the near future, cholesterol crystals could be grown in the presence of bilirubin, glucoronic acid and most of the components of the bile extract, in order to see whether these crystals will show the same crystallographic faces or not. These crystals could be re-dissolved in alcoholic solutions and the proteins extracted. These proteins could be purified and characterized from the cholesterol crystals in order to check their similarity to the AAG protein.

Finally, the AAG protein should be co-crystallized in the presence of bilirubin-cholesterol searching for the specific bonding-sites into the crystallographic structure either by X-ray Crystallographic methods or by NMR, as performed in the crystallographic projects nowadays. In future investigations we might be looking for inhibitors for the AAG Protein, like the promising cholesterol antinucleating 120kDa glycoprotein found by Ohya et al. [29]; or look for strategies to turn the genes off related to the over-expression of this AAG protein in vivo to control the cholesterol crystallization.

4. Biosensor's design

From these electroanalytical results, the most plausible way of producing an ad hoc choles-terol-AAG/BR biosensor based on biological macromolecules like AAG, should follow the steps from A to D as shown in the box-diagram of Figure 6. The first step (A) corresponds to the gold electrode. The AAG protein would then be deposited on the gold electrode by us-ing a thin film of the protein, this is the second step called (B). In the following step, (C) the protein should be complexed with bilirubin either by electrochemical interaction, or by Langmuir-Blogget isotherms as proposed by Xie et al., [30]. Finally, (D) the amperometric biosensor could be tested in the presence of different concentrations of alcoholic solutions of cholesterol, based on the strategy described in the experimental set up on this contribution.

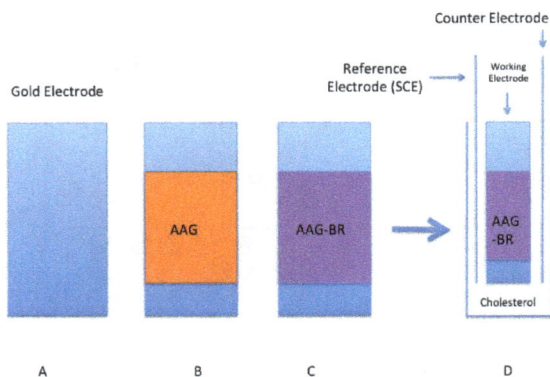

Figure 6. It shows different steps in order to produce an ad hoc Cholesterol-Bilirubin-AAG Biosensor.

The electrochemical response using this biosensor will be used to optimize the detection limits to check cholesterol in different media either in solution or in gel (this latter which would emulate the tissue texture or the mucin usually found in the bile).

5. Conclusions

We can observe from this research, that there is no a direct interaction between α_1-Acid Pro-tein (AAG) and cholesterol molecules. The perfect biosensor for the evaluation of this chemi-cal interaction should take into account that bilirubin is the intermediate molecule between cholesterol, and the AAG protein. This protein works as promoter, but needs some intermediate molecules like bilirubin to solubilize cholesterol in aqueous solutions when precipitating gallstones. Future investigations should be focused on co-crystallizing these proteins in the presence of bilirubin/cholesterol. The crystallographic 3D structure of the native glycoprotein can give answers about the specific sites (amino acids), where cholesterol molecules could be attached to the protein molecule working as nucleation precursor.

Acknowledgements

Two of the authors acknowledge to DGAPA-UNAM Projects PAPIIT No. IN201811 (for AM) and No. IN202011 (for BF-U) the financial support for this research. The authors (A.M. and M.E.M.) thank to CONACYT for the Consolidación de Grupos Initiative at the Instituto de Física, Benemérita Universidad Autónoma de Puebla (BUAP). The support and sponsorship for one of the authors (M.J. R-A) as a postdoctoral fellow is acknowledged to the Mexican Softmatter Network (CONACYT). The authors also acknowledge the technical support from I.Q. Leonel San Roman from the (IFBUAP).

Author details

José Manuel Bravo-Arredondo[1], María Josefina Robles-Aguila[2], María Liliana Marín-García[3], Bernardo A. Frontana-Uribe[4,5], Abel Moreno[2] and María Eugenia Mendoza[1]

1 Instituto de Física, Benemérita Universidad Autónoma de Puebla, Puebla, Mexico

2 Instituto de Química, Universidad Nacional Autónoma de México, México, México

3 Centro Interdisciplinario de Investigaciones y Estudios sobre Medio Ambiente y Desarrollo (CIIEMAD), Departamento de Sociedad y Politica Ambiental. Instituto Politécnico Nacional, México, D.F., México

4 Centro Conjunto de Investigación en Química Sustentable UAEMex-UNAM, Toluca, México, México

5 Permanent position at the Instituto de Química, Universidad Nacional Autonoma de Mexico, Mexico, Mexico

References

[1] Mann S. Biomineralization. Principles and Concepts in Bioinorganic Materials Chemistry, Oxford University Press, ISBN 0-19-850882-4, Oxford, UK. 2001:p1-5.

[2] Lowenstam HA. Minerals Formed by Organisms. Science 1981;211:1126 –31.

[3] Shaffer EA. Gallstone Disease: Epidemiology of Gallbladder Stone Disease. Best. Pract. Res. Clin. Gastroenterol. 2006;20: 981-96.

[4] Admirand WH, Small DM. The Physicochemical Basis of Cholesterol Gallstone Formation in Man. J. Clin. Invest. 1968;47:1043–52.

[5] Stolk MF, J Van Erpecum KJ, Peeters TL, Samson M, Smout AJPM, Akkermans LMA, Vanberge-Henegouwen GP. Interdigestive Gallbladder Emptying, Antroduodenal Motility, and Motilin Release Patterns are Altered in Cholesterol Gallstone Patients. Dig. Dis. Sci. 2001;46:13 28-34.

[6] Eggins RB. Chemical Sensors and Biosensors, John Wiley & Sons, London, 2002, p: 1-9.

[7] Morrison DWG, Dokmeci MR, Demirci U, Khademhosseini A. Clinical Applications of Micro- and Nanoscale Biosensors. In Gonsalves KE, Laurencin CL, Halberstadt, Nair LS. (Ed.) Biomedical Nanostructures, John Wiley & Sons 2008:p1-10.

[8] He J-B, Jin G-P, Chen Q-Z, Wang Y. A Quercetin-modified Biosensor for Amperometric Determination of Uric Acid in The Presence of Ascorbic Acid. Analytica Chimica Acta 2007;585:337-43.

[9] Chen P, Liu X, Guang B, Chen G, Wang P. A Biomimetic Taste Receptor Cell-based Biosensor for Electrophysiology Recording and Acidic Sensation. Sensors and Actuators 2009;B139:576-83.

[10] Apetrei C, Ramos-Fernandes EG, Zucolotto V, Oliveira ON. A Biomimetic Biosensor Based on Lipidic Layers Containing Tyrosinase and Lutetium Bisphthalocyanine for The Detection of Antioxidants. Biosensors & Bioelectronics 2011;26:2513-19.

[11] Wu C, Du L, Wang D, Zhao L, Wang P. A Biomimetic Olfatory-Based Biosensor with High Efficiency Immobilization of Molecular Detectors. Biosensors & Bioelectronics 2012;31:44-48.

[12] Marín-García ML, Frontana-Uribe BA, Reyes-Grajeda JP, Stojanoff V, Serrano-Posada HJ, Moreno A. Chemical Recognition of Carbonates Ions in Biomineralization Processes and Their Influence on Calcite Crystal Growth. Crystal Growth and Design 2008;8:1340-45.

[13] Hyogo H, Tazuma S, Cohen DE. Cholesterol gallstones. Current Opinion in Gastroenterology 2002;18: 366-71.

[14] Ko CW, Lee SP. Gallstone Formation: Local Factors. Gastroenterology Clinics of North America 1999;28:99-115.

[15] Fred H, Killian E. What Genes and Genomes Tell Us about Calcium Carbonate Biomineralization. In: Sigel A, Sigel H, Sigel KO. (Ed.) Metal Ions in Life Sciences, John Wiley & Sons, 2008:p37-69.

[16] Veis A. Crystals and Life: An Introduction. In: Sigel A, Sigel H, Sigel KO. (Ed.) Metal Ions in Life Sciences, John Wiley & Sons, 2008:p1-35.

[17] Abei M, Schwarzendrube J, Nuutinen H, Broughan TA, Kawczak P, Williams C, Holzbach RT. Cholesterol Crystallization-promoters in Human Bile: Comparative Potencies of Immunoglobulines, α_1-Acid Glycoprotein, Phospholipase C, and Aminopeptidase N. Journal of Lipids Research 1993:34:1141-48.

[18] McPherson A, Friedman ML, Halsall HB. Crystallization of α-1-Acid Glycoprotein. Biochemical and Biophysical Research Communications 1984:124(2):619-24.

[19] Shiyan SD, Bovin NV. Carbohydrate Composition and Immunomodulatory Activity of Different Glycoforms of α_1-Acid Glycoprotein. Glycoconjugated Journal 1997;14:631-38.

[20] Schoenfel DL, Ravelli RBG, Mueller U, Skerra A. The 1.8-Å Crystal Structure of α1-Acid Glycoprotein (Orosomucoid) Solved by UV RIP Reveals the Broad Drug-Binding Activity of This Human Plasma Lipocalin. J. Mol. Biol. 2008;384:393-405.

[21] García-Ruiz JM, Gonzales-Ramirez LA, Gavira JA, Otalora F.. Granada Crystallization Box: a New Device for Protein Crystallization by Counter-diffusion. Acta Cryst. 2002;D58:1638-42.

[22] Gavira JA, García-Ruiz JM. Agarose as Crystallisation Media for Protein: II. Trapping of Gel Fibers into The Crystals. Acta Cryst. 2002;D58:1653-56.

[23] Lorber B, Sauter C, Théobald-Dietrich A, Moreno A, Schellenberger P, Robert MC, Capelle B, Sanglier B, Potier N, and Giegé R. Crystal Growth of Proteins, Nucleic Acids, and Viruses in Gels". Progress in Biophysics and Molecular Biology 2009;101:13-25.

[24] Schid, K. The Plasma Proteins: Academic Press New York, F. Putman (Ed.) 1975:p183-228.

[25] O'Leary DP. Biliary Cholesterol Transport and The Nucleation Defect in Cholesterol Gallstone Formation. Journal of Hepatology 1995;22:239-46.

[26] Konicoff FM, Chung DS, Donovar JM, Small DM, Carey MC. Filamentous, Helical and Tubular Microstructures During Cholesterol Crystallization from Bile. J. Clin. Invest. 1992;90: 1155-60.

[27] Loomis CR, Shipley GG, Small DM. The Phase Behavior of Hydrate Cholesterol. J. Lipid Res. 1979;20:525-35.

[28] Garti N, Karpuj L, Sarig S. Correlation Between Crystal Habit and The Composition of Solvated and Nonsolvated Cholesterol Crystals. J. Lipid Res. 1981;22:785-91.

[29] Ohya T, Schwarzendrube J, Busch N, Gresky S, Chandler, K, Tabayashi A, Igimi H, Egami K, Holzbach RT. Isolation of a Human Biliary Glycoprotein Inhibitor of Cholesterol Crystallization, Gastroenterology 1993;104:527-38.

[30] Xie A, Shen Y, Chen C, Han C, Tang Y, Zhang L. Glycoprotein Adsorption into Bilir-rubin/cholesterol Mixed Monolayers at The Air-water Interface. Colloid Journal 2006;68:390-93.

Permissions

The contributors of this book come from diverse backgrounds, making this book a truly international effort. This book will bring forth new frontiers with its revolutionizing research information and detailed analysis of the nascent developments around the world.

We would like to thank Dr. Toonika Rinken, for lending his expertise to make the book truly unique. He has played a crucial role in the development of this book. Without his invaluable contribution this book wouldn't have been possible. He has made vital efforts to compile up to date information on the varied aspects of this subject to make this book a valuable addition to the collection of many professionals and students.

This book was conceptualized with the vision of imparting up-to-date information and advanced data in this field. To ensure the same, a matchless editorial board was set up. Every individual on the board went through rigorous rounds of assessment to prove their worth. After which they invested a large part of their time researching and compiling the most relevant data for our readers. Conferences and sessions were held from time to time between the editorial board and the contributing authors to present the data in the most comprehensible form. The editorial team has worked tirelessly to provide valuable and valid information to help people across the globe.

Every chapter published in this book has been scrutinized by our experts. Their significance has been extensively debated. The topics covered herein carry significant findings which will fuel the growth of the discipline. They may even be implemented as practical applications or may be referred to as a beginning point for another development. Chapters in this book were first published by InTech; hereby published with permission under the Creative Commons Attribution License or equivalent.

The editorial board has been involved in producing this book since its inception. They have spent rigorous hours researching and exploring the diverse topics which have resulted in the successful publishing of this book. They have passed on their knowledge of decades through this book. To expedite this challenging task, the publisher supported the team at every step. A small team of assistant editors was also appointed to further simplify the editing procedure and attain best results for the readers.

Our editorial team has been hand-picked from every corner of the world. Their multi-ethnicity adds dynamic inputs to the discussions which result in innovative outcomes. These outcomes are then further discussed with the researchers and contributors who give their valuable feedback and opinion regarding the same. The feedback is then collaborated with the researches and they are edited in a comprehensive manner to aid the understanding of the subject.

Apart from the editorial board, the designing team has also invested a significant amount of their time in understanding the subject and creating the most relevant covers. They scrutinized every image to scout for the most suitable representation of the subject and create an appropriate cover for the book.

The publishing team has been involved in this book since its early stages. They were actively engaged in every process, be it collecting the data, connecting with the contributors or procuring relevant information. The team has been an ardent support to the editorial, designing and production team. Their endless efforts to recruit the best for this project, has resulted in the accomplishment of this book. They are a veteran in the field of academics and their pool of knowledge is as vast as their experience in printing. Their expertise and guidance has proved useful at every step. Their uncompromising quality standards have made this book an exceptional effort. Their encouragement from time to time has been an inspiration for everyone.

The publisher and the editorial board hope that this book will prove to be a valuable piece of knowledge for researchers, students, practitioners and scholars across the globe.

List of Contributors

Donald M. Cropek
U.S. Army Corps of Engineers, Engineer Research and Development Center – Construction Engineering Research Laboratory, Champaign, IL, USA
Department of Veterinary Biosciences. University of Illinois at Urbana-Champaign, Urbana, IL, USA

Jill M. Grimme
U.S. Army Corps of Engineers, Engineer Research and Development Center – Construction Engineering Research Laboratory, Champaign, IL, USA

A. Reshetilov and T. Reshetilova
Federal State Budgetary Institution of Science, G.K. Skryabin Institute of Biochemistry and Physiology of Microorganisms, Russian Academy of Sciences, 5 Pr. Nauki, Push-chino, Moscow Region, Russia

V. Arlyapov and V. Alferov
Federal State Budgetary Educational Institution of Higher Professional Education, Tula State University, Tula, Russia

Carmenza Duque, Edisson Tello, Leonardo Castellanos and Catalina Arévalo-Ferro
Departamento de Química, Universidad Nacional de Colombia, Colombia

Miguel Fernández
Departamento de Biología, Universidad Nacional de Colombia, Colombia

Feng Long
School of Environment and Natural Resources, People's University of China; State Key Joint Laboratory of ESPC, School of Environment, Tsinghua University, Beijing, China

Anna Zhu
Research Institute of Chemical Defence, Beijing, China

Chunmei Gu and Hanchang Shi
State Key Joint Laboratory of ESPC, School of Environment, Tsinghua University, Beijing, China

Saloua Helali
Research and Technology Centre of Energy, Hammam Lif, Tunisia

V. Somerset
Natural Resources and the Environment (NRE), Council for Scientific and Industrial Research (CSIR), Stellenbosch, South Africa

E. Iwuoha
SensorLab, Department of Chemistry, University of the Western Cape, Bellville, South Africa

B. Silwana and C. van der Horst
Natural Resources and the Environment (NRE), Council for Scientific and Industrial Research (CSIR), Stellenbosch, South Africa
SensorLab, Department of Chemistry, University of the Western Cape, Bellville, South Africa

Mingqiang Zou, Xiaofang Zhang, Xiaohua Qi and Feng Liu
Chinese Academy of Inspection and Quarantine, China

Yanfei Wang and Mingqiang Zou
Chinese Academy of Inspection and Quarantine, China

Ping Yao
The People's Hospital of Juxian, China

Yuande Xu
COFCO Corporation, China

Lívia Maria da Costa Silva, Vânia Paula Salviano dos Santos and Andrea Medeiros Salgado
Laboratory of Biological Sensors, Biochemical Engineering Department, Chemistry School,
Technology Center, Federal University of Rio de Janeiro, Ilha do Fundão, Rio de Janeiro, Brazil

Karen Signori Pereira
Laboratory of Food Microbiology, Biochemical Engineering Department, Chemistry School, Technology Center, Federal University of Rio de Janeiro, Ilha do Fundão, Rio de Janeiro, Brazil

Magdalena Stobiecka
Department of Biophysics, Warsaw University of Life Sciences SGGW, Warsaw, Poland

Antonio Aparecido Pupim Ferreira, Carolina Venturini Uliana, Michelle de Souza Castilho, Naira Canaverolo Pesquero, Marcos Vinicius Foguel, Glauco Pilon dos Santos, Cecílio Sadao Fugivara, Assis Vicente Benedetti and Hideko Yamanaka
Instituto de Química, UNESP - Univ Estadual Paulista, Brazil

Stephen J. Pearton
Department of Materials Science and Engineering, University of Florida, Gainesville, FL, USA

Fan Ren and Byung Hwan Chu
Department of Chemical Engineering, University of Florida, Gainesville, FL USA

Florina Brânzoi
Institute of Physical Chemistry, Bucharest, Romania

Viorel Brânzoi
Department of Applied Physical Chemistry and Electrochemistry, University Politehnica of Bucharest, Romania

José Manuel Bravo-Arredondo and María Eugenia Mendoza
Instituto de Física, Benemérita Universidad Autónoma de Puebla, Puebla, Mexico

María Josefina Robles-Aguila and Abel Moreno
Instituto de Química, Universidad Nacional Autónoma de México, México, México

María Liliana Marín-García
Centro Interdisciplinario de Investigaciones y Estudios sobre Medio Ambiente y Desarrollo (CIIEMAD), Departamento de Sociedad y Politica Ambiental, Instituto Politécnico Nacional, México, D.F., México

Bernardo A. Frontana-Uribe
Centro Conjunto de Investigación en Química Sustentable UAEM ex-UNAM, Toluca, México, México
Permanent position at the Instituto de Química, Universidad Nacional Autonoma de Mexico, Mexico